Selected Writings

Edited by Stefanos Geroulanos and Todd Meyers

TRANSLATORS

Colin Anderson
Danielle Dubois
John Duda
Gina Fisch
Oleg Gelikman
Daniela Ginsburg
Vincent Guillin
Jaron Paktor
Nils F. Schott
Lenn Schramm
Cara Weber
Henri Atlan

Selected Writings

On Self-Organization, Philosophy, Bioethics, and Judaism

Henri Atlan

Edited by Stefanos Geroulanos and Todd Meyers

FORDHAM UNIVERSITY PRESS

NEW YORK 2011

Cet ouvrage publié dans le cadre du programme d'aide à la publication bénéficie du soutien du Ministère des Affaires Etrangères et du Service Culturel de l'Ambassade de France représenté aux Etats-Unis.

This work received support from the French Ministry of Foreign Affairs and the Cultural Services of the French Embassy in the United States through their publishing assistance program.

Atlan, Henri.
    Selected writings on self-organization, philosophy, bioethics, and Judaism / Henri Atlan ; edited by Stefanos Geroulanos and Todd Meyers.—1st ed.
        p.   cm.— (Forms of living)
    English translations from French originals.
    Includes bibliographical references and index.
    ISBN 978-0-8232-3181-2 (cloth : alk. paper)
    ISBN 978-0-8232-3182-9 (pbk. : alk. paper)
    1. Medical ethics—Philosophy.   2. Bioethics—Philosophy.
    3. Self-organizing systems.   4. Complexity (Philosophy)
I. Geroulanos, Stephanos.   II. Meyers, Todd.   III. Title.
R724.A78   2011
174.2—dc23

                                                        2011017246

Printed in the United States of America
13   12   11   5   4   3   2   1
First edition

ACKNOWLEDGMENTS

The editors would like to thank the Office of the Dean of the Humanities at New York University, as well as Ruth Leys and the Humanities Center at Johns Hopkins University, for generous grants that supported the publication of this volume. We would also like to thank, at Fordham University Press, Helen Tartar for proposing this project to us and shepherding it to completion with truly extraordinary meticulousness, and Tom Lay for going through the difficult process of acquiring the rights for all the texts included here. We are grateful to Elliot Wolfson for checking Hebrew transliterations. At the concluding stage of revisions, Ari Edmundson, Isabel Gabel, Jacob Krell, James Philips, and Simon Taylor assisted greatly in bringing this volume together, and for this assistance we are grateful.

We met Henri Atlan at Johns Hopkins in 2006 and then followed his seminar on self-organization and Spinoza's thought in late 2007. M. Atlan provided very valuable help both at the drafting stage and during editing; this help has been generous and very consequential to the improvement of this book.

Chapter 1, "Is Science Inhuman?" is a complete translation of the book *La science est-elle inhumaine?* (Paris: Bayard, 2002). © Bayard 2002, 3–5 Rue Bayard, 75008 Paris.

Chapter 2, "Intentional Self-Organization: Emergence and Reduction," was published in English in *Thesis 11* 52 (February 1998):5–34. An alternative version of this text was published in French in *Les étincelles de hasard*, vol. 2, *Athéisme de l'écriture* (Paris: Seuil, 2003), chapter 7, "Auto-organisation intentionnelle: Vers une théorie physique de l'intentionnalité," 237–76. Reproduced by permission of SAGE Publications Ltd., London, Los Angeles, New Delhi, Singapore, and Washington, D.C. © Sage, 1998.

Chapter 3, "Noise as a Principle of Self-Organization," translates "Du bruit comme principe d'autoorganisation," in *Entre le cristal et la fumée* (Paris: Seuil, 1979), 39–60. An earlier version was published in *Communications* 18 (1972): 21–36.

Chapter 4, "The Intuition of the Complex and Its Theorizations," is a translation of "L'intuition du complexe et ses théorisations," in *Les théories de la complexité: Colloque de Cérisy autour de l'oeuvre d'Henri Atlan*, ed. Françoise Fogelman Soulié (Paris: Seuil, 1991), 9–41. It appears here courtesy of the Centre Culturel International de Cérisy.

Chapter 5, "The Genetic Program," is a translation of "Le programme génétique," part 1 of *La fin du "tout génétique"? Vers de nouveaux paradigmes en biologie*, 13–51. © INRA, 1999. It is reproduced by permission of Éditions Quae.

Chapter 6, "Underdetermination of Theories by Facts," is a translation of "Sous-détérminations des theories par les faits," in *Tout, non, peut-etre: Education et verité* (Paris: Seuil, 1991), 130–37.

Chapter 7, "Internal Purposes, Vitalism, and Complex Systems," is a translation of two sections from *Tout non peut-être*, namely, "Finalité interne et vitalisme: L'héritage de la philosophie critique," 55–71, and "Systèmes complexes: Nouveaux aspects de la finalité formelle," 71–76.

Chapter 8. "Ectogenesis and Reproductive Cloning," is a translation of "Ectogenese et clonage reproductif," from *L'uterus artificiel* (Paris: Seuil, 2005), 79–93.

Chapter 9, "Weak Reductionism," appeared in English in *Enlightenment to Enlightenment: Intercritique of Science and Myth*, Trans. Lenn J. Schramm (Albany: State University of New York Press, 1993), 61–63. Reprinted by permission; all rights reserved.

Chapter 10, "The Spinoza Path," is a translation of "La voie Spinoza," part of an interview with Roger-Pol Droit, published in Henri Atlan and Roger-Pol Droit, *Chemins qui mènent ailleurs : Dialogues philosophiques* (Paris: Stock, 2005), 51–63.

Chapter 11, "Immanent Causality: A Spinozist Viewpoint on Evolution and the Theory of Action," was published in English in *Evolutionary Systems: Biological and Epistemological Perspectives on Selection and Self-Organization*, ed. G. Van de Vijver et al. (Dordrecht: Kluwer Academic Publishers, 1998), 215–31. A French version was published in *Les étincelles de hasard*, vol.2: *Athéisme de l'écriture*, chapter 6, "Un point de vue spinoziste sur l'évolution et la théorie de l'action: De la philosophie analytique à Spinoza," 201–35. The essay appears here with the permission of Springer Science and Business Media.

Chapter 12, "Spinozist Neurophysiology" reprints a short section, entitled "Philosophical Interpretation: A Non-Mentalist Model of Intentional Actions," from Yoram Louzoun and Henri Atlan, "The Emergence of Goals in a Self-Organizing Network," *Neural Networks* 20 (2007): 156–71. © 2007, Elsevier. Reprinted with permission.

Chapter 13, "Knowledge, Glory, and 'On Human Dignity,'" first appeared in *Diogenes* 54, no. 3 (2007): 11–17. An expanded version of this text can be found in *Les étincelles de hasard*, vol. 2, *Athéisme de l'écriture*, chapter 4, "La réalité, la perfection et la gloire," 101–53. Reproduced by permission of Sage Publications Ltd., London, Los Angeles, New Delhi, Singapore, and Washington, D.C. © Sage, 1998.

Chapter 14, "Sparks of Randomness," is a translation of *Les étincelles de hasard*, vol. 1, *Connaissance spermatique* (Paris: Seuil, 1999), "Introduction," 11–16. It appears in Henri Atlan, *Sparks of Randomness*, vol. 1, *Spermatic Knowledge*, trans. Lenn Schramm (Stanford, Calif.: Stanford University Press, 2011), 1–10, and is reprinted courtesy of Stanford University Press.

Chapter 15, "Nature's Ultimate Trick: The Parable of the Divine Intrigues (*'Alilot*)" renders the section "Ruse ultime de la nature: La parabole des intrigues ou *'alilot* divines," of *Les étincelles de hasard*, vol. 1, *Connaissance spermatique*, chapter 2, 124–29. This translation appears in *Sparks of Randomness*, 96–101, and is reprinted courtesy of Stanford University Press.

Chapter 16, "Mysticism and Rationality," includes two parts of the chapter "Mysticism and Rationality," from *Enlightenment to Enlightenment: Intercritique of Science and Myth*, trans. Lenn J. Schramm (Albany: State University of New York Press, 1993), 93–96, 105–13. Reprinted by permission; all rights reserved.

Chapter 17, "Souls and Body in Genesis," appeared in English in *Koroth (The Israel Institute of Medical History, Jerusalem)* 8, no. 5–6 (1982): 115–21.

Chapter 18, "Israel in Question," first appeared in French in *Entre le cristal et la fumée*, 235–58.

Chapter 19, "The Self, the Person, the 'I,'" translates "Le soi, la personne, le 'je': Perspectives hébraïques: Mise en dialogue avec un modèle mécanique d'auto-organisations mémorisées," from a special issue, *Le soi dans tous ses états*, of the Canadian journal *Théologiques* 12. nos. 1–2 (2004): p.71–93. The original article can be downloaded for free at http://www.erudit.org.

Chapter 20, "Golems," translates part of the chapter "Golems" from *Les étincelles de hasard*, vol. 1, *Connaissance spermatique*, chapter 1, 39–50. This translation appears in *Sparks of Randomness*, 28-34, and is reprinted courtesy of Stanford University Press.

Chapter 21, "The Mother Machine," is translated from "La 'mère machine,'" *L'uterus artificiel*, 143–59.

Chapter 22, "Human Cloning: Biological Possibilities, Social Impossibilities," is a translation of Atlan's contribution, "Possibilités biologiques, impossibilités sociales," to Henri Atlan, Marc Augé, Mireille Delmas-Marty, Roger-Pol Droit, and Nadine Fresco, *Le clonage humain* (Paris: Seuil, 1999), 17–42.

Chapter 23, "Possibility in Development," renders two distinct and non-sequential sections from *Tout non peut-être*, "Négation, totalité et possible dans le développement de l'individu," 38–42, and "Possibilités de personnes ou potentialités?" 126–30.

Chapter 24, "Does Life Exist?" is a translation of part 2 of *La fin du "tout génétique?"* 51–63.

Chapter 25, "The Knowledge of Ignorance," renders into English "La connaissance de l'ignorance," *Tout non peut-être*, 139–46.

Chapter 26, "The Fraternal Utopia," translates "L'utopie fraternelle," from *L'uterus artificiel*, 161–93.

Chapter 27, "To Teach Virtue," translates "Le relativisme relatif: La mesure des choses qui ne sont pas" and "Le bien-bon et le mal-mauvais," from *Tout non peut-être*, 245–54.

Chapter 28, "The Center of the Universe and the Domain of Ethics," renders into English "Le centre de l'univers et le domaine de l'éthique," from *Tout non peut-être*, 176–84.

Chapter 29, "Pleasure, Pain, and the Levels of Ethics," translates "Le plaisir, la douleur, et les niveaux de l'éthique," *International Journal of Bioethics* 6 (1995): 53–68. An extended version of this article was published as "Les niveaux de l'éthique: Pour une généalogie" in *Les étincelles de hasard*, vol. 2, *Athéisme de l'écriture*, chapter 2, 31–67.

Throughout, earlier translations have been reworked as necessary to fit into the context of this volume. Also throughout, we have used Atlan's French renditions as the basis for quotations from Spinoza, sometimes modifying them via the Shirley translation (Spinoza 2002a and b).

Selected Writings

An Introduction to Complexity   *Stefanos Geroulanos and Todd Meyers*

> It seems that the idea of the simple is already to be found contained in that
> of the complex and in the idea of analysis, and in such a way that we come
> to this idea quite apart from any examples of simple objects, or of
> propositions which mention them, and we realize the existence of the simple
> object—*a priori*—as a logical necessity. So it looks as if the existence of the
> simple objects were related to that of the complex ones as the sense of ~p
> is to the sense of p: the *simple* object is *prejudged* in the complex.
>
> —LUDWIG WITTGENSTEIN, *Notebooks 1914–1916*

The present volume aims to introduce a wide readership to the thought of
Henri Atlan, who is best known as a biophysicist and whose impressive philosophical, ethical, and political contributions have yet to receive the full attention of Anglo-American audiences.[1] Atlan has published at length and in
detail on matters as seemingly distinct as complexity and the theory of self-
organization, artificial intelligence, parables from the Hebrew Bible, the
faults of the "genetic program" theory in genetics, cloning and the possibility

---

1. Atlan's thought has been discussed and debated in scientific circles, and it has also
been evoked by a number of critics and humanists (notably in matters of literature and
science and network culture), including Mark C. Taylor (2001), William R. Paulson (1988),
Michel Serres (2007), N. Katherine Hayles (1991), and others. Historian of cognitive science Jean-Pierre Dupuy has contributed an excellent overview ("Henri Atlan, 1931–")
to the *Columbia History of Twentieth-Century French Thought*, (Kritzman 2007: 393–96).
Nevertheless, largely due to the absence in translation of his books (except for *Enlightenment to Enlightenment*), discussion among humanists of Atlan's writing has usually been
restricted to vague praises of their implications and scope.

of an artificial uterus, Spinoza, scientific mysticism, biotechnology, the figure of the Golem, and the idea of freedom. His contributions to biology and biological thought are marked throughout by a philosophical and humanistic sophistication that allows the "practice of science" and the "philosophy of science" to become joined in conversation while retaining their distinct characters; conversely, his work in self-organization, genetics, and systems theory links up with his work in philosophy and religious studies—deeply motivating his understanding of Spinoza's theory of freedom, for example, while refusing to offer a strict hierarchy or in any way reduce the scientific to the philosophical, or vice versa.

In other words, an introduction to the work of Henri Atlan is much like an introduction to complexity itself. It can offer potential directions for reading but cannot summarize Atlan's thought. This introduction, like the book as a whole, attempts to show how the different elements of this complexity allow for a picture that is irreducible to the sum of its parts, yet nevertheless reveal a trajectory in Atlan's work, one that cannot be reduced to a few keywords but rather links its interweaved elements. We start out here from a number of points, none of which can be granted strict priority: Atlan's approach to the theory of self-organization; his reading of Spinoza as at once mirroring self-organization's theory of nature and offering an understanding of the way determinism in nature organizes and allows for our sense of freedom and ethics; his writing on Judaism, on the ethics of an artificial alteration if not fabrication of the living, and on questions of ethics and freedom; his historicization and critique of biological concepts and practices.

Life and Thought

Henri Atlan was born into a Jewish family in Blida, a town in what was then French Algeria, in 1931. He studied biophysics and medicine in Paris and eventually in the United States. Among the figures who defined his studies and early research were Heinz von Foerster and Aharon Katchalsky, whom he met at Berkeley and cites as a decisive influence.[2] His first book, *L'organisation biologique et la théorie de l'information* (*Biological Organization and Information Theory*), appeared in 1972 and attempted to link the question of

---

2. For an autobiographical essay, see Atlan 2008.

organization in biology with the theoretical potential of computer programming. In 1979, Atlan published *Entre le cristal et la fumée: Essai sur l'organisation du vivant* (*Between Crystal and Smoke: An Essay on the Organization of the Living*), which elaborated his understanding of self-organization, while aiming for (and finding) a far wider audience.[3] In the 1980s, thanks in part to a careful study of Spinoza, Atlan moved to develop the philosophical consequences of his studies. In 1986, he published *À tort et à raison: Intercritique de la science et du mythe* (published in English as *Enlightenment to Enlightenment: Intercritique of Science and Myth*), in which he argues against scientists' tendencies to extrapolate metaphysical and philosophical principles from specialized kinds of research. Starting in 1983, Atlan served on the newly founded Comité Consultatif National d'Éthique pour les Sciences de la Vie et de la Santé (National Advisory Committee on Ethics in the Life and Health Sciences)—a committee charged with drafting reports for the president of the French Republic on matters of new technologies and practices, their ethical import, feasibility, and desirability, the most famous of its concerns being cloning.

His balancing of scientific and philosophical research further led to *Tout, non, peut-être: Education et vérité* (*All, No, Perhaps: Education and Truth*, 1991), and, above all, his magnum opus, written from 1999 to 2003, the two-volume *Les etincelles de hasard* (*The Sparks of Randomness*). Several of the major strains in his thinking have appeared in short book form. Already in *Between Crystal and Smoke*, Atlan is critical of facile reductions of biology to the simple idea of a genetic program, an argument he elaborated on in *La fin du "tout génétique"?* (*The End of "Everthing Is Genetic"?*, 1999; translated in full as Chapters 9 and 24, below). Similarly, *La science est-elle inhumaine? Essai sur la libre nécessité* (*Is Science Inhuman? An Essay on Free Necessity*, 2002; translated as Chapter 1, below) again addresses the status of science in modern human inquiry, while *L'utérus artificiel* (*The Artificial Uterus*, 2005) recalls the historical, philosophical, and cultural questions concerning the artificial creation of new beings, notably new human beings. Two books of interviews, *Questions de vie, entre le savoir et l'opinion* (*Questions of Life: Between Knowledge and Opinion*, 1994) and *Chemins qui mènent ailleurs* (*Paths That Lead Elsewhere*, 2005) also contributed to his influence by facilitating a broader understanding of his positions. What distinguishes *The Sparks of Randomness* is its tremendous

---

3. As Douglas Morrey notes in his book on Godard's films, Godard repeatedly cites and quotes from *Entre le cristal et la fumée* (Atlan 1979) in his films *Sauve qui peut* (1980) and *JLG/JLG: Autoportrait de décembre* (1995). See Morrey 2005: 156–58.

scope: in its pages Atlan moves freely from his studies on self-organization to work on Jewish thought, from Spinoza to the ethical import of scientific thought. Today, Atlan divides his time between France and Israel, where he is a Director of Research at the École des Hautes Études en Sciences Sociales in Paris, Professor Emeritus of Biophysics at the Universities of Paris VI and Jerusalem, and Director of the Human Biology Research Center of the Hadassah University Hospital in Jerusalem.

Self-Organization, Emergence, Chance, and Randomness

In his classic and paradigm-defining study of the conceptual and experimental history of biology, *La logique du vivant* (1970; translated as *The Logic of Life*, 1993), François Jacob presents *life* and *memory* as the two major concerns of biology after the discovery of DNA. The barrier between life and the nonliving and the emergence and operation of an apparatus of storing experience were problems that had defined the science of biology, problems that had to be addressed as much in everyday as in experimental terms. Although these problems were not new, Jacob explains how they came to be addressed together through the examination of DNA, whose discovery had drastically altered the landscape and utility of biology, establishing molecular biology, biophysics, and other subdisciplines as essential to the understanding (at once scientific and philosophical) of living and knowing.

Atlan began to develop his theory of self-organization shortly thereafter, during the turn into the 1970s, using the notion of *self-organization* to address the problems of *life* and *knowledge* (the latter of which, as we shall see, is for him grounded in memory). *Self-organizing processes* are processes in which a dynamic relation between different elements allows these to self-organize, without any sort of external aid or cause, raising them as a system to a level of higher complexity. At the level of the complex system emerge new properties that are not the result of any force transcendent to the system; at the same time, information contained within the system is rendered meaningful thanks to codes that pertain to the system itself rather than the different elements. "These models [of the self-organization of matter] show us how global properties of a complex system made of many simple elements are qualitatively different from the properties of these elements themselves" (Atlan 2003: 9). Self-organization and complexity theories exploded in the

1970s and 1980s, with various strands of research (by no means restricted to biology) involving such diverse and influential scientists as Stuart Kauffman, Ilya Prigogine, John Holland, Edward Lorenz, and Atlan himself. For some thinkers, like Jacques Monod, who had played a role in the earlier stages of the revolution of molecular biology, the theorization of self-organization and complexity in the 1970s came to offer a solution to major problems raised by the transcription of protein sequences from DNA; for others, like John von Neumann, it came to serve as the major question of the twentieth century.

Central to Atlan's approach was the utility of information theory, which complemented the birth of molecular biology, particularly in an effort to understand the role of noise and the transmission and correction of errors in a message. In his first book, *Biological Organization and Information Theory*, as well as his crucial essay "Noise as a Principle of Self-Organization" (Chapter 3, below), Atlan took up Heinz von Foerster's critique of Erwin Schrödinger's thinking that order derives necessarily from order. Demonstrating that this was neither necessary for nor capable of generating any sort of nonpreformed organization, Atlan raised the central question of how noise, randomly introduced by an organism's environment or thanks to errors in a system or a message (e.g., chance mutations in genes), affects the complexity of this system, adding variation, threatening the reliability of the "original" message, but at the same time making possible new information and new interpretations (see Atlan 1972, chaps. 3–4). For the principle of noise, Atlan relied on Claude Shannon's theory of probability, showing (contra Shannon, see Chapter 3) that in a channel of communication noise does not merely increase the ambiguity of an original message, distorting it, but also allows this message, thanks to the distortion, to augment in complexity. Noise's production of "errors" affects a system's (or organism's) relation to its environment by affecting its response:

> it depends on the subsequent reaction of the system in relation to which, a posteriori, these factors are recognized as either random or as part of the organizing process. A priori, they are in effect random, if one defines randomness as the intersection of two independent chains of causality: the causes of their occurrence have nothing to do with the chain of phenomena that has constituted the prior history of the system until then. It is thus that their occurrence and their encounter with the system can constitute noise from the viewpoint of the exchanges of information in the system, where these encounters are susceptible to producing only errors. But from the moment the system is capable of responding to these "errors," not just so that it does not disappear, but rather so that the system uses them to

modify itself in a way that benefits it or that at least ensures its subsequent survival—in other words, from the moment the system is capable of integrating these
errors into its own organization—then these errors lose, a posteriori, a little of
their character as error. (Chapter 3)

The systems at stake may be dynamic computer programs as much as evolving organisms, neural networks, the physics of disordered and complex systems, ecosystems, or immune networks (Atlan and Cohen 1989). The
bridging of information theory and biology, a form of reductionism to which
we will return, allows for the elaboration of self-organizing systems across a
series of different realms of inquiry (though Atlan refuses any identity
between such realms, as a problematic kind of reductionism; Atlan 1994a:
72). In any case, at the broadest level, if order can emerge from noise thanks
to a dynamic it develops, then we can begin to approach how matter, in self-
organizing, can make possible forms that would seem otherwise irreducible
to it—forms such as life and thought.

Neural networks, and more generally automata networks, exhibit self-organizing
properties of different types. In particular, large, random Boolean networks can be
studied as models of functional self-organization in which both structure and function emerge without an explicit program. A network microstructure, together with
its initial conditions, is set up randomly. Then, under some conditions of connec-
tivity, network dynamics almost always lead to a robust attractor with a
macroscopic spatiotemporal structure characterized by a division into subnets,
some of which are stable, while others oscillate with a relatively short, nontrivial
limit cycle. Furthermore, it can be shown that the network, after having reached
its attractor, behaves like *a machine to recognize patterns on the basis of self-generated
criteria*. After the network has been stabilized, it reacts to incoming temporal
binary sequences by changing its state differently according to the temporal pattern of the sequence, thus defining a class of patterns that are "recognized."
Therefore, what emerges from the dynamics of the network is not only a macroscopic structure but also some functions in the form of classification procedures
capable of differentiating between classes of input sequences. This work shows
that networks are able to organize themselves so that they *behave in a meaningful
fashion*—at least to our eyes as human observers—*without having been programmed
to do so*. This kind of functional self-organization can serve as a generic model to
understand how the emergence of function and meaning can take place in nature
without having to resort to planning that would involve some sort of conscious
intentional design. This is what I have called . . . self-organization in a *strong* sense.
Self-organization in a *weak* sense differs in that the origin of meaning, as a goal to

be achieved, is set up from outside by a designer, as for learning machines. A third kind of self-organization, in an even stronger sense, can be called "intentional" if the emerging behavior not only appears to be a meaningful function to the observer's eyes but also is known to be so by the system itself, capable of self-observation. (Atlan 1994a: 71–72)

In other words, at stake in these self-organizing processes is not only an emergence of complex forms out of seemingly meaningless matter that does not postulate either a necessary *creatio ex nihilo*, a *creatio continua*, or even an external impulse and movement to aid matter in its organization; at stake is (perhaps above all) the creation of *meaning* and *intentionality*. In his later work, finding Shannon's probabilistic information theory insufficient to address the problem of meaning, Atlan would turn toward the approach and language of Boolean networks. For him, central to the development of an adequate theory of self-organization was the question of how complex systems offer evidence not merely of complexification but of the development of certain sorts of internal discernment and agency. Meaning would emerge thanks to nature itself: not because of any animal or human agency, but because the emergence of complex systems out of simple elements would of itself necessitate the systems' "choices" between, say, useful and useless changes occurring within them (not only in the case of *intentional* self-organization, but also in that of *strong* self-organization, where the system would not be consciously choosing but would do so nonetheless).

Atlan emphasizes these questions of meaning and intentionality. In his essay "The Self-Creation of Meaning," he writes:

The creation of meaning in natural systems is one of the most difficult problems both in linguistics and in biology. It is at the root of the known difficulties in simulating natural languages, where apparently an unlimited number of new meanings can be created and understood, although no syntactic structures are sufficient to program their appearance in a predictable way. In biology, meaning is not experienced from within, as in our process of understanding human natural languages. However, biological information has a meaning that is manifest in physiological functions: for example, the meaning of genetic information is to be looked for (and sometimes found) in phenotypic features, both morphological and functional. (Atlan 1987a: 563)

Indeed, throughout much of his career Atlan has sought to offer models for this self-creation of meaning as well as approaches to the problem of intentionality. (His arguments about natural language and frequent discussions of

Wittgenstein also seek to address this problem, though from a specifically philosophical perspective). Because "the appreciation of meaning and intentionality takes place at the interface between a self-organizing system and its environment," Atlan's emphasis on the presence of self-organizing behavior both in evolution and in cognition aims precisely to understand the conditions of possibility for subjectivity and meaning. The problem of intentionality—Atlan divides self-organizing systems between weak (i.e., receiving external support), strong, and intentional (i.e., not just strong and thus unsupported from the outside but capable of intentionality)—quickly comes to concern networks capable of self-observation (Chapter 7).

Central to self-observation are the chances of *memory* and the subject's perception and subsumption of them. Atlan notes that "memory devices keeping record of previous histories of such interactions are essential ingredients for models of what appears to us as both conscious and unconscious intentional behavior" (Chapter 2). In other words, memory is a, if not *the*, fundamental determinant of consciousness and intentionality:

> Conscious will and intentionality should be analyzed as secondary to the interaction between two different primary processes: consciousness as memory of the past on the one hand, and unconscious self-organizing processes on the other. Basically, consciousness would be mere recording, storage, and retrieving of past events, whereas innovation and adaptation to newness with the creation of new structures and functions would be produced by unconscious self-organizing processes. On top of these two basic primary processes, and secondary to their interactions, conscious will and creative intentionality would be interpretations of memorized products of previous self-organization, in such a way that previous unplanned consequences of actions are transformed into possible causes for the repetition of similar—though not necessarily identical—sequences of actions. (Atlan 2000: 135–36)

To the extent that the problem of consciousness and the biological function of un- or nonconscious self-organizing processes that occur in the body with growth and learning (i.e., with adaptation to, and efforts toward the mastery of, the external milieu), memory comes to serve as the biological grounding of consciousness. This is not a static ground, however: self-organization guarantees that consciousness retains a dynamic character that relates this ground to the world it needs to adapt to and aims to control—in other words, its *intentionality*. Memory becomes the ground of a being's sense of its world and its ties to the world.

Concepts of Modern Biology and Their Philosophical Implications

Atlan's writings are filled with references to the history of physics, chemistry, and the life sciences, and it would not be an exaggeration to claim that, in his thought, self-organization has its own conceptual history, a history with direct implications for the philosophical claims made for it. Atlan's critical treatment of DNA, complexity, and biotechnology is firmly grounded in a theoretical understanding of their conceptual and cultural history and is pertinent in part thanks to the remarkable extent to which that history is tied to the historical rise of the major concepts and problems of biology. His critique of the genetic paradigm—the *tout génétique*, or "everything is genetic"—attempts both to interpret biology's development in the last fifty years and to rethink more traditional and long-term problems of *mechanism* (since Descartes), *finalism* (particularly in Kant), and *vitalism*, as well as their historical and contemporary stakes. His discussion of the artificial womb and cloning similarly speaks not only to contemporary biotechnological possibilities but also to the conceptual and cultural traditions surrounding (and, for better or for worse, frequently framing) these possibilities. The very term *self-organization* reaches back to questions of the organization of the living (back to Cuvier, if not to Aristotle), of evolution and mutation, not to mention the question of *who* does the organizing. Similarly, Atlan's thoughts about the traditions of empiricism and determinism motivate a thinking of the links between the theory of science and its history, as well as his approach to reductionism and the "underdetermination of theories by facts." In these we begin to glimpse a theory not only of modernity but also of humans' relations to their own being and world, amid the efforts of human scientific inquiry.

Atlan's understanding of modern science is organized around the notion that two scientific revolutions determine modernity: the first, in the seventeenth century, concerns physics, and the second, which has dominated the last half-century or so, concerns biology (Atlan 2003: 18). Concerning the former, Atlan highlights the relationship between physics and biology: either the objects of the latter were reducible to those of the former or they were so patently superior in their complexity as to suggest a rigid and absolute divide between the two sciences. This led to the rise of two competing attitudes concerning living organisms: *mechanism*, which postulated that organisms were merely complex machines whose exact laws would eventually be discovered by physics and mechanics, and *vitalism*, which postulated a vital principle or impetus thanks to which living beings differed from matter. In Atlan's genealogy, what granted vitalism force was its alliance with *finalism* (Chapter 7), the belief that nature has particular purposes and goals, that organs function to serve particular goals, and that organisms not only exhibit this existence of purposes and goals but carry in themselves internal purposes. All of these attitudes pose problems. Atlan objects to classical mechanism's crudeness, and he rejects both vitalism (roundly discredited by biology since the discovery of DNA) and finalism, finding in them some of the major dangers and misunderstandings of modern biological thought.[4] He premises his approach to self-organizing systems both on the lack of goals within nature and on the lack of external forces guaranteeing the direction or reliability of evolution and organismic functions:

> Most of the researchers who were Kant's contemporaries held that biology obeyed vitalist presuppositions, principles, or theories. Living beings could ontologically, or at least epistemologically, be distinguished from nonliving beings. Thus Kant could think living organisms through the internal purposiveness that they display and contrast them with others determined by causal mechanisms alone. This a priori purposiveness has practically disappeared from the discourse of contemporary biology. Vitalism is finished, and so is what so solidly grounded the difference between living beings and others. Every day, molecular biology shows us that organisms, far from obeying an internal purpose, are regulated by physicochemical mechanisms. Today, biology and the neurosciences reveal a continuity between the nonliving and the living, between the world without consciousness and the world of human consciousness. We have definitively left behind a period that could be described as "prebiological," when the existence of the soul cut the world in two, distinguishing animate beings from inanimate beings and man from every other living being. Now, the soul exists only for philosophers and poets. All previous visions, from Plato to Descartes, of the soul governing the body like a charioteer his chariot have collapsed. (Chapter 1)

By contrast, the second scientific revolution of the twentieth century—or rather, the two revolutions, in molecular biology in the 1960s and postgenomics, synthetic biology, and biocomplexity in the 1990s—sets up a decidedly *biological* period and forces a conceptual and philosophical reorientation on the grounds that biology can be reconciled with mechanics in a particular way. In "Noise as a Principle of Self-Organization," Atlan endorses the cybernetic revolution, as well as W. R. Ashby's claim that computers have

---

4. See also the account of mechanism in Atlan 1999b: 126–33.

allowed us to understand medium-complexity machines that come between organisms and basic machines (Chapter 3). Elsewhere, Atlan expresses his stance more precisely: what matters is not an adoption of modern mechanism *tout court*, but a recognition that the living, natural machine transforms itself in a fashion that is not available to artificial machines (Chapter 3). That is to say, Atlan objects to the strict and overly reductive mechanism that would identify the genetic code (DNA) with a *program* that is merely played out in the organism. By contrast, the alternative he favors is premised on self-organization, which allows for internal dynamism. He sees the mechanistic approach as out of date and tainted by scientific faults, both of which have gradually undermined its status as (according to Atlan) the central dogma of biology since the 1960s. Moreover, for him, it suffers from a conceptual muddle because it depends on a dated notion of the computer program and thereby reconstructs a kind of preformationism (the eighteenth-century theory that saw an organism as being wholly contained in the genetic material from which it emerged; Chapter 24) and is also bound to a neo-finalism of "mechanistic purposefulness" (Chapter 5). By contrast, Atlan works from the perspective of a new sort of epigenetics, which articulates DNA as a mixture of program and data,[5] which deploys itself both on its own and in response to its environment.

Importantly, at stake here is not only the conceptual development of modern biology but also the kind of mechanism and relation to information theory that it will develop. In his largely epistemological argument, Atlan fashions a history of biology that is divided between the central dogma of genetics (based on the hypothesis "one gene—one enzyme—one function or characteristic" and a univectoral flow of determination from the first to the last) and a more complex model, one that sees the organism as influenced by its environment and that is unwilling to equate code with program.[6] In his early work, Atlan already utilized self-organization to object to theories that treat the genome as a preestablished program:

> where the evolution of species is concerned, no mechanism is conceivable outside of those suggested by theories in which random events (chance mutations) are responsible for evolution toward greater complexity and greater richness of organization. Where the development and maturation of individuals is concerned, it is

5. See Chapter 5 and, for the rising public awareness of epigenetics and recent engagements with it, see Stephen S. Hall, "Beyond the Book of Life," in *Newsweek* (July 13, 2009).

6. Atlan already considers the problem of genetic coding in 1972: 74–79.

strongly possible that these mechanisms also play a non-negligible role, especially if one includes the phenomena of nondirected adaptive learning, where the individual adapts to a radically new situation for which is it difficult to call upon a preestablished program. In any case, applying the notion of a preestablished program to organisms is a debatable move, insofar as this concerns programs of "internal origin," created by the organisms themselves and modified in the course of their development. To the extent that the genome is produced from the outside (by its parents), one often compares it to a computer program, but this comparison seems to us to be totally abusive. If a cybernetic metaphor can be used to describe the role of the genome, that of data stored in memory [*memoire*] seems to us more adequate than that of program, since the latter implies mechanisms of regulation that are not present in the genome itself. Otherwise, one cannot avoid the paradox of a program that needs the products of its own execution in order to be read and executed in the first place. By contrast, theories of self-organization permit us to understand the logical nature of systems in which what fills the role of program modifies itself unceasingly, in ways that are not preestablished, as a result of "random" factors in the environment, which produce "errors" in the system. (Chapter 3)

The model of DNA as program is mechanistic in a traditional sense—it reduces the living organism to a prefabricated machine whose essence is contained in the gene (hence the accusation of preformationism Atlan levels at it). It is not, however, mechanistic in a sense that would allow for an immanent dynamism, or more specifically, it precludes any real self-organization.[7] In more recent work, Atlan has similarly emphasized how techniques in cloning, advances in epigenetics, and the utilization of embryonic stem cells have demonstrated the insufficiency of the "genetic paradigm." Phenomena of epigenetic heredity—including the three-dimensional structures of proteins (Chapter 5), how cell nuclei function in a growing embryo and how adult cell genomes can be used for cloning (as in the case of Dolly the sheep; Chapter 22), and techniques of reverse transcription and RNA splicing in human virology (Chapter 5)—demonstrate that the genetic paradigm is insufficient to account for the uniqueness of individual organisms, and thus to theorize the overall role of DNA. For Atlan, an emphasis on epigenetics—or rather, a rejection of the self-sufficiency of genetic code—is not vitalism: "life" exists neither in DNA nor outside of it (Chapters 1 and 24). Because mechanism and determinism act in relation to one another, epigenetics discredits the

---

7. In this regard, see Atlan's reference to Richard Lewontin in Chapter 24.

"program"-based mechanism Atlan sees as dominant and lays the ground for a research program at once biological, epistemological, and ontological.

Atlan's outline of a conceptual history of biology, hinging on the birth of molecular biology, the discovery of DNA, and the association with information theory, allows him to address a number of further problems. One concerns scientific reductionism, and another concerns what Atlan calls the *underdetermination of theories by facts.* The question of reductionism has been crucial to the philosophy of science since the rise of positivism in the nineteenth century. Traditionally, reductionism has been used to account for analogies drawn from one domain of inquiry and applied to another that amount to a transcription of the former onto the latter—analogies that explain psychology by reducing it to biology, for example, or—as in the early mechanist example—that reduce biology to physics. For Atlan, the stakes involved in reductionism are substantial. They concern the metaphysical premises of scientific experimentation, the disunity of scientific knowledge, and the status of reason (or, rather, of reasons; see Chapter 9 and, more generally, Atlan 1986, chap. 2). Atlan opposes the classical kind of reductionism, which imposes a materialist metaphysics that reduces biology and chemistry to physics, and celebrates the failure of new "strong" reductionisms (see the appendix to Chapter 9). Following the advent of double-helix biology, this problem has been transformed, largely because the new kind of reduction of biology to physics involves a profound transformation of physics by biology, particularly a transformation in the meanings of the terms (*program, code,* etc.) used by biology.[8] Atlan's work on self-organization, by contrast, concerns the emergence of systems more complex than the elements on which they are based without allowing a classical reduction of these systems to those elements. However, his understanding of emergence is mechanistic and in some sense even more reductionist than classical reductionism, since it is based on mechanisms of self-organization and not on properties of life, as was claimed by vitalist emergentism in the nineteenth century.

He proposes a distinction between weak and strong reductionism: weak reductionism is "indispensable to scientific practice," whereas strong is "the result of the belief in this axiom" (Atlan 1993: 44). That is to say, reduction is inescapable, indeed, a necessary tool in the movement between different scientific languages and methods (e.g., cybernetics and biology in Atlan's own

---

8. Jean-Pierre Dupuy, in Kritzman 2007: 394.

work). Denying all reductionism would amount to negating "the unity of an organism," he writes (ibid. 38); however, the practice of systematizing reductionism quickly amounts to a "metascientific philosophy" that envisions a rigid materialism and denies whatever it cannot reduce (in other words, that denies the very possibility of complexity; ibid. 43). Though one must not systematize reductionism, weak reductionism is of considerable importance:

> this pragmatic reductionism circumscribes science's domain of legitimacy; it indicates the limits of the scientific process, which can progress only by *forcing itself* to be reductionist, only by "playing the reductionist game," whereas "believing in it" would be evidence of great naïveté—the naïveté of believing in the objective truth, in some fashion or other, of the "fact" of the reducibility of the real to some unique (ultimate) reality, whether material or not, on the basis of scientific theories. (Chapter 9)

In other words, weak reductionism guards against the overreach of a scientific rationality convinced of its self-sufficiency and autonomy. Despite, or rather because, it paradoxically demands that science play a game without believing in it, it allows for the possibility of research that is not premised on restrictive and overly defined reductionist theoretical postulates. Most importantly (and to this problem we will return), it guards against a mystification of scientific reasoning.

"Underdetermination of theories by facts" concerns the relation between experimental findings or observations and the theoretical models that aim to explain them. Atlan's question here is epistemological: What do we do when we have findings that can be explained by very different models? On what basis is it possible to judge between such alternatives—and to what degree is it possible to be satisfied with a theory that cannot determine or predict, strictly and with far higher success than do its alternatives? Atlan here addresses much of twentieth-century philosophy of science. Versions of the problem of underdetermination have been present in analytic philosophy since Willard V. O. Quine's work in the 1950s, including his use of Pierre Duhem's 1906 *The Aim and Structure of Physical Theory* (Duhem 1914),[9] and in France since Duhem and especially in the wake of the epistemological thinking of Gaston Bachelard, Alexandre Koyré, and Georges Canguilhem, who influenced thinkers as different as Thomas Kuhn, François Jacob, and Michel Foucault. Bachelard's studies of a "scientific unconscious" that influences interpretations of experimental results, Canguilhem's discussion of the

---

9. See Quine 1960, and Duhem 1914.

relations between concept and life, and Koyré's efforts to show the dependence of early modern experimental results on metaphysical premises broadened the problem beyond Duhem's critique of positivism by asking where theories acquire their explanatory value and how the facts they establish or the observations they make interplay with nonscientific presuppositions to give theories their particular character and aim.[10] Atlan approaches the problem from a somewhat different perspective, noting that, in discussions of automata and neural networks, the complexity of the interpretive models required to explain observations can lead to their vastly outnumbering the basic elements being studied, and that a certain indetermination or apparent redundancy is inescapable. Thus Atlan argues that in such networks only a limited application of Ockham's razor is reasonable or useful, that *"an underdetermination of theories by facts does not necessarily imply a useless redundancy in the theory itself"* (Chapter 6). Put affirmatively: "The more complex and singular a phenomenon is, the more underdetermined any theory giving some account of this phenomenon will be" (Chapter 25). Such an affirmation comes to concern not only the capacity to engender adequate models but the very complexity of complexity, the possibility and meaning of norms, the limits of theoretical activity, and the production of scientific objects. Underdetermination further points toward an image of the status and goals of philosophy:

> Thus, contrary to the ideal of the neo-positivist philosophies that tried to imitate the logical-mathematical shape of physics, the role of philosophy is to speak of what cannot be formalized, to use a natural language with its metaphors, analogies, and all the vagueness that comes with them, yet without giving up rationality. This means not giving up on distinguishing good analogies from bad ones, enriching metaphors from misleading ones, the vagueness [*le vague en moins*] that conceals what should be said from the vagueness [*du vague en plus*] that stands for the potential of creation. For the sake of this new ideal of philosophy, one must envision how this philosophy based on natural language—nonformalized, but rational and rigorous nonetheless—distinguishes itself from scientific language, and does so without confounding itself with either mythology or ideology. (Chapter 25)

This joining of philosophy, biological research, and the history of science from the angle offered by self-organization further suggests that concepts cannot be understood in static or essentialist terms. Because one cannot

---

10. Atlan discusses Canguilhem in 1999b: 41–42.

appeal to science in general (or to particular scientific discoveries more specifically) in order to offer a single unifying interpretation or model of facts or observations, because definitions need to be flexible enough to operate across fields, and because self-organization undermines a sense that what is given is produced as such, it is in general impossible to offer concepts based on self-identity. Writing of the embryo, for example, in relation to advances in potential therapeutic applications of uses of stem cells, Atlan says:

> We are forced to consider new discoveries in biology that modify traditionally accepted definitions of life and human nature, along with the representations associated with them. Essentialist definitions must be abandoned and replaced by evolving definitions: what is not alive, for example, molecules, can be alive at some degree of organization; what was not human life has become human life in the process of evolution; what is not conscious can become conscious in the process of development, etc. (Atlan 2007: 3)

This is thus one way in which issues of reductionism and underdetermination direct Atlan toward one of the crucial openings in his scientific writing: not just to questions of the humanities but also to matters of mythology, religion, ethics, and Judaism. Understanding the reductionism that Atlan endorses allows us to grasp how central tenets of his thinking (e.g., determinism) are carefully joined to a forceful refusal of scientistic tendencies toward all-encompassing and thus mystical theories, yet also to a willingness to identify affinities and analogies between his own postulates and philosophical and religious approaches to similar problems. As he argues in "Mysticism and Rationality" (Chapter 16) and, more generally, in *Enlightenment to Enlightenment*, claims to reason and scientific rationality have often been linked to concepts of reason that do not much differ from mythology—both because of myth's inherent rationalities but also because of reason's own tendency to determine a whole world. At stake is the possibility of balancing different rationalities, with the goal of usefully linking—through differences rather than through analogies—recent neurobiological languages with philosophical and religious ones. Chief among these, in Atlan's work since 1980, is the thought of Baruch de Spinoza.

Spinoza

In "Is Science Inhuman?" (Chapter 1), Atlan presents his turn to Spinoza in terms of a discovery that finally made possible a reconciliation between the determinism he saw as ineluctable in scientific inquiry and the moral, political, and metaphysical concepts of freedom and agency that one experiences every day:

> My scientific experience led me to defend in theory the position of absolute determinism. The sheer fact of searching for causes implied positing determinism as a postulate. But at the very same time as I upheld this position, I fought to obtain concrete things, made certain decisions, chose one strategy over another. . . . A determinist position changed nothing in my everyday life, in the relationships I formed, including within my laboratory. In sum, I lived fully the contradiction between theory and experience. It was in this uncomfortable position that I discovered the philosophies of absolute determinism: first the Stoics, then Spinoza. It thus became possible to reconcile these two experiences . . . (Chapter 1)

Out of all the philosophers with whom Atlan engages, Spinoza has certainly exercised the greatest influence on his thought—indeed, Spinoza's *Ethics* offers a conceptual system that addresses not only the paradox of freedom within determination but, just as importantly, the status of nature and the mind.[11] In him can be found the philosophical terms of Atlan's refusal of finalism and also of his support for a new and different kind of "mechanism," opposed to the paradigm of the genetic program and open to continual and disordered self-transformation.

Atlan works with Spinoza in two main areas of his writing. The first involves his ideas of complexity and emergence in self-organization, which parallel Spinoza's rendition of nature as at once *natura naturans* ("naturing nature") and *natura naturata* ("natured nature"). The main question here for Atlan is the capacity of nature to operate without either external impetus or internal purpose: "From a philosophical point of view, [the mechanistic experience of emergence] is more akin to the Spinozist notion of *causa sui*, cause of itself, used to describe the power of Nature under its two aspects of *Natura naturans* and *Natura naturata*, producing in itself an infinite multiplicity of forms and beings" (Atlan 2003: 19). The dual character of Spinoza's Nature mirrors self-organization's postulation of nature as at once ordered and dynamic; his monism erases the need for an impetus or force that would be external to matter and would organize it from without. Atlan is concerned to identify self-organization with a specific sort of mechanism: Spinoza, he

---

11. Atlan is careful not to grant either Spinoza's thought or his own scientific interests priority over one another. While he seeks support in each for interpretations involving the other, he does not suggest that either has explicative force for the other.

writes, cannot be understood as a vitalist, because there is no external motivation for nature's actions, and he eludes the finalist problem by allowing for a
dynamism and, more importantly an *internal causality* in nature that is
unbound by the ordered states we usually experience. *Natura naturans* can
thus be understood as nature in the process of self-organization; *natura naturata*, "natured nature," reflects these ordered states. The mechanist experience of emergence thus allows the subject to perceive nature as in some way
ordered, even while preventing this very subject from claiming any sort of
control or complete understanding of nature, or from pretending that the
order experienced is in any sense static or given:

> [Spinoza's] cause of itself, *causa sui*, which pertains to substance, is distributed in
> the modes through their essences or *conatus*, although the modes are produced by
> one another, come into existence, and are destroyed in their infinite chain of effi-
> cient causes. Given such a notion of immanent causality, evolution can be seen as
> the unfolding of a dynamic system or a process of complexification and self-organi-
> zation of matter, produced as a necessary outcome of the laws of physics and
> chemistry. In this process, new species come into existence one after the other as
> effects of mutations with stabilizing conditions working as their efficient causes,
> whereas their particular organizations are instances of the whole process. This
> view of evolution is compatible with the idea of a dynamic evolutionary landscape
> with peaks of local stability. (Chapter 11)

The *conatus* here implies a "presence of self-transforming infinite power in
every finite thing"—the possibility and presence of self-organization across
different organisms and at different levels of organization in nature.

The second subsumption of Spinoza occurs when Atlan brings together
neurophysiology and the mind/body problem, and it involves the reliability of
our notions of freedom and agency. Atlan refers to Spinoza's understanding of
the mind as the *idea* of the body, as nothing external to it, merely a different
attribute. Here he finds a central parallel to his writing on self-organization as
a process that determines living and knowing, a dynamic immanence capable
of accounting for the difference between mind and body without at all separating the two. He frequently quotes the second proposition of part 3 of the
Ethics: "The body cannot determine the mind to think, nor can the mind determine the body to motion or rest, or to anything else (if there is anything
else)" (pt. 3, prop. 2; 2002a: 279), arguing that Spinoza's mind/body monism
is essential for an understanding of body and consciousness through and thanks
to memory, as well as for the establishment of meaning and intentionality in a

biological system that is self-generated and self-generating without any external spark or force (Chapter 12). At stake here is at once the grounding of consciousness in memory and desire, the establishment of mental and bodily events as identical and expressed in different aspects of human being, and the very possibility of meaning and decision: How can decisions and free will appear if their author is matter? If they do so through self-organization, how can we explain away the presumed independence of thought? What meaning could they, and their freedom, have?

In this argument, Atlan's main targets in his turn to and citation of Spinoza are Kant and, in a less direct sense, Edmund Husserl. He aims three critiques at Kant. First, because he understands thought to be a self-organizing process founded on memory, Atlan criticizes and rejects Kant's dependence on a mind/body dualism that can be read as a relationship in which the mind influences or commands the body's movements: "Today, it is in relation to biological discoveries that Kant's philosophy has been thrown off balance" (Chapter 1). Second, Atlan targets Kant's understanding of nature— especially his finalism, which imbues nature with an internal purposiveness (see Chapters 1 and 16).[12] And third, Kant's understanding of free will is incommensurate with the determinism—the absolute determinism—that emerges from physical biology and information theory.

As for Husserl, Atlan refuses his pure ego and prioritization of consciousness and intentionality over the natural sciences as basic to our horizon of apprehension and our understanding of nature. As a result of his immanent approach, Atlan finds no place for such a separate consciousness or an intention that stands on its own apart from the world of matter and life, and only then engages with it. Atlan writes of phenomenology:

> Husserlian and post-Husserlian phenomenology constitutes a powerful attempt to understand what it is to understand and think, based on analysis of the flows of our consciousness as the origin of all meanings. However, even for Husserl, the ascendancy of vitalism made the idea of physical biology impossible, whereas, this is precisely the type of biology that we practice today. That is why we must consider a symmetrical relationship between *two kinds* of reductionism. On the one hand, the *phenomenological reduction*, Husserl's famous *epochē*, brackets the "naturalistic attitude," that is, puts aside the natural sciences, to be explained and founded later by transcendental phenomenology (Husserl 1947, 1950). On the other hand, the *physico-chemical, materialist reduction* brackets consciousness, to be explained later by the natural sciences. (Chapter 2)

---

12. See Kant 2007, §§5–8 of the introduction.

For Atlan, phenomenology's postulate of a pure ego is neither necessary nor helpful in understanding the operations of consciousness and indeed disregards the very possibility of thought's emergence in evolution. The problem with the transcendental reduction is that it relies on a separation between the biological conditions for consciousness and the operation of consciousness and its perception of the world. This grants consciousness autonomy and gives an inferior epistemological status to biological processes, thus raising the twin specters of a vitalist interpretation of the mind (targeting the body and the physical world) and a subjectivist interpretation and reduction of the reach and meaning of biology.

To these thinkers, Spinoza offers a striking counter, a conception of nature and freedom that surprises us by its modernity and its compatibility with biological research. Atlan's turn to Spinoza echoes the renewed attention paid to him in both analytic and continental thought in recent years. Atlan converses chiefly with the analytic and postanalytic tradition, in whose discussions of language and mind he is particularly well versed; he cites Hilary Putnam, Jerry Fodor, and G. E. M. Anscombe (Chapter 12) and takes on Donald Davidson for his rapprochement between Spinoza and Kant (Chapter 11). But he does so from a somewhat more complex perspective than that offered by analytic philosophy and in some respects engages in bridging between it and concerns in European thought. This is evident less in his occasional alliance with Antonio Damasio than in his echoing of postwar French philosophy's resolute suspicion of Kantianism. Though Atlan is far from allied with the readings and uses of Spinoza by Gilles Deleuze or by Michael Hardt and Antonio Negri, through Spinoza Atlan engages with priorities of late-twentieth-century continental thought.[13] Specifically, Spinoza allows Atlan to argue for a decided antianthropocentrism and antisubjectivism, which would deny human authorship of actions while admitting that an illusion of free will exists in our experience of them (Chapter 1). Here we find the reasons behind his engagement with Foucault (Chapter 28) concerning the transformation of man, as well as questions concerning the status of freedom.

---

13. Recent debates on Spinoza have raged along a series of axes. For a few of them, see: on Spinozism and the history of the Enlightenment, Israel 2001, 2009, critiqued by Moyn 2010 and Lilti 2009; for contemporary political appropriations of Spinoza, Hardt and Negri 2000, and also Connolly 2002; for a Spinozist neurology, Damasio 2003. For an excellent account of the importance of Spinoza in twentieth-century French thought, including his paramount influence on the oeuvres of Gilles Deleuze and Louis Althusser, see Peden 2009.

The problem of freedom thus poses the third and crucial question where Spinoza offers Atlan a philosophical hand. Atlan cites Spinoza's definition 7 of part 1 of the *Ethics*: "That thing is said to be free which exists solely from the necessity of its own nature, and is determined to action by itself alone" (2002a: 217). Atlan treats determinism as defining human experience, when he writes that:

> We must begin by accepting what we encounter every day in our science of determinisms: that is, that our subjective consciousness of free choice is increasingly refuted by our objective knowledge of causes and of the impersonal laws that determine these choices and show that they are not free as we believe. Instead of first positing our daily and affective experience of free will and then trying to accommodate it to our cognitive experience—that is, to our wider and wider understandings of mechanisms—let us do the opposite. Let us forget our experiences of free will for a moment and start from the postulate that today we know only part of an infinity of determinisms, and let us postulate a totally determined world. In other words, let us stop debating the gaps in determinism and assume that there no longer are any. (Chapter 1)

He appeals to determinism precisely in the sense allowed by Spinoza—one that does not contrast determinism and freedom, indeed that interprets classical free will as an illusion, given that the subject experiencing it ignores the reasons behind it. At stake then, as per Spinoza's definition 7, is the freedom that Atlan calls "not being determined by anything other than one's own law" (Chapter 1), a freedom that thoroughly depends on knowledge of causes or determinisms, a freedom to which one aspires without ever being able to achieve it:

> Our subjective experience of voluntary acts—what Spinoza designates as what *we* call the decree of free will from the viewpoint of the attribute of thinking—is not more illusory than our objective experience of the physiological mechanisms that we discover in these acts. It is a consequence of our ignorance of a part—a very large part—of their determinations. And we have no means *of acting as if* this ignorance did not exist; *we have no means of knowing what our behavior would be like if we had a total knowledge of the determinisms in nature.* (Chapter 25)

Ignorance plays an important part in this theory of freedom. While it demonstrates a profound lack of freedom and allows the persistence of a grand delusion of free will, it also buttresses against the interpolation of grand schemes arising from specific scientific results—a tendency Atlan identifies among scientists as much as among laymen and that he targets as mystical in a pejorative sense:

Every scientific system rests on the postulate or faith in the ability of reason to disclose an order beneath the complexity and apparent chaos of our experience of the world. From this point of view, symbolic and interpretive thought goes as far as possible down the path of faith in the possibilities of reason: nothing is accepted as being devoid of meaning. Every phenomenon, even the most fortuitous and confused, every myth, even the strangest and most enigmatic, finds an explanation that renders it rational by means of a change of level—the symbolic explanation—in which reason appears not only in what is said overtly but also symbolically, "between quotation marks." (Chapter 16)

It is when reason is charged with seeking to offer complete explanations of the complexity and chaos confronting science (and because mystical systems of thought also convey a strong sense of rationality and are thus by no means simply opposed to reason) that one must emphasize the limits of the human capacity to know and believe. Spinoza offers not only a philosophical way of understanding the possibility and hopes of freedom (setting up a crucial link between freedom and happiness), but a way of guarding against easy pretenses to having achieved it (Atlan 2003: 23).

This understanding of freedom as tied to perfect knowledge yet also bound up with our ignorance of total determination offers three further openings into Atlan's ethical and anthropological thought: his treatment of traditions of Judaism and its continuing importance and pertinence; his understanding of ethics, especially an ethics of life; and his treatment of the motif of the creation of life, in both historical and biotechnological terms.

On Judaism, Mysticism, and Mythology

Atlan's approach to self-organization is not scientistic; he does not seek to ground in biological premises the study of cultural, philosophical, and religious traditions. Nonetheless, he does not hesitate to offer arguments concerning the autonomy and legitimacy of philosophical and ethical inquiry, as well as the independence of different kinds of rationality. The biological antianthropocentrism of self-organization does not make totalizing claims about the forms and possibilities of culture—rather, it allows for their difference and for their own forms of order and rationality. This openness is clear in Atlan's frequent engagement with Wittgenstein's thought (particularly about natural language, intentional descriptions, and the role of philosophy). A far more complex case involves Atlan's extensive studies of Judaism and the

rationalities of myth: Judaism allows Atlan a space in which to extend principal concerns of his scientific research into a humanistic arena, a space whose
legends and problems can be brought into conversation with those of a particular scientific inquiry he undertakes, while remaining a different realm of
human inquiry.

Before tackling Atlan's approach to the Kabbalah, it is useful to recall his
understanding of mysticism, which has two central elements. First is his frequent critique, in *Enlightenment to Enlightenment*, of totalizing and mystical
trends in scientific thought—notably the trend, born of "strong reductionism," to attempt a complete explanation of life while ignoring metaphysical
elements of the concept or mistakenly identifying them with experimental
results that favor the reductionist's own position. Second is his attention to
Jewish mystical thought and its particular rationality, one that need be neither at odds nor in agreement with modern scientific thought. Mystical traditions "do not stop at the ecstatic experience of the ineffable but call on the
intellect and on discursive reason. They did not wait for the development of
Western science and philosophy to make use of the human rational faculties,
which were nourished by them and which carried them forward and renewed
them over the centuries" (Chapter 16). As he writes in *Enlightenment to
Enlightenment*—in passages that could retrospectively be read in conversation
with post–World War II thinkers such as Adorno, Foucault, and Cavell—this
two-sided relationship between rationality and mysticism is essential to an
understanding of either concept. The persistence of similar questions across
different mystical traditions—questions concerning reality, finitude, individuality, and so on—as well as the effort made in each of them to divulge an
order of the world that can be commensurate with sense (Chapter 16).
Among these traditions, Atlan works in particular on Jewish myth, the Kabbalah coming to serve as a realm that can be approached in a naturalist and
rational fashion, not as a spiritualist tradition valuable for its extra-rational
character.

In his autobiographical essay "From French Algeria to Jerusalem," Atlan
writes that his early upbringing was secular and indeed involved little reference to Judaism; it was the 1940 anti-Semitic laws barring Jewish children
from school in Vichy France and its possessions that made of him "a perfect
illustration of Jean-Paul Sartre's thesis that 'the anti-Semite is the one who
produces the Jew'" (Atlan 2008: 340). Following the liberation, Atlan studied

for a year at the École Gilbert Bloch in Orsay (a school founded by Robert Gamzon for the purpose of giving Jewish adolescents a space to consider their identity and heritage, as well as their status as Jews, after the experience of occupation and the Shoah). There he began a life-long engagement with the Hebrew Bible. Atlan writes that it was thanks to his studies at the École Gilbert Bloch that he began to work on the rationality of Jewish mysticism:

> I realized that, if we study these texts not as mere expressions of more or less dogmatic religious beliefs based on articles of faith, then we can find in them a kind of formal rationality associated with their mythical contents. Thus, in the end, these teachings appeared to us, in the light of our twentieth-century critical experience of reason, more rational than most texts of Jewish philosophy and theology, including Maimonidian ones, classically supposed to express the rational path in Judaism, as opposed to so-called Jewish mysticism. . . . Rather than an opposition between rational and irrational trends in Judaism, we are dealing here with an opposition between Jewish Aristotelian theology and other philosophical schools more related to the Stoics and Neoplatonists. (Ibid., 342)

Atlan deploys his reading of Hebrew texts, notably of the speculative ("rationalist") Kabbalah, throughout his work. "Israel in Question," for example, Chapter 18 of the present volume, is an account of Israel following the destruction of the Temple. It dates to his 1979 *Between Crystal and Smoke*, from which we also take Chapter 3, "Noise as a Principle of Self-Organization." This aspect of his work culminates in *The Sparks of Randomness*, an interdisciplinary reading of Jewish myth and law in which he brings these constantly into conversation with biological theory and contemporary biotechnological and ethical debates. At the core of this work is the notion of a seminal reason or "spermatic knowledge," which suggests a rethinking of the relations between life and knowledge. It refers to the Jewish legend that Adam and Eve were separated for 130 years between the birth of Cain and that of Seth (Genesis 4:25, 5:3). During that time, according to legend, Adam emitted drops of semen, whether thanks to commerce with demons or whether due to his involuntary nightly torment, that became demons or spirits. These drops were called *nitsoutsot keri*, translated literally as "sparks of randomness." Atlan treats this legend

> not as a story with a moral, but as an album of images of various and contrasting aspects of the human condition, associated in particular with the protracted period of childhood and maturation that follows birth and with the long interval between sexual maturity and intellectual and emotional maturity: the Tree of Knowledge is

assimilated before the Tree of Life (although we can easily imagine how the inverse chronology might have had better consequences). (Chapter 14)

He invokes the "ambivalence" of the "sparks of randomness" legend to discuss the novelty and chance involved in birth and in the creation of a new being, showing that the radical determinism he advocates in biology, in his reading of Spinoza, and, as we shall see, in his reading of the Kabbalah (which he calls "expanded Spinozism"; Atlan 1999b: 335–46) does not simply deny individuality or hopes for autonomy. Seth is born in Adam's likeness, in his image (Genesis 5:3), a status that rabbinical commentaries accord neither to Cain nor to Abel (let alone the demons born of Adam's semen drops; Atlan 199b 50). Atlan extrapolates from this the chance element involved in conception and birth, and expands further into a broader interpretation of the ethical considerations surrounding the episode. In Atlan's retelling, Seth comes to hold a privileged status: while his birth in Adam's likeness does not mean that he recaptures the unity and transparency lost with the fall, he reopens humanity to the possibility of such a recapturing. At the same time, he denotes the expulsion of the serpent's influence—and therefore the potential of a human innocence separated from an external reality dominated by the demons similarly expelled or generated from Adam's drops of semen (Atlan 1999b: 107–8).

This subject/world contrast marks the construction of a flawed subjective interiority distinct from the "animal knowledge of the serpent," a knowledge increasingly separate from life and eventually related to it (and to its possible redemption) only through the five books of the Torah (Atlan 1999b: 54).

Why emphasize this legend, though, and this particular reading? Atlan has two further interrelated problems in mind: first, that of chance or randomness; second, that of the motif of the creation, manufacture, or fabrication of life. Through these he deploys not a classical narrative of damnation and redemption (the narrative that he traces across the traditions he examines) but an anthropological questioning concerning man's sense of his world as hostile to, and constantly overwhelming, his efforts to control it. By the same token, the legend points to man's perseverance in this world and his efforts at control: "What is the status of the randomness of birth, chance, and the ignorance of causes that we call 'destiny' in a world that we are increasingly able to control, where uncertainty itself seems to be programmed by probabilistic estimates of risk?" (Chapter 16).

Atlan is particularly concerned with the role granted to chance, especially in traditions attentive to deterministic explanations. (The French *hasard* encompasses both "chance" and "randomness.") The legend of the sparks of randomness leads to the question of possibility and determination in that it balances the figure of Seth, an offspring who successfully bears Adam's likeness, against the demons begotten of Adam's nocturnal emissions (Atlan 1999a: 107). As in his biological writings, chance and determinism are not contradictory alternatives. Atlan sides with traditions of rabbinical and philosophical commentary that emphasize determinism and the lack of control over human action.[14] Somewhat like the effects of noise in mediating or mobilizing a self-organizing process, chance mediates for humans the ineluctable order of the world set forth by God/nature, which is unintelligible to them. In an essay on "divine intrigues" or *'alilot* (Chapter 15), Atlan rereads classical references of rabbinical and philosophical commentary that are tied closely to the question of free agency. To Maimonidean resolution, which highlights the possibility of free will (a resolution that Atlan believes can lead to refusing the possibility of a kind of reason offered by kabbalist mysticism), Atlan contrasts other schools of rabbinical interpretation, notably those of the Saragossa philosopher and rabbi Hasdai Crescas, who lived at the turn of the fifteenth century and is best known for his book *Or Adonai* (*The Light of the Lord*; Chapter 1), and Shlomo Eliaschow (Chapter 15). In a reading of the claim of *Pirkei Avot* (*Ethics of the Fathers*) "Everything is foreseen and the possibility of choice is granted" (3.15), Atlan, with Crescas and Eliaschow, highlights the deterministic first part of this sentence, which focuses on the "conduct of the world, which seems to play with men, who must believe that they are guilty of their crimes, even though all their doings are the results of an eternal decision from the Infinite" (Atlan 2008: 351). Free will then becomes an illusory, subjective experience and a trick or intrigue that the world's forces play with humans in cunning demonstrations of their responsibility for the good and bad they are determined to do (Chapter 15).

The second element Atlan evokes in his reading of the Adamic myth concerns creation—the role played by (or granted to) human beings in the creation of life, the discourses involved in understanding such creation, and the kind of control over the future and over nature that creation promises. Atlan here addresses two principal elements: the relation between soul and body in the

14. Significantly, these traditions at once locate Spinoza's thought in a broader context and open out from Spinoza's reading of the Torah in the *Theologico-Political Treatise* toward a broader tradition of commentary that similarly emphasizes the idea of human choices as foreordained. For Atlan's reading of Spinoza's treatment of philosophical versus religious texts, see Atlan 1999b: 26.

creation of man in Genesis (Chapter 17) and the tradition of the Golem. In the first, he seeks to emphasize that "the soul was not always 'spiritual' in the sense of incorporeal" (Chapter 15), that it can be understood instead as enveloping the body and not, in a Neoplatonist sense, as opposed to it. Here, Atlan explicitly links readings of Genesis with his blending of neurobiological insights and Spinoza, both to demonstrate the noncontradiction between these realms of inquiry and to open a discussion of the potential stakes and openness to interpretation of seemingly internalist and "dated" systems of thought.

Atlan's treatment of the figure of the Golem opens his study of Judaism to bioethical concerns that have arisen with the development of technologies for cloning, stem-cell research, and control over conception and procreation—of biotechnology more generally. As Atlan notes, biotechnology has frequently led bioethicists to engage with traditions of creation (Chapter 26): "Here we propose . . . to analyze origin myths, whose anachronistic juxtaposition with the new situations created by these technologies can yield new insights. These make it possible to project diverse (and perhaps contradictory) meanings onto the moral and legal positions we are tempted to take in a given situation" (Chapter 20). For Atlan, the Golem poses a particularly forceful reference (in comparison with other literatures of a new man or a man-playing-God—most famously that of Faust) because its creation, regardless of the Golem's destructive consequences, does not result in summary judgments about the futility or destructiveness of human creativity. On the contrary, "the creation of a golem, far from being sacrilegious, was seen as the fulfillment, by means of the secrets of the Torah, of a long and difficult ascent toward wisdom and holiness" (Chapter 20), thus confirming for commentators both the capacity of the Torah to raise human knowledge and consciousness and the human failure to perfect and fully control knowledge and its promise. The very ambiguity of the Golem, Atlan notes, continues to haunt ethical debates today—in matters of gender roles (Chapters 21 and 26), in confusion and worry about the kind of "construct" that a new man created by man alone would be (Chapter 22), and in the very status of the human (Chapters 22 and 23).

Practices, Possibilities, and Ethics

As they define a questioning of the new man, the problems of knowledge and possibility become crucial for Atlan's thinking of the intricate and ever-changing relationship forged between human biology, scientific knowledge (and practice), and everyday life. The continual reworking of this relationship brings us to the final section of our introduction and to an important conceptual domain—not to mention popular moniker—for moral philosophical concerns that reside within science and human experience: bioethics. Over the past several decades, the term *bioethics* has come to signify a seemingly novel set of disciplinary practices involving biology, philosophy, psychology, theology, and jurisprudence—and with this new discipline comes the assumption that equally new ethical concerns (as well as phantasms of thought; Chapter 8) have arisen out of and in the face of scientific advancement (see Bosk 1999). As a number of historians of science and medicine have suggested, what remains puzzling about this apparent renaissance of ethics as a means to gauge scientific value is the recognition that technical advancement brings with it increasingly difficult dilemmas, and thus bioethics as a discipline is hard put to extricate itself from the aura of future threat and the figure of crisis (Rosenberg 1999; see also Fox 1999). (As Alasdair MacIntyre has suggested, the problems and concerns that arise in bioethical debate seem always just beyond the reach of an answer.[15]) Atlan does not take lightly questions such as "Where do bioethical problems come from?" and "Where do they lead?" (Chapter 22). In his writings, however, whether or not moral philosophical challenges arise from (or run alongside) scientific and technological advancement, how biological theory is transmitted and taken up within social, political, and economic domains is an entirely human endeavor, science itself. Therefore, it is often hard to understand *bio-* in the term *bioethics*. Atlan states:

> [Bioethics] implies the existence of a new discipline derived from biology, rather like biophysics or biochemistry, and suggests that via bioethics it may be possible to find, within biology itself, the answers to the ethical problems posed by biology. But this is not at all the case. Biology poses problems of an ethical nature, but it does not solve them, even if it is clear that we cannot begin to tackle them without an understanding of their biological roots. Solutions will result from different ethical approaches that vary from person to person and culture to culture. These problems are so new that people belonging to the same tradition often respond spontaneously to them in totally different ways.[16]

Atlan refuses to assert a scientific morality, that is to say, an ethics that would rest entirely within science itself. He states: "In this field the progress of

---

15. See MacIntyre 1975.
16. Interview with Atlan, conducted by Geraldine Schimmel, in the *UNESCO Courier* (Schimmel and Atlan 1996).

science and technology has created problems—of legitimacy, of good and evil, of what is permitted and what is not—that did not exist before and that neither biology nor medicine alone can solve."[17] Nowhere do we see this more clearly than in the polemical debates about reproductive cloning and ectogenesis—birth following artificial and not human gestation.

Ectogenesis, as it applies to the human production and reproduction of life outside the uterus, and reproductive cloning through somatic cell nuclear transfer (the implantation of the nucleus of a somatic cell into an enucleated egg cell, which then multiplies, replicating the implanted rather than its own nucleus) are two technologies in which biological possibility has become (or is at present) socially untenable (Chapter 26). Even technologies that seek to "correct" certain conditions (i.e., infertility and dangerous pregnancy) have incurred societal wrath because they are imagined to denature life (see the discussion of the 1992 Clothier report in Chapter 24). Atlan writes:

> each of these two techniques constitutes the crossing of a qualitative threshold in the denaturalization of reproduction. Debates about the acceptability of crossing the threshold of asexual reproduction, seen in relation to techniques of medically assisted procreation already employed, will very probably change their nature in societies where the other threshold, that of ectogenesis, has already been crossed. Once ectogenesis were to become a normal mode of gestation that would make it possible to do without a mother's uterus in giving birth to a child, the question of what an embryo is and of its legal status would be posed in a new way, even more dramatically. (Chapter 8)

The problem here is that the nature of reproduction is seen as reflecting the nature of life as such. So long as the genome is treated as containing the "essence of life," manipulating it means destroying life and its supposed sacredness. The opposition to these technologies derives from what Atlan has described as a "fetishism of the gene" (Chapter 8), asserting that the genome contains all the instructions for the development and growth of the organism, now and in the future. Here, the diffusion of biological theory into a social (and ethical?) critique comes to absorb issues such as a woman's right to make choices about her own body and possible new models of kinship, as well as the value of biological life and how that value is determined. Atlan's own approach is more careful. On the one hand, he emphasizes the problems involved in speaking of "life" in an essentialist fashion. He repeatedly quotes Albert Szent-Györgyi's claim that "Life as such does not exist" (Chapters 1,

---

17. Ibid.

14, and 24), interpreting it to mean that "life does not exist as an explicative notion of organic properties . . . life does not exist as an object of biological research" (Chapter 1). This is also to say, in the context of bioethics, that any thinking of the living cannot maintain its rigor if it postulates life as an essentialist notion. On the other hand, he emphasizes the need (regardless of Szent-Györgyi's "good news") for localized and precise responses to problems of biomedical ethics: "This is why, if one is preoccupied with questions in biomedical ethics and, as we are here, with the precautions to be taken in this domain, it is necessary to try to define these precautions in a pragmatic fashion, to analyze each situation in its particularity—that is, for each disease, each technique, and so on to analyze the specific potentially dangerous or undesirable effects, without launching into notions as woolly as 'the essence of life'" (Chapter 24). Thus Atlan suggests that human ectogenesis poses ethical, social, and political problems very different from those of reproductive cloning (Chapter 8): the technology actualizes biological concerns in a way that involves much wider social issues.[18] Similarly, the essay on cloning included here (Chapter 22) centers on an elaboration of different practices and techniques that commonly fall under the moniker "cloning" and supports the Comité Consultatif's negative response to reproductive cloning of human beings (to be distinguished from the nonreproductive cloning of cells) as a biotechnological possibility on the grounds of an ethics of anticipation, claiming the problems of kinship and moral confusion that would result render cloning socially impossible at present.

An "ethics of anticipation": we use this term heuristically, in part to complement Atlan's explicit attention to ethics and moral judgment—an attention that begins with Spinoza's thought concerning pain and pleasure as the first and subjectively determinant level of ethics and extends to normative and meta-ethical concerns (Chapter 29). As Atlan repeatedly notes, what has come to dominate ethical questions is *possibility*: "what is not an embryo can become an embryo, what is not a human person can, under certain conditions, become a human person. The ethical debate . . . shifts to the conditions of such becomings" (Chapter 1). The ambiguity of biomedical ethics at present, for Atlan, concerns the question of addressing the relationship between things that are not (but might come to be) and things as they ought to be: in other words, not only the relation between things that are and things that are not, nor only that between things that are and those that ought to be. Rather,

18. See esp. Atlan's recent (2007) work with Mylène Botbol-Baum.

the effort to think and organize becomings, potentialities (Chapters 8, 23, and 28)—the effort to think the future—becomes pedagogically, scientifically, and ethically the basis of a persistent self-questioning grounded in a quasi-subjective experience and thought. Thanks to these ethical emphases, in his work Atlan is able to expose the analytical problems we face when solutions to moral philosophical systems fail to take hold in the everyday realities of individuals' lives.

❀

In conclusion, let us return to the practical aspects of organizing a volume that encompasses such a far-reaching body of work. The chapters of this book are arranged to cohere around substantive aspects of Atlan's work. It goes without saying that there remain topics and concerns beyond the scope of a single volume. This book should not be mistaken for a reductive or single theoretical perspective (or suggest a single norm). Writing about the different levels of ethics, Atlan states: "In particular, if a norm must be erected, rightly or wrongly, on the basis of a theory (either by deduction and predictive projection of what must be on the basis of what is or merely by indicating the constraints that limit the possibilities), then each theory will permit us to erect very different norms" (Chapter 25). In this volume, we hope readers will find not the limit but the opening of possibility.

ONE

Is Science Inhuman? An Essay on Free Necessity

(2002)

The inhuman concerns only the human species. Only human beings can be inhuman or be confronted with the inhuman. Minerals, vegetables, and animals can only be "nonhuman." And it is precisely because science is one of the most characteristic activities of the human species that the question of the human or inhuman character of its productions arises today. So, is science inhuman?

Since the invention of fire, science and technology have always at once fascinated and terrified, for they have increased men's power over nature and over one another, including in their inhumanity. But is science, as knowledge, fundamentally inhuman—through its search for mechanisms, causal explanations, and predictive laws, which eat away at the field of free will? Under the influence of a certain humanist tradition, man's humanity has long been associated with what has been believed to be the exercise of his free will. The affirmation of determinism to which scientific discoveries increasingly lead thus inevitably seems to affirm their inhumanity.

This question calls forth another, one that seems crucial and even urgent to me, yet that rarely attracts our attention. It arises today in a relatively new way: "What is our freedom?" Although this question has become unavoidable for me, it is difficult to remember at what precise moment I first asked it. Perhaps this is simply because there was no such precise moment. The reflections that follow are not the fruit of a sudden illumination but rather the result of a daily practice, biological research, and of philosophical reflection on that practice. When one spends a great deal of one's time searching for the causes of phenomena, of determinations, and sometimes even of laws, pursuing this ideal of scientific research day after day, the question cannot fail to arise. Even as I was devoting so much time to this research, in my private life I continued to behave like anyone else. I made plans for the future, I experienced conflicts; in short, I behaved like an agent who chooses what he wants to do. My scientific experience led me to defend in theory the position of absolute determinism. The sheer fact of searching for causes implied positing determinism as a postulate. But at the very same time as I upheld this position, I fought to obtain concrete things, made certain decisions, chose one strategy over another . . . Determinism changed nothing in my everyday life, in the relationships I formed, including within my laboratory. In sum, I lived fully the contradiction between theory and experience. It was in this uncomfortable position that I discovered the philosophies of absolute determinism: first the Stoics, then Spinoza. It thus became possible to reconcile these two experiences . . .

If inhumanity consists in demystifying as much as possible alienating passions and human illusions, including those that science itself has contributed to maintaining, then yes, science is inhuman. But if inhumanity consists in enslaving bodies and minds in suffering, powerlessness, and ignorance, then science, to the contrary, is an irreplaceable factor of humanity.

Until the second half of the twentieth century, we lived according to the idea that we were free agents, not only politically, but above all metaphysically. We decide freely to undertake certain actions, aimed toward ends that are just as freely chosen, according to our judgment of what is good or bad. It is always on the basis of this idea—"we are endowed with free will"—that we are judged responsible. This idea is at the root of moral philosophies, at least the best known and most common ones, among them the one that first established the foundations of our free condition: Kantian philosophy. Many thinkers have pointed out that Kant's philosophy has been compromised as a result of the development of the sciences. At the beginning of the century, his epistemology was examined closely in the light of then relatively recent developments in physics. The influence of an outmoded Newtonian physics was detected in his conception of time and space. But the upheaval we are currently witnessing is just as important, even more so. Today, it is in relation to biological discoveries that Kant's philosophy has been compromised, for a reason of the same order.

In order to gain an understanding of the current crisis, we must first take a look at recent advances in biology. Most of the researchers who were Kant's contemporaries held that biology obeyed vitalist presuppositions, principles, or theories. Living beings could ontologically, or at least epistemologically, be distinguished from nonliving beings. Thus Kant could think living organisms through the internal purposiveness they display and contrast them with others determined by causal mechanisms alone. This a priori, purposiveness, has practically disappeared from the discourse of contemporary biology. Vitalism is finished, and so is what so solidly grounded the difference between living beings and others. Every day, molecular biology shows us that organisms, far from obeying an internal purpose, are regulated by physicochemical mechanisms. Today, biology and the neurosciences reveal a continuity between the nonliving and the living, between the world without consciousness and the world of human consciousness. We have definitively left behind a period that could be described as "prebiological," when the existence of the soul cut the world in two, distinguishing animate beings from inanimate beings and man from every other living being. Now, the soul exists only for philosophers and poets. All previous visions, from Plato to Descartes, of the soul governing the body like a charioteer his chariot have collapsed.

Is this to say that all difference has vanished, that we no longer distinguish between a dog and a cloud? Quite obviously not. The difference remains, but its nature has changed. What separates dogs from clouds are the questions we ask about them. When we observe a cloud, we ask questions about structure, or perhaps about causality—how it turns into rain, for example—but never questions about function. Who would maintain that the function of a cloud is to make rain? In other words, we ask about function only when faced with living beings. And even then we no longer expect this function to explain the organism's structure. Just like what appeared to be internal purposiveness, function must be explained mechanistically. This is the task of molecular biology today. The difference is no longer a difference in nature, and in this sense one may speak of a continuity between the physical world and the living world.

The same laws apply; only the properties vary—a stone does not breathe, an amoeba does not think . . .

For a long time, the capacity for reproduction and the accompanying transmission of hereditary characteristics was considered the phenomenon most specific to life and was thought totally irreducible to physicochemical laws. Similarly, metabolism and the organism's capacity to adapt seemed proof of the existence of vital forces. Until the twentieth century, it was thought that within organisms there existed so-called organic molecules, which could be observed and analyzed but not synthesized, and that this secret contained the specificity of life. The biological revolution of the twentieth century has consisted in explaining, at least in rough form, behavior claimed to be characteristic of life on the basis of the physical and chemical properties of molecules whose structure is known and which can be artificially synthesized in the laboratory.[1] The first blow came at the beginning of the twentieth century with the synthesis of urea, a molecule of organic origin. Biochemistry continued to synthesize new organic molecules (proteins during the sixties), increasingly reducing the domain of vitalism. Today, the only specificity of life stems from the complexity of its organization and of the activities that accompany it. This activity, to which we refer in continuing to speak of functions, in no way corresponds to a finality. Radioactive substances also display activity, but they do not thereby have either function or purpose. The discovery of programmed cellular death—apoptosis—finally upset our certainties about life. We know today that certain cells have the capacity to kill themselves. Once again, it is not a matter of purpose, of a sort of consciousness these cells might have of their own life and death, but of enzymatic mechanisms latent in the cell that, when activated, bring about its destruction. This mechanism, far from being exceptional, characterizes almost all

---

1. From the synthesis of urea until the discovery of the genetic code in the sixties, those who thought they could explain the living being via the mechanist method stood in opposition to those who maintained that this could only ever be a partial success. The great discoveries of molecular biology took place during the sixties: the discovery of the molecular substratum of genes in the form of DNA, on the one hand, and that of mechanisms by which these genes direct protein synthesis, on the other. The physico-chemical nature of genes and the mechanism of protein synthesis (proteins being molecules only observed in living beings) were the two great classical problems of biology that the biochemical method had not yet succeeded in resolving, thus leaving some space for a nonmechanist conception of life. These discoveries put an end to vitalism.

living beings. The leaves of a tree would not fall from its branches were it not for the death of a certain number of cells located where the leaves are attached to the branches. Moreover, this cellular death plays a role in embryonic development. How do the forms of the hands and feet visible on an ultrasound develop into fingers and toes? Precisely by the death of a certain number of cells located in what are to become spaces. This process is very common and is found in the development of the central nervous system and the immune system, as well as elsewhere. By 1979, it could already be observed that life, which Xavier Bichat defined as "the ensemble of phenomena that resist death" (Bichat 1827: 10; trans. modified), is, rather, "the ensemble of phenomena capable of using death" (Atlan 1979: 278).

The biologist Albert Szent-Györgyi's expression "Life as such does not exist" seems to sum up the matter.[2] When he made this somewhat abrupt declaration, in all likelihood he was not doubting his daily experience. I do not think I betray him in thus clarifying his words: life does not exist as an explicative notion of organic properties. In other words, life does not exist as an object of biological research. This is exactly what François Jacob meant when he wrote: "One no longer studies life in the laboratory" (Jacob 1993: 299; trans. modified). The question "What is life?" no longer belongs to biologists. It may preoccupy philosophers or any person uneasy with all these discoveries, but it is no longer a biological question. This in no way changes our daily lived experience [*vécu*] and the enormous difference that remains for us between the life and death of someone dear to us. But this unwavering difference partakes of our lived experience, our language, and the exchanges we have with others, whether human beings or animals. Life is an undeniable experience, but never more than an experience.

The gap between scientific advances and our sense of daily existence thus becomes wider and wider. These discoveries point in a direction whose very possibility Kant rejected. He was convinced of the impossibility of a biological Newton one day explaining living beings purely mechanically. For Kant, all living beings, man first and foremost, express purposiveness. The human capacity to set one's own ends grounds the possibility of freedom. It was not, however, one Newton but rather an army of little Newtons who demolished finalism and discovered the mechanisms that explain the behavior of the simplest organisms, as well as our own. The mechanisms observed today at the

---

2. Albert Szent-Györgyi was one of the first great biochemists of the century and the discoverer of vitamin C.

endocrine level show us how certain of our behaviors, certain of our thoughts, and a fortiori certain of our sentiments and our passions are determined by biological phenomena of all sorts (not to mention the social, psychological, or linguistic mechanisms with which they interact).[3] The idea that we have the capacity to decide our actions freely is thus necessarily upset. When it is discovered that a decision has in fact been determined by a hormone imbalance, a genetic predisposition, or a social or cultural influence, this idea becomes very difficult to defend. Our increasing knowledge of physicochemical mechanisms inevitably leads to a conception of determinism that leaves little if no room for free will. It even leads us to consider our subjective and social experiences of free will as an illusion created by our imagination. In this way, biology seems to complete absolute determinism's conquest and, in consequence, to eliminate entirely our experience of freedom, conceived as the capacity for efficient free choice.

The crisis is without precedent. Someone who has not had the experience of scientific research will not be very sensitive to it. Such a person may be fascinated or horrified by the horizons opened by certain techniques, such as cloning, but will not thereby call his or her free will into question. By contrast, scientific researchers who combine their research with the experience of everyday life live this dilemma to the fullest. Through their work or interests, they experience the determinisms that our current knowledge reveals, all the while continuing to live and make choices in their daily lives. Thus they live the gap from within, divided between their cognitive experience and their affective experiences.

Confronted with this situation, a person can adopt one of two possible attitudes. The first advocates resistance. Yes, there does still exist a domain in which the subject is free and not determined: the famous Kantian supersensible realm. The argument wielded by those who resist is forceful: if we give up this position, everything collapses. Morality and responsibility crumble along with free will. And who would dare accept that there is no longer either morality or responsibility? It is thus necessary, in spite of everything, to affirm our free will. It is necessary to oppose all the discoveries—not just

---

3. We know, for example, that the amount of adrenaline in our bodies makes us more or less capable of resisting anger. Certain imbalances in sexual hormones can also lead to uncontrolled acts, even to crimes. Other factors, of course, play a role in performing an action, but today it is impossible to deny the influence of hormones. The efficacy of hormonal treatments to decrease libido offers one proof of this. Certain sexual criminals have themselves asked for this treatment on leaving prison out of fear of repeating their offenses.

scientific, but also psychic, psychoanalytic, and sociological—that lead to the death of the subject. According to these discoveries, what I believe I decide or will by myself as a free subject, master of my destiny and behavior, is determined by a set of factors I know little about. The free subject becomes an illusion and disappears, like the face of man in the sand described by Foucault (Foucault 1994: 387). These philosophies of suspicion—which I do not espouse[4]—are at the origin of a veritable social crisis. Everyday morality and the law itself are based on the existence of agents who are responsible for their acts. The gap continues to grow between what is socially agreed upon and the new knowledge we acquire.

From here it is but a step to seeing the development of knowledge and the sciences in particular as a menace to the moral and juridical stability of society. More science seems to mean less humanity. There are multiple responses to this increasing unease; the institution of ethical committees is probably one of the most convenient, if not most rational.[5]

But this habitual way of resisting resembles denial: closing one's eyes or refusing to draw the consequences of the accumulation of knowledge; continuing to believe in the efficacy of our free choices in order to continue being social and juridical agents. When this denegation is proclaimed, it becomes negation pure and simple. Science is then accused of draining the foundations of morality and society, and its place is taken by the celebration of certain forms of the irrational, within both institutionalized religions and sects.

Another danger—nihilism—lurks, though it does not necessarily lie where one would first think. It is present in a certain determinist attitude in an

---

4. Foucault is an heir of the philosophers of suspicion: Marx, Nietzsche, and Freud. All three contributed to the death of the subject by demonstrating that behavior thought to be free is in fact determined by: for Marx, social factors; for Nietzsche, biological factors; for Freud, unconscious factors. The Stoic or Spinozist philosophies, to which I am close, are precisely not philosophies of the death of the subject but, on the contrary, ways of thinking in which the subject is constructed through his or her own determinisms.

5. Science and technology today pose ethical problems that they cannot solve on their own. The usefulness of these committees is to create the conditions for the widest possible public debate on these questions, which sometimes have quite complicated technical aspects and are therefore poorly understood by the public—but which are not reducible to these aspects. When ethical questions are involved, it is unthinkable to leave the task of resolving these problems to scientific experts alone. The task of ethics committees is to get the debate started before it is brought to the public forum and, as sometimes happens, turned into law.

obvious way: If my actions are determined, then in the name of what principles could I be condemned? But is this danger, so coarse, greater than another, more insidious one hidden under the attitude of resistance, among those who, in the name of the defense of the subject, make free will into a supreme value? Perhaps in reaction to the revelation of all these mechanisms, a kind of morality of sovereign desire is often put into place. The spontaneity of our desire would be the truest expression of our freedom, and there would be no reason to oppose it. But what world would this be, if not a world of the war of all against all? Even if certain political barriers prevented actual war, it would already be a preoccupation. This is indeed a practical nihilism, which threatens to create a society in which I would not recognize any values other than my own desires. Perhaps I would respect the law that juxtaposes and adjusts the desires of all, but the only value that would guide me would be my own desire. A certain philosophy of the free subject thus arrives at a nihilism just as dangerous as the one we reach via a poorly understood determinism.

I refuse this position of resistance. Today it is difficult to prove that such a position is inadequate, since not all behavior has as yet been explained mechanically, and thus one can always dream that something will remain sheltered from the increasingly sophisticated advances of mechanism, which includes those advances that, through the calculation of probabilities, make use of statistics and have "domesticated" chance [*hasard*], as Nietzsche put it.[6] The trouble, nevertheless, is that the domain of free will continues to shrink. So why not instead assume the opposite attitude and fictively extend current discoveries? Indeed, I think it more economical to let everything go at once. Let us suppose that one day we will manage to explain mechanistically all the behavior and choices we feel are free. Would this be the end of morality, responsibility, and all social life? I claim not. It is possible to construct an existence and a philosophy that would be no less felicitous and moral. We must learn to reconsider how we are responsible for what we are and what we do, independent of a metaphysical belief in free will, by accepting that we cannot escape universal determinism and that we are determined

---

6. Chance is not at all in opposition to absolute determinism here. Through the calculation of probabilities, it becomes, on the contrary, a way of reducing our ignorance even in the explicit absence of knowledge. When we think that we are incapable of provisionally or definitively knowing causes, chance, formalized through the calculation of probabilities, allows us to reduce our ignorance.

to do as we do—even when we think we choose freely, and although we do not feel constrained or forced. To conceive this requires an intellectual procedure that is certainly somewhat difficult, but one that is far from new. Before us, the Stoics, the Epicureans, and more recently Spinoza have undertaken this exercise.[7] The philosophies of determinism, which see free will as an illusion linked to our lack of knowledge about the causes of our wills, never stopped searching for an ethics of responsibility and freedom. Physics can help us in this procedure. It prepared me, for one, to accept that certain of my experiences are illusory; if our immediate experiences of space and time are shown to be inadequate through the observation of elementary particles (which can be in two places at the same time), why not other experiences along with them? Thus, one could conceive that certain of our experiences of daily life are illusory without doing away with them . . .

This is just what Spinoza, in his day, said: free will is an illusion linked to our ignorance of true causes. Far from thinking he was putting an end to all morality, he entitled his central work the *Ethics*. Through him, we can find another way of thinking freedom, one more in accord with the current advances of biology and the human sciences.

❖

It is therefore urgent to return to this philosophical question: In an entirely determined world, can freedom, morality, and life in society still exist? This question is sometimes formulated differently by the Stoics, Spinoza, and certain Kabbalists: the world is perfect, for it could not have been other than what it is, and yet it is perfectible; the search for an ever-greater perfection is even the specific vocation of the human species. The basic intuition of a given perfection is accompanied by an exhortation to walk in the path of the righteous—that is, by a quest for salvation in the form of a perfection greater than or different from the one initially given to us. The same paradox consists in affirming an absolute determinism that leaves no place for free will other than that of an illusion, while positing the reality of freedom. For this, freedom must be considered in a way radically different from our immediate experience of free will, from the experience we have when, confronted with a choice, we choose one path rather than another. Let us be clear: we should

---

7. They are echoed by certain thinkers of absolute determinism in other philosophical traditions, such as Buddhism and Islam, or, in the Jewish tradition, by Hasdai Crescas (in the fourteenth century) and Schlomo Hetkhil Eliaschow (in the twentieth).

not content ourselves with regarding free will as a naïve and degraded form of liberty insofar as it is not informed by universal practical reason and our judgment remains obscured by passions and opinion.

For the majority of philosophers after Kant, only those choices that reason develops and that obey the moral imperative are considered free, while the rest are seen as being determined by the force of drives or the pursuit of interests. But it is this very distinction between reasoned choice and impulsive choices, if it is thought to be radical and absolute as regards causal determinations, that the cognitive sciences—neurology as well as psychology—reveal to be partially illusory. The exercise of reason is not disembodied. It requires the body and its passions. Furthermore, in what sense is the choice of an end in a process of rational decision free? Isn't it too dictated by the laws of desire and its biological and psychosocial determinations—such as mimetism and want [*envie*]? Some have held that the knowledge we have acquired of these determinations enables us to free ourselves from them. In any case, it gives us the feeling of being able to act freely, spurred by the causes and motives of our desires and our will. But how could we fail to see that there is an infinite regress here—if we do not accept, as Kantian and post-Kantian philosophies do, a "supersensible" domain of freedom, outside of the world and its physical determinisms? According to them, the free subject is an absolute origin, a first agent capable of creating first causes, who initiates new causal chains ex nihilo, independent of the rest of the world. This is the idea of man as an "empire within an empire," at which Spinoza already scoffed at a time when discovery of the mechanisms by which living and nonliving bodies are moved had only just begun.

This post-Kantian morality is the foundation of humanist morality. And, from a certain point of view, what I am proposing here is antihumanist. I assert this openly. But I refuse to think, like Peter Sloterdijk, for example, that if we can no longer be humanist, we must accept immorality. I identify with the denunciation of certain aspects of humanism as a Western and oppressive bourgeois morality. Classical humanism occasioned the moral horrors of colonialism and war. But beyond that, I refuse to accord to man the place of an empire within an empire, for he instead appears to me to be determined like every other natural being. I simply do not see who the man of humanism, Man with a capital *M*, is. This Man does not exist; there are only individuals and subjects in the process of becoming subjects [*sujets en devenir*]. Today we hear one sentence constantly repeated: the future is uncertain and Man must choose. The first part of this sentence is undeniable, but the second is insane: Who is this subject who could choose? The capacity to choose is at the very least subject to discussion, as we are in the process of seeing, and this Man does not exist. Humanism put man in the place of God. Today, it suffers from the same flaws as theology. And it seems to me that we can rid ourselves of this humanism, as we have of theology, without thereby denying ethics and the recognition of the subject in everyone.

The revolution we must undertake is far more total. We must begin by accepting what we encounter every day in our science of determinisms: that is, that our subjective consciousness of free choice is increasingly refuted by our objective knowledge of causes and of the impersonal laws that determine these choices and show that they are not free, as we believe. Instead of first positing our daily and affective experience of free will and then trying to accommodate it to our cognitive experience—that is, to our wider and wider understandings of mechanisms—let us do the opposite. Let us forget our experiences of free will for a moment and start from the postulate that today we know only part of an infinity of determinisms, and let us postulate a totally determined world. In other words, let us stop debating the gaps in determinism and assume that there no longer are any.

This reversal immediately implies another. Spinoza will once again serve here as our guide. The freedom whose reality he posits at the beginning of his ontology is identified with the free necessity that characterizes the work of God—that is to say, of Nature—and, at the end of his *Ethics*, with human freedom. True freedom is that of God, understood as the infinite nature that for us expresses itself under two aspects, extension and thought. Its reality corresponds to its self-producing nature. Absolute freedom, that of God, coincides with the self-production of what exists in nature by nature and with the infinite knowledge of its absolute determinism. "That thing is said to be free [*liber*] which exists solely from the necessity of its own nature, and is determined to action by itself alone. A thing is said to be necessary [*necessaries*] or rather, constrained [*coactus*], if it is determined by another thing to exist and to act in a definite and determinate way" (Spinoza 2002, Part I, Definition 7, p. 217). Far from being the capacity for arbitrary choice, freedom here depends on not being determined by anything other than one's own law. Human liberty thus lies at the end of a path: the philosopher gradually learns to free himself from the passive servitude in which unthinking submission to affects and external causes keeps him and learns to determine himself more and more as he reaches an adequate knowledge of things and of himself.[8] The more a man is determined to act out of the necessity of his nature alone, not influenced by the other parts of nature on which his existence and affects depend, the freer he is. Thus, infinite knowledge of determinism would coincide with total freedom. Our being would then merge with our knowledge, and we would be able to perceive in ourselves the power and efficiency of a "self-cause [*causa sui*]." In other words, thanks to this infinite knowledge, everyone can be his or her own self-cause, and, in this sense, can be a truly free agent—because produced by his or her own determination. It is obvious that here we are not speaking of our effective knowledge, since we never have access to this infinite knowledge. It is a horizon, and no one, neither Spinoza nor Freud, has claimed to have arrived at the end of this search. We are only ever en route to this infinite knowledge and this total freedom. Human existence, which unfolds in time, can be the occasion for an ever-greater search for perfection, for a history of salvation and freedom in which the highest demands of ethics join together with the experience and knowledge of the laws of Nature.

And this postulate is necessary to every enterprise of knowledge. The simple fact of undertaking a search for causes, reasons, or natural laws obviously implies that we suppose causes, reasons or natural laws to exist.[9]

---

8. For Spinoza, human perfection, like that of the rest of nature, is given not as a possibility but as a reality—more precisely, as a part of reality. This part grows or diminishes according to whether adequate knowledge of the causes of affects increases or decreases, for this knowledge increases the number of adequate ideas produced by and contained in the mind. By this very fact, such knowledge increases the power to order the affects in accordance with the order of reasons—that is, the power to act, rather than simply being subjected to affects produced by exterior causes that inadequate knowledge does not allow one to interiorize. Rational knowledge allows one to encompass the causes of phenomena by comprehending them. Intuitive knowledge, the so-called third kind of knowledge, which cannot be obtained without rational knowledge—that is, without first knowing, via reason, the universal laws of nature—goes even further in that it understands particular things.

9. More and more, searching for causes is found to be inadequate and tends to be replaced by a search for laws. Mathematical physics has exploded the notion of cause. Today, physical phenomena are described with the help of mathematical laws, that is, equations that allow one to predict from a given state of a body what subsequent states will be. The classical notion of cause has disappeared from the description of these phenomena. For Wittgenstein, belief in the reality of a cause-and-effect relation experienced in time was only a superstition. In physics, it is difficult to contest this development, because it has been formalized, mathematized. In the other sciences—biology, the human sciences, medicine—it is difficult to push the denunciation of causality to this point, and causes are

The initial postulate is thus absolute determinism. The hypothesis of absolute determinism is simultaneously a methodological postulate preceding any experience of knowledge and an ontological postulate grounding the possibility of our freedom. We can neither access nor possess this infinite knowledge. But that is enough to reverse entirely the relation between knowledge and freedom. True freedom is asymptotic. It is projected onto the horizon of an infinite knowledge of things, others, and oneself. From this point of view, we can perceive our knowledge and experience our existence as projections and partial recoupings in which the infinity of things and of thought expresses and produces itself in its finite modes. Though it is but a horizon, this projection nevertheless illuminates our choices—not in that they are less determined, but in that they are inscribed within the free necessity of infinite power, self-cause within each of us—what Spinoza calls our *conatus*—and they are lived as such.

The experience of this freedom is not, for all that, resignation and the passive acceptance of a determinism fallen upon us from above. To know one's nature is above all to know what determines one in the same way as any human being, and then to see how these common determinisms are singularized in oneself. Though we cannot modify things arbitrarily by whim, we are conscious of our acts as we effect them; we understand them. The experience of this free necessity is that of an intense activity of our minds and bodies.

---

still sought. It would thus be prudent to apply the Spinozist distinction between adequate causes and inadequate causes, between partial causes and total causes. In biology and medicine, there is unfortunately still a great deal of confusion, whose effects are harmful and even catastrophic. A very sad example of this occurred during the fifties. It had been observed that women who repeatedly spontaneously aborted had a lower than normal level of urinary estrogen. On the basis of this weak statistical correlation, it was deduced that the abortions were caused by this deficiency, and these women were prescribed a synthetic estrogen, diethylstilbestrol. Hormones were the fashion at the time. Twenty years later, a large number of girls born from these pregnancies developed serious cancers of the vagina and ovaries. The scandal is that a statistical study had only observed a correlation. Another study would have consisted in studying the percentage of women with lower than normal levels of estrogen who aborted. But this second study, more difficult and more costly, was not performed. When it was done later, it showed that the correlation was a fact of causality in the opposite direction: repeated abortions cause a decrease in estrogen levels. A relation of causality had been deduced without reason. Two errors were compounded: a confusion first between statistical correlation and causality, and second between partial cause and total cause. There is indeed a form of superstition in the search for causes at any costs and in the inability to accept that an event may exist without our knowing the cause.

We can experience this freedom at certain privileged moments: when, for example, we understand something. It is when I am active that I experience being the subject of what I am and what I do. I am not a subject like an empire within an empire, escaping determinism, but rather in understanding and knowing the determinisms of nature that act in me and make me act. And it is within this activity that I constitute myself as a subject.

We are, quite simply, not accustomed to such a thought. Moreover, it belies our affective experience. In Spinoza's time, scientific knowledge was extremely limited, and to hypothesize absolute determinism was much more daring an extrapolation. The price he paid for this audacity was to incur the most serious accusation of his time, that of atheism. For him, God's freedom was no longer to be understood as an absolutely free will, because that would imply the arbitrariness of such a will.

Our temporal existence and our daily choices remain. Knowledge, the understanding we can have of the laws of nature, in a certain way brings us in into a timeless reality. There is nothing mystical or vague here. It is the experience of any physicist or mathematician. The mathematical expression of a law of nature in which time is a variable has as an immediate consequence the disappearance of time. As soon as a future event can be predicted by a law, this event already exists, in a certain way, in the knowledge we have of it, and the future will bring nothing more. Here again there is a difference of level between, on the one hand, our existence, which is spatial and temporal and which we experience through our sense perceptions and memory, and, on the other, this abstract and rational knowledge. From a certain point of view, it is when a scientific prediction is realized that we should be surprised. The most surprising thing, in fact, is that sensible experience accords with what had been predicted abstractly. In our daily life, the new still exists because we can never get out of time; the future always brings something that did not exist yesterday, even if we were able to predict it. Theoretical certitude does not suppress the curiosity with which we await the result of an observation. When our predictions are confirmed, we are of course satisfied, but at the same time always a bit astonished that "it worked." The general affirmation that "everything is foreseen" or that "there is nothing new under the sun" in itself does not do away with our experience of the passing of time and the novelty that accompanies it. The infinite knowledge of determinism, necessarily timeless, is not in opposition to this freedom as a horizon but, to the contrary, constitutes it.

There is a simpler and more common way to try out this form of freedom. Compare a child's behavior with that of an adult. When a child wants something, when it chooses one thing instead of another, it is never difficult for an adult to see that in fact the child is not really choosing but is influenced by a desire, a suggestion, or a habit. Children feel themselves to be free insofar as they often have the impression that they are doing what they want. Their desires are nonetheless seen as caprices, as arbitrary decisions and not the exercise of real freedom. Adults, by contrast, have the feeling that they exercise their freedom in full knowledge of the facts. And they are not entirely mistaken, for an adult has much wider knowledge of his or her own determinisms and the determinisms of the world than an infant does. In California in the sixties, hippies also thought that freedom expressed itself completely in the spontaneity of desire, in the blossoming of all its potentialities. When these young people were asked what freedom meant, they answered, "To do what I want," and to the question "What do you want?" their answer was circular and caricatural: "To be free." This is the experience of the child, who has the impression it is free. And yet we know that it is determined—all the more so in that we ourselves can drive it to act in a certain way. Thus there is indeed a difference between this illusory infantile experience and the experience we ourselves have when we have acquired a certain knowledge and try to determine our behavior through rational analysis of its eventual consequences. But to the degree that our knowledge is itself partial and limited, this experience is necessarily illusory in relation to the total freedom we imagine—even if it is less so than the child's. Here we have a remarkable reversal: contrary to the received idea, a better knowledge of the determinisms that govern us enables us to experience greater freedom. The same principle applies in law, where only a person capable of judgment is considered to be responsible. However, within that conception of things, knowledge is thought merely to give the means to choose, while the choice remains completely undetermined. In the end, we would set our choices. Within the perspective I would like to adopt, choices are not arbitrarily separated from knowledge but instead are determined by factors that this knowledge can sometimes allow us to discover. The fact remains that the common attitude toward childish choices shows us that another vision of freedom is not so strange to us.

If, through an unaccustomed intellectual effort, we can conceive of a freedom that would be free necessity, for the moment this in no way solves the problem of our everyday experience. Though from time to time we can have a timeless experience, as we do with mathematical knowledge, we can never escape the experience of our own duration. Our experience of time and of duration varies incessantly over the course of our lives and sometimes even over the course of a single day. An hour of boredom is never the equivalent of an hour devoted to an enthralling activity. Our experience of time is not an absolute experience. If we are capable of considering our experience of time in its relativity, why do we not look at the choices we make in the same way? Isn't this made easier by the fact that we have proof of this relativity every time the discovery of a determinism contradicts our impression of having chosen freely?

There are two ways of seeing things here: minimally or maximally, one could say. Minimally, things must be separated. There is the cognitive world in which we can approach total freedom, the experience of free necessity, through the development of our intellect and our understanding. And then there is the world in which we live, made up of partial knowledges, in which we are submerged by passions and sentiments that unfold in time and give us our experiences of free choice. In other words, there are two domains, one of which is more noble than the other. When we reflect on the causes that make us act, we realize that we are not free agents, but in our personal experience and in our social relations, we continue to act "as if" we were and to consider ourselves responsible for what we do. This experience is irreducible, despite any eventual knowledge of our determinations. Although I was able to predict the future, I still had an experience of newness. Even if I know in theory that I am determined, I nevertheless experience free choice.

Let us try to widen our perspective. Can't our experiences of free choice allow us to approach free necessity? Why not see them as signs or symbols of true freedom? These experiences, illusory though they may be, give us an idea of free determination. In a certain way, they prefigure the limit experience of freedom we would have if we had access to infinite knowledge of natural determinisms. One can thus see these experiences as impoverished, blurry, and inadequate images—but images nevertheless—of free necessity. We recognize this illusory freedom we are given as a mirage when we observe a child. It is as if in growing up we forget that it remains illusory and convince ourselves we have arrived at an age at which it no longer is. The double sense of the term *subject* in relation to the possible may be clarifying here.[10] On the

---

10. I am borrowing this distinction between the conditions of the existence and nonexistence of the possible from Hasdai Crescas, *Or Hachem*, bk. 2, section 5 (Crescas 1998).

one hand, there is the active subject, "subject of," subject of his actions and history; on the other hand, there is the subject "subject to," subjected to what happens to him, to his history and what he does, or rather, to what is done through his actions. For the subject as agent—rather, for the agent who perceives himself as a subject—the possible is real, as the possibility of acting or not acting, of acting in one way or in another. And it is in experiencing the reality of this possibility deep within ourselves that, in a first moment—let us call it the moment of childhood—we constitute ourselves as responsible subjects. But in reality the possible does not exist, for the same subject is subject to what happens to him and to the causal necessity that determines him to choose this rather than that. Thus, in a second moment—that of adulthood and of lucidity—when we acquire objective knowledge we find ourselves torn between what this knowledge teaches us and our subjectivity, which has just as much reality. But at the same time we can acquire experiences of this other freedom through the development of the intellect. Our experiences of free choice are then more than images. They allow us to go from ordinary experience, where we think we choose without being determined, to other experiences in which, at the very moment we are choosing, we can be aware of the causes that drive us to choose. We then become aware of everything that determines us to make a choice, even if not in detail. And as we little by little gain access to these determinisms more precisely, our feeling of freedom changes. From the infantile sentiment of being able to make arbitrary choices, we gradually move to acquiesce in what is done by itself within us [*ce qui se fait en nous*]. It is this acquiescence that opens onto the experience of true freedom. The more knowledge we have, the more the experience we have of our free choice asymptotically approaches the experience of true freedom. The gap between lived freedom and theoretical freedom is thus gradually filled in, thanks to progress in the knowledge of causes. Our knowledge, always finite and limited, in a certain sense creates a space that is illusory if we believe it to be real but that is real in the experience we have of it.

Spinoza and Hasdai Crescas, each in his own way, help us understand how true liberty is found in an increasingly active consciousness of this determinism within oneself and in the "intellectual love" that results.[11] It seems that something escapes determination by the infinite chain of causes: the way it is understood and used, conscious adhesion to the questioning it sets in motion,

---

11. See Atlan 2003.

its deepening as the beginning of wisdom, and, finally, the joy that this process itself brings. In fact, an external determinism—by exterior causes—is replaced by internal determinism,, "virtue" in the Spinozistic sense, which is defined as determinism by adequate ideas. An example of this can be found in chains of violence and counter-violence, or in social or political situations that produce crimes followed by repression that leads to other crimes, and so forth. The existence of causes does not do away with perpetrators' responsibility, as we shall see. Understanding and appealing to these causes cannot serve as a justification. What is more, these causal chains can sometimes be stopped by some saving awareness, a self-examination, and sometimes by forgiveness. Forgiveness and self-examination are also determined, of course, but within a different causal chain—internal, differently ordered, near to what Spinoza calls "the order of reasons." This is what is stated explicitly in the first propositions of the fifth part of the *Ethics*, where the association between the causal order of natural things and the order of ideas of these things is turned in such a manner that the sequence of ideas becomes predominant: "The affections of the body, that is, the images of things, are arranged and connected in the body in exactly the same way as thoughts and the ideas of things are connected in the mind." (Spinoza 2002, Part V, Proposition 1, p. 365). The experience of determinism that becomes dominant is then the active experience of conscious reflection on oneself and on things, and the joyous assent that accompanies it.

Another image will allow us to better understand how we can live these different experiences: that of games. The same person can play both tennis and basketball without thereby being a schizophrenic; this person simply applies in each case the rules of the game he or she is playing. No one would think to play one game following the rules of another or to invent a sort of "meta-game" that would combine all the rules. The synthesis between games here takes place at the level of the subject who plays them, who would be different if he or she devoted him or herself to playing only one of them, for example. Far from being theoretical, the synthesis of these different experiences is of the order of lived experience. In a similar way, we sometimes live in a domain ruled by passions, sometimes in a domain in which these passions are integrated into intellection.

There is an expression that captures this reality in a striking manner. In the "Tractate of the Fathers,"[12] one of the fundamental texts of Judaic

---

12. The "Tractate of the Fathers" is a moral treatise within the Talmud. It is sometimes called the "Tractate of Principles," because in Hebrew the same word signifies "father" and "principle."

morality, it is written: "All is foreseen and permission—or possibility—is given."[13] One can take the first part of the sentence as the affirmation of an absolute determinism. But it is immediately specified that permission itself is part of this.[14]

Free will is an illusion from the viewpoint of knowledge of causes, but it is no less a real experience for every finite human. It is part of our reality, even if our freedom also consists in becoming aware that it is an illusion to believe we determine things. The absolute determinism in which "all is foreseen" is carried out through our choices themselves, which are given to us as possibilities. But these choices cannot change anything in the chain of causes. It is possible for us to play at being free, though all the while we must be aware that this is an illusion. It is in this space of play that ethics and politics develop. Even if it is only as a misunderstanding that we know ourselves to be free, we must act as if we were so truly.

As for political freedom, it is just as different from free choice as it is from active assent. Here one is not concerned with whether an individual is aware of the causes that determine him or her, but only with whether he or she is compelled to act by his or her consciousness and not by force or coercion by other individuals. In other words, the claim to political freedom in no way contradicts the affirmation of an absolute determinism of nature, and the defense of democracy is not weakened by awareness of the causes of our behaviors. Political freedom is in opposition to coercion, to certain individuals' power over others.

❁

How can the question of responsibility be avoided once we have affirmed the double possibility of absolute determinism and freedom without free will? The entire moral tradition we have inherited maintains the opposite: if there is no free will, then there is no freedom, no responsibility, and thus no morality. Just as it is possible to conceive of a freedom different from free will, it is possible to think of responsibility and morality within determinism. The

---

13. [*Pirke Aboth*, 3.19: "All is foreseen and permission—or possibility—is given" is Atlan's direct translation from Hebrew (see also the next note, on "possibility."—Trans.]

14. Thus, Crescas comments on the second part of the sentence: "And when it is said 'possibility is given,' this is a testament to the secret of choice and of the will, that is, that the possibility of choosing is given to every human being from his own point of view, such that commands—the commandments of the law—do not fall upon someone constrained and forced at any rate."

more one discovers that men are driven to act by multiple causes, whether they know or are unaware of them, the more difficult it becomes to define responsibility in the traditional manner. Much new knowledge calls the classical conception of responsibility into question, in theory certainly but also in practice: for example, in the juridical domain.

The most flagrant current example is the French jurisdiction relating to crimes perpetrated by the mentally ill. Traditionally, punishment is aimed at criminals recognized to have acted not only with knowledge of the facts but freely, in the classical sense of the term. It thus corresponds to a supposed use of free will. If a psychiatrist, by contrast, can demonstrate that an individual has acted under the influence of a drive, of a change in state that has rendered him or her unaware of what he or she was doing, the individual will be declared not responsible under the law. He or she will not even be judged, but will be considered a sick person in need of treatment. Treatment then substitutes for punishment. The case of sexual criminals is one example among many showing the difficulty of establishing this distinction firmly. Often, one cannot consider these criminals to have acted without knowledge of what they were doing, since many premeditate their acts down to the smallest details. But we also know that they are compelled to act by drives that are more difficult for them than for others to resist; by a hormonal or psychological imbalance. They are thus subject to punishment, to being condemned in court, but are also in need of treatment. No psychiatrist would say that this criminal has acted without judgment. He or she would even maintain the contrary, but would specify that at the moment of acting the criminal was determined by an ensemble of factors, some of which are known and others of which are not. The 1994 legislation was designed to facilitate a psychiatrist's task.[15] An intermediary category was created, between the category of fully responsible persons and persons recognized as totally irresponsible because they acted without having judgment: a category for individuals

---

15. The new penal code (article 122–1) introduces a distinction between an "abolition of judgment" and an "alteration of judgment" in order to permit attenuated conviction. The result is that psychiatric experts diagnose an "abolition of judgment" less and less frequently and prefer to limit themselves to the diagnosis of altered judgment. This distinction has led to a drastic decline in the number of persons benefiting from nonprosecution on grounds of penal irresponsibility, which dropped from 611 in 1989 to 190 in 1997. For some ten years now, more and more individuals judged "borderline" by specialists have been sent before tribunals before receiving prison sentences. In fact, this modification of the penal code only aggravates the problem, for it maintains the same

having acted under altered judgment. Suffice it to say, many psychiatrists at present opt for this nuanced decision. The result is sad and well known: there are more and more mentally ill prisoners in our prisons, sick persons who do in fact have some degree of awareness of their actions and who are thus judged, condemned, and imprisoned like other criminals.[16] Recently, this jurisdiction underwent a veritable revolution: the institution of a "penalty with therapeutic follow-up." This proposition, which joins punishment and therapy, is a medical and juridical heresy. For all that, it is perfectly justified. It states just this: the individual is considered legally and criminally responsible even though his or her actions were determined.

As for the question of the individual's guilt and an eventual penalty, it must be adjusted and separated from that of responsibility. Responsibility is not the same as guilt. Georgina Dufoix's declaration during the infected blood scandal that she felt "responsible but not guilty" emphasized just this distinction, and did not merit the bursts of laughter it provoked.[17] Responsibility does not necessarily imply guilt. To be persuaded of this, we need only list all the things we are led to do that do not imply any guilt but for which we may be answerable. This essential distinction is, moreover, accepted in cases of civil responsibility but seems to be forgotten when it comes to penal responsibility.

In fact, we must learn to distinguish between two forms of responsibility. The first can be described as ontological, a priori responsibility. It means: I am responsible because I am in charge of something. It is not necessarily a question of a crime here. I must simply answer for what I do, even if it is not

---

procedural framework: psychiatrists are still asked to define states of consciousness that would imply an absence of responsibility.

16. After the terrible events in Tours (a man opened fire on passersby, killing four and wounding seven), Jean-Michel Dumay wrote in *Le Monde*: "What is striking is that society, constrained to a binary choice between care and punishment, by virtue of its tendency to favor prison for the mentally ill in the end denies or minimizes the fact that they are ill. (At Auxerre, no one asked the slightest question regarding the paroxysmic suffering of a paranoiac living like a 'lab rat.') What are we to think of a society that no longer recognizes its ill?" (Dumay 2001).

17. [France's "contaminated blood scandal" erupted in 1991, when it was revealed that the National Center for Blood Transfusion (*Centre National de Tranfusion Sanguine*) had been giving HIV-infected blood to patients between 1984 and 1985. Georgina Dufoix was Minister of Social Affairs and National Solidarity at the time, and in 1999 she was tried for manslaughter and acquitted. She made this statement as part of a televised defense in 1991.—Trans.]

a crime, just as I must answer for what I am. This absolute and unconditional responsibility is connected with human nature and its capacities for representation, all of which depend neither on the nature of the decision nor on its execution and eventual effects. It is, ultimately, a maximal claim of human dignity: I remain a human being, and thus responsible, no matter what my actions. Certain criminals declared irresponsible have testified to having been overtaken by a feeling of indignity and have said they would have preferred to have been judged. Responsibility is not reducible to a juridical category nor to a feeling on which moral judgment could be based. It is first and foremost a given of the human condition. To be in charge of something implies the possibility of having to answer for its being carried out, to answer for its success or failure and their possible consequences, including details that had not been imagined, much less chosen. This demand for an answer is the corollary of the power to act: the "permission" or "possibility" to constitute oneself as a subject that is given to every human being.

Responsibility after the fact [*la responsabilité après coup*], by contrast, is situated in the domain of our contingent existence, where we are brought to exercise the capacity of responsibility proper to every human being. This second form of responsibility comes into play when I am implicated as one cause among others of an event. I am then to answer for a certain situation, and for this I am submitted to a moral and/or juridical judgment. The question takes on a dramatic form when one of the consequences of my actions has been detrimental to others. It is this responsibility after the fact that involves the question of guilt and that is today made problematic by the discovery of the many determinisms that govern us. If these determinisms do not free us of all responsibility, they show us the limits of too close a connection between responsibility and free will. We are all responsible for what we do, including things done through us that we have not chosen to do. We are, for example, responsible for the actions of our children, even if we did not will those actions. In the same way, I am responsible, by definition, for the actions of persons placed under my responsibility.

Among the characteristic properties of the human species is the human body's consciousness of itself, or rather, the different states of self-consciousness that accompany its existence by virtue of the complexity of the brain. To at least the same degree as upright posture and articulated language, these states contribute to organizing human societies—unlike even the most evolved animal societies—in a specific way. The ethical and legal domains and the notions of responsibility with which they are linked are produced by the different levels of complexity that characterize the organization of the human body, its different states of consciousness, and the social life proper to the human species. This is why, if a total and unconditioned responsibility is attached, a priori, to human nature, a responsibility after the fact can be defined in a different way, relative to the states of consciousness that may have preceded, accompanied, and followed an action.

It could be a matter of a reflex action, "independent of my will." At the other extreme, it could be a premeditated act, developed slowly. The deliberations preceding this action lead to a feeling of free choice, and no philosophy of absolute determinism could suppress this sentiment. If free will does not determine the sequence of causes, an awareness of choice and a certain degree of assent accompany that sequence. The feeling of freedom that accompanies voluntary action is indeed real as a state of consciousness, even if it coincides with an erroneous representation of the mechanism of action. Even if free will is an appearance, subjects' assent to what their will—determined by internal and external causes—makes them do is quite real.[18] This assent is also an irreducible given of human nature in its finitude and must be taken into account, even in a totally determined world. The whole richness of human existence stems from the infinity we conceive and the finitude we live. The notion of person in the juridical, moral, and grammatical senses is closely linked to this state of consciousness. Taking individuals' sentiment of free will into account and proceeding as if free choice presided over their actions has allowed for the constitution of societies of morally responsible subjects. But here again the existence of a different freedom can be detected. When deliberation is short, we always have the impression that we could have chosen something else. When, on the other hand, it is long and rational, our feeling of free choice is both weakened and strengthened: weakened because we perceive the reasons that have pushed us to choose, to the point that we sometimes even go so far as to claim "not to have a choice"; strengthened by the feeling that it is no longer a matter of an arbitrary decree of our will.

Does this mean that, in a roundabout way, we have arrived back at the usual notions of responsibility and irresponsibility linked to capacities for judgment? No, there is a great difference. The question has been shifted, and it

---

18. Through this notion of assent, Crescas reconciles human choice with absolute determinism. The extent to which choice affects results remains illusory, but the awareness of effort and will that accompanies our behavior when we do not feel ourselves con-

is no longer a matter of all or nothing—responsible and thus guilty and to be condemned, or irresponsible and innocent. We have seen that psychiatric expertise can answer this question less and less and that it would be better to stop asking it. In the absolute, the criminal madman who kills during an episode of delirium is as responsible as Henri Landru,[19] who plans his crimes and how he will hide them, or the young delinquent who kills while robbing a bank. But the punishment or treatment will not be the same. The question will center on the gradation of states of consciousness and on accessibility to sanction and treatment. In another domain, if a business head, a minister, or a general makes a complex decision, perhaps counseled by advisors, which on account of unpredicted factors or a failure in its execution leads to a catastrophe, the relevant question is not who is responsible. If, in order to find the fault and condemn the guilty party, it is necessary to find the person responsible in the usual sense of the one who has been the cause of harm, failure is certain, since each person has played only the role of partial cause. Responsibility in this sense has been diluted to the point of disappearing. But when one recognizes from the outset that all were responsible (in relation to their office and hierarchy), the question is shifted onto the circumstances and manner in which each person exercised this responsibility.

In the same way, doctors and nurses are obviously always responsible for their activity, no matter what the circumstances in which they exercise it. Today, the obligation to inform patients is creating a tendency toward an ideal of shared decision, which would imply shared responsibility. This in no way means that the doctor's responsibility diminishes in inverse proportion to that of the patient, who now participates in the decision. The risk of removing the accountability of doctors is, however, quite real if one has a contractual vision of the doctor-patient relation and if one insists on looking for a responsible—in the sense of guilty—party. In cases of medical malpractice or of harm to a patient, it is not a question of knowing who is responsible, since all the implicated agents are, a priori, by the simple fact of having been there and participated in the action. It is, rather, a question of investigating whether the conditions under which the doctor or nurse exercised his or her responsibility bring to light any fault or guilt, considering his or her state of

---

strained—or, to the contrary, the awareness of passivity and even an opposition to our will when we do feel constrained—is real.

19. [A French serial killer famous for planning his murders.—Trans.]

consciousness—in light of medical knowledge possessed and evaluation of risks.[20]

Finally, researchers whose discoveries lead to initially unpredictable criminal applications are clearly responsible for their work and its consequences. Technological developments and their social, political, and economic repercussions mobilize a number of actors, not only experts but all those who express and help shape the desires of public opinion. In instances where things skid out of control—the Manhattan Project and the dropping of the bomb on Hiroshima and Nagasaki in spite of the opening of negotiations, or criminal applications of genetics—the search for one or more guilty party cannot be restricted to identifying those responsible for the initial discoveries. The fact remains that the notion of scientific responsibility has imposed itself along with awareness of scientists' a priori responsibility, which is linked to the exercise of their occupation regardless of its consequences. This responsibility does not imply guilt automatically. But it obliges researchers and other social actors to engage in ethical, juridical, and political reflection on the exercise of this a priori responsibility. Ethics committees are the direct outcome of this awareness.

When one is looking for a guilty party, the question being asked has nothing to do with free choice or with the first form of responsibility. It concerns an analysis of causes. The question is all the more delicate in that very often I am not the only person involved. Can one, for all that, speak of shared responsibility? Is the agent's responsibility then to be considered proportional to the probability—necessarily subjective—that he or she was the cause of the damage? This difficult question is constantly being asked now that more and more decisions are collectively made and applied in situations where risks are uncertain.

The well-known principle of precaution, so often brandished these days, would appear to be the only guide. I see it as nothing more than a wager. The desire to take precautions can and must accompany and moderate action, but it certainly must not serve as a principle from which the right decisions could be deduced systematically, for in certain cases its effects can be catastrophic. We refer to this principle in situations of uncertainty, when there is

---

20. Because the right to health entails an institutional dimension that exceeds individual members of the medical profession, society can sometimes assume responsibility for responsibility for reparation of damages. The notion of responsibility without fault can then be applied collectively.

a perceived risk that can be neither precisely evaluated nor quantified. (Otherwise it would no longer be a matter of precaution properly speaking but of prevention, based on the management and evaluation of risk.) Put back into this context, it is troubling to see that the principle of precaution self-destructs as a principle. It is supposed to avoid the worst, but since the worst cannot be predicted, there is a risk that the decision dictated by this principle might have worse consequences than those one would have imagined without it. The principle of precaution thus demands not to be applied: "When in doubt, abstain," does not always prevent the worst. Thus it cannot be invoked to justify a decision or an incrimination. All one can and must do, in ethical domains where a rule cannot give a precise determination, is decide and act with the prudence Aristotle recommends (Aristotle 1998, 5.10.1137b, p. 133). Prudence is in no way a principle. It is a virtue accompanying action, and it is at least as difficult to quantify as intelligence and good sense. Exercising prudence consists, among other things, in progressing in small steps, remaining on the watch for new signals, and always being prepared to change course.

In France, the affair of the contaminated blood is associated with awareness of the need to apply such a principle. But looking at things a bit more carefully, one realizes that the malfunctions at issue there could have been avoided without any recourse to this principle. The heads of the National Center for Blood Transfusion deliberately put economic imperatives ahead of attention to the safety of their products and the well-being of patients. This is a case of serious professional malpractice, reprehensible and to be condemned without reference to the principle of precaution. The number of contaminations also resulted from not applying the ministerial circular suggesting that blood donations from persons at risk, such as homosexuals and drug addicts, no longer be used. At the time, there was strong social pressure, shared by patients and the medical corps alike, not to stigmatize certain groups, especially within the French context of the practice of free donation, symbol of solidarity. Due to the moral rehabilitation attached to this gratuitous gift, prisons remained privileged sites for blood drives, despite the fact that their inmates constituted populations at risk. Here there was a fatal error of judgment in the evaluation of risks: the risk of contamination was underevaluated in relation to that of exclusion. It is a matter of the comparative evaluation of risks and not of obvious consequences that could have been deduced from implementing the principle of precaution as a principle of action.

If there is a great lesson to be drawn from this national tragedy, it is the need to take ignorance into account. The continuous progress of biology and medicine since the beginning of the century had produced a certain arrogance. What was not known simply did not exist. AIDS and the succession of errors, faults, and crimes committed by health professionals set off a veritable earthquake. But invoking a principle of precaution, which would have averted these tragedies had it been respected, is also a myth. It is simply deceitful to claim that the principle of precaution is adequate to take into account the portion of ignorance inherent in scientific knowledge. Claiming to follow such a principle can have additional detrimental consequences: avoiding responsibility by invoking its consensual application. The recent examples of mad-cow disease and foot-and-mouth disease unfortunately demonstrate this consequence. The decision to massacre all those animals seemed irreproachable, for the simple reason that the government sheltered itself behind the principle of precaution. It is indeed politicians' fear of being judged "responsible and guilty" that motivates such decisions.

This uneasiness and the way it is handled stem once again from confusing the two senses of responsibility. The prudence that leads to acting with precaution independent of a risk that cannot be evaluated involves total, situational responsibility. Precaution, falsely erected into a principle, claims to be able to avoid harmful consequences. Yet here one situates oneself at the level of responsibility after the fact and of possible guilt. At the risk of injustice, determining responsibility involves answering the following question: "Who is responsible for what and before whom?" The answer will be very different depending on whether one asks it with respect to situational responsibility or responsibility after the fact. In the latter case, a very precise answer must be given: the responsible party is the specific person or group of persons who, under certain specific conditions, made a decision that brought about an action executed in a certain fashion. And those responsible are responsible precisely before the persons who were harmfully affected by their action. By contrast, in the case of responsibility a priori, the answer to the question "Who?" is "everyone," insofar as he or she is in a position to act; abstention from action is still action. To the question "For what?" the answer is "For all his or her existence and the effects of his or her acts"; to the question "Before whom?" it is "Before everyone, including oneself, potentially exposed to these effects." Setting up precaution as a guiding principle of action amounts to applying what stems from responsibility a priori to responsibility after the fact.

For me, this whole discussion highlights the importance of renewing and deepening philosophical reflection as a practice of thought and of life. We can no longer content ourselves with repeating what we have already been repeating for centuries in schools as if nothing has changed.

On the one hand, it is necessary to recapture things as they were before the upheavals of humanity's great pivotal periods. I mean the departure out of the ancient world and its commerce with the world of the gods, and then the scientific revolution of the seventeenth century. We are today living in a new period of that kind, and no one knows what new humanity will come out of it.

But we must, on the other hand, enter into the details of situations and not content ourselves with general concepts, which it is certainly necessary—though not sufficient—to think clearly and rigorously. In this way, one can kill two birds with one stone. The first advantage of this procedure is that it allows us to discover that a knowledge of the details of determinism leads to deepening morality, contrary to what has long been believed.

The second and certainly not the least advantage is to keep us from purely verbal debates. The ethical problems posed by biotechnologies make this attitude especially necessary. Greater and greater knowledge of natural determinisms, ones potentially replaced by technology [*technique*], is indispensable in order to assume the responsibilities of the new situations created by the development of this technology. Too often, global accusations—like that of eugenics, for example—are slung. But it is important to go into the detail of techniques [*techniques*] that put into play very precise mechanisms, just in order to know what one is talking about, before then analyzing as broadly as possible what may be at stake, including at the symbolic level, in their utilization.

A recent example concerns the ill-named nonreproductive or therapeutic cloning. The debate over this question is often presented as an alternative between, on the one hand, reifying or instrumentalizing a human embryo, and thus offending human dignity, and, on the other, impeding research and thus slowing down possible therapeutic, scientific, or economic progress. Formulated in these terms, the ethical debate seems closed; it opposes a moral concern to a quest for utility and profit. But this way of posing the problem is in fact the result of multiple confusions linked to abusive uses of the terms *cloning* and *embryo*. Let us simply recall here that cloning is a matter of transferring a nucleus from an adult tissue into an egg whose nucleus has been removed. The cell thus obtained is a pure artifact, existing nowhere in nature. We know, however, that this modified egg can be stimulated, can divide and develop in a laboratory culture, and thus can produce cells that have the properties of embryonic cells, although there has been no fertilization. Better—or worse—yet, since the birth of Dolly, we now know that, if this cellular artifact is implanted into a uterus, it can under certain conditions that have not yet been mastered entirely and with a still very small probability, develop as an embryo and produce an adult organism. Must it, for all that, be called an "embryo" if there is no question of implanting it into a female uterus and if the technique used boils down to laboratory engineering that will never be able to create anything more than cell lines? It is this question of definition that should be debated before all else. Instead, it is considered settled, and people butt heads over the instrumentalization of embryos. The trouble is precisely that there is none. I would like to defend the argument that this is indeed a case of instrumentalization, but of cellular artifacts produced without fertilization and not of embryos—even though these artifacts can under certain conditions display properties in common with embryonic cells in the course of their development. This distinction seems to me very important, especially if one respects the position of those who, following the teachings of the Catholic Church, consider a human embryo a person from the moment of fertilization. So why speak of an embryo and, what is more, of a human person when there has been no fertilization, unless it is out of a strained concern for classification? Why want to define at any cost what a human being *is*, what a person *is*, what an embryo *is*, with words fixed once and for all, when one has to do here with evolving processes [*des processus évolutifs*] and beings in becoming [*des êtres en devenir*].

Before the birth of Dolly, it would not have occurred to anyone to consider the product of transferring a somatic nucleus into an enucleated egg to be an embryo. Indeed, (almost) all biologists agreed in considering the development of such a cell into an adult organism to be impossible. Since Dolly and the more or less normal calves, mice, and pigs that have succeeded her, an egg with a replaced nucleus is considered an embryo. At the very least, we are told, even if it is not exact from a strictly biological point of view, one must "ethically" consider it an embryo, for it can become one. Yet nothing is more obscure than this notion of potentiality. A bud can become a tree. Seeds can produce a harvest. Is this to say that ethically a bud is a tree or seeds a standing crop? To the contrary, I do not see what would be ethical about putting aside the efforts of nature (and of man) that transform buds and seeds into trees and harvests. In the case we are concerned with here, it is even worse: it is a matter of the potentiality of an embryo—of the potentiality of a potentiality. If an embryo is considered a person, or even as the potentiality of a person from the moment of fertilization, it would be much more coherent to oppose the utilization of supernumerary embryos—true embryos produced by fertilization—than that of cells produced by the transfer of somatic nuclei into eggs.[21]

These technical processes demonstrate to what degree the attempt at simple definitions allowing one to label once and for all what is an embryo, what is a human person, and so on fails as soon as one considers evolving processes [*processus évolutifs*], where what is not something or someone can become that thing or that person. In other words, the attempt to find a definition by appealing to what would be the immutable essence of a thing, an animal, or a human being fails before the unity of nature viewed in its evolution. Certainly, our techniques and the fabrication of living artifacts contribute to thus exploding essentialist definitions, but techniques can succeed only insofar as they obey the laws of nature—even if this is in order to transform it. There is a lesson to be taken from this: let us renounce essentialist definitions and look instead for evolving definitions. To return to our problem: what is not an embryo can become an embryo, what is not a human person can, under certain conditions, become a human person. The ethical debate then shifts

---

21. Several possibilities exist for preparing embryonic stem cells. Currently, one way in particular is being explored: the utilization of stem cells with the properties of embryonic cells normally present in small numbers in adult tissue. If this technique proves applicable, it would probably be the best from the viewpoint of therapeutic applications and certainly the most consensual from an ethical point of view. Let us note that attempts to transfer human somatic nuclei into cow eggs have thus far resulted in failure. What would happen if a better understanding of the mechanisms for the reprogramming of adult nuclei by egg cytoplasm were to make it possible to cultivate embryonic stem cells from such artificial cells? The genes would be exactly identical to those of the donor being treated, and the cytoplasmic factors would be of animal origin. These artificial cells would never lead to the development of any organism, either human or animal, and it really is not clear how one could ever be describe them as embryos, even if they had several properties in common with embryonic cells and thus could be used for therapy or simply research. Some protest against transgressing the boundaries between species. But insofar as these techniques concern only cells and molecules, this boundary (which then is not one), is transgressed every day by the use of cells of animal, bacterial, or hybrid origin for producing medications or prostheses, without this posing the least ethical problem.

to the conditions of such becomings. Here, whether or not the embryo has been implanted in a female uterus is not a "detail" of the technique having no bearing on the data of the matter.

Today we face ethical, social, and political problems that technology on its own does not give use the means to resolve. It is crucial to know what is at stake in order to understand the problem and, a fortiori, to propose a solution. One of the most important tasks is thus to achieve the greatest possible diffusion of knowledge, so that debates are not reserved for experts alone. But the diffusion of knowledge and philosophical reflection about this knowledge are quite difficult tasks, for they involve the co-operation of scientists, the media—the intermediaries between scientists, the public, and the politicians who must decide—and the public, who must make the effort to receive in a critical fashion what the media and scientists say. I have proposed (Atlan 1991a) that in this domain democracy is expressed in a division of the powers of speech into three: politics, science, and the media. And these three powers must not only be kept separate, they must above all critique one another. This is certainly one of the challenges to be taken up today.

In this context, as we have seen, the most ancient philosophical questions, such as that of determinism and freedom, are posed again in a new way in the light of events that make it necessary to reexamine certain seemingly obvious facts.

Translated by Daniela Ginsburg

PART ONE

Self-Organization

TWO

Intentional Self-Organization: Emergence and Reduction, Toward a Physical Theory of Intentionality

(1998)

The object of this essay is to propose a mechanistic model of intentional behavior, in which intentionality is not assumed from the beginning but is an emerging property of local causal constraints. This will be done by building on previous results of computer simulations of structural and functional self-organization. These results will be revisited and expanded by analyzing several aspects of what it is for behavior to be meaningful.

The mechanisms of the emergence of structure and meaning have become a legitimate problem for biological and physical sciences in the framework of current works on self-organization (See, e.g., Atlan 1972, 1979, 1987a, 1994a; Nicolis and Prigogine 1977; Eigen and Schuster 1979; Kauffman 1984, 1993; Salthe 1993). Following this trend, a successful theory of self-organization

NOTE: First published in *Thesis Eleven* 52 (1998): 5–34. This paper was given in a condensed version at the Boston Colloquium for Philosophy of Science on Emergence and Reduction, in 1995. I am indebted to critical comments and encouragement from Fred Tauber, Irun Cohen, Evelyn Fox Keller, and Gertrudis Van de Vijver. This work was supported by an Ishaiah Horowitz Scholarship in Residence in Philosophy and Ethics of Biology.

should be able to explain the emergence of meaningful intentional behavior from networks of nonintentional physical units defined by local laws only. This approach obviously departs from idealist phenomenology, Husserlian and post-Husserlian, where consciousness and intentionality are given as basic premises at the foundation of natural sciences. Therefore, I will start by analyzing some aspects of the encounter between phenomenology and the natural sciences on the question of intentionality and meaning.

Then, I shall present current works attempting to give an operational content to intuitive aspects of meaningful complexity. The classical algorithmic complexity of a task to be achieved by a computer program does not take into account the meaning of the program. The achievement of the task is assigned from the outside as the goal to be reached, and this provides the meaning of what the computer program is doing. In the functioning of natural machines, where an obvious goal is not set up from the beginning, the meaning of what the machines are doing must be assessed in a different way. The notion of sophistication as a measure of meaningful complexity adapted to the analysis of natural systems will be presented. As an example of application, the usual metaphor of DNA working like a computer program (the "genetic program") will be questioned, and an alternative metaphor of DNA as data will be discussed.

Relatively simple networks of boolean automata are able to classify and recognize patterns of binary strings on the basis of nonprogrammed, self-generated criteria (Atlan, Ben Ezra, et al. 1986; Atlan 1987a). However, what is missing in this model for behavior to be really intentional is a capacity for self-observation and interpretation. Therefore, the basic model of functional self-organization presented in section 5 will be extended and used as a basis to analyze two difficult problems that a naturalistic account of intentionality must solve. *First*, there must be a plausible mechanism for the apparent time-reversal of causality in the definition and achievement of a project. What is defined as the effect at the end of a causal process seems to come first, as the intended goal of the process. Since no final causes are assumed, the emergence of a goal in a machine must imply a mechanism to account for this apparent time-reversal. *Second*, the appreciation of meaning and intentionality takes place at the interface between a self-organizing system and its environment. For this reason, it will be shown that memory devices keeping record of previous histories of such interactions are essential ingredients for models of what appears to us as both conscious and unconscious intentional behavior.

Our seemingly infinite capacity for new interpretations requires machines with infinite sophistication. Such machines, although they can be formally conceived, by definition cannot be programmed. In practice, this behavior cannot be distinguished from that of a partly random, partly determined system, where emerging new meanings cannot be predicted but are recognized after the fact (Atlan 1987a). Thus, the notion of infinite sophistication provides a way of reconciling the idea of possible truly self-organizing systems in nature with physical determinism. The aforementioned models will be used to explain unexpected neurophysiological data on the time course of neural events in the onset of voluntary movement and to suggest, along the lines of Spinoza's philosophy, that the voluntary aspect of decision making is merely a memorization of functional self-organizing processes rather than an ex nihilo creation of a causal series. In conclusion, the notion of "underdetermination" will be presented as a general property of network dynamics that imposes an intrinsic limitation on models of complex natural systems but that makes possible mutual understanding and intersubjectivity, in spite of, and thanks to, what Wittgenstein called the "vagueness" of natural languages.

1. Emergence and Reduction

A first point must be clarified concerning the idea of "emergence," in view of the usual opposition between "emergentism" and "reductionism," the former being often assimilated to vitalism of one sort or another. The kind of emergence dealt with in physical and computer models of self-organization departs from this tradition, and it will be shown further that our approach here is both emergentist *and* reductionist. For a long time, emergence was viewed as an irreducible phenomenon of life, manifested in the occurrence of structural and functional properties at a global level in organized beings, when these properties could not be observed and could not be accounted for at the level of the constitutive elements. In that sense, at a first glance, chemical properties of molecules, such as chemical affinities and others, could also have been considered as emergent in relation to the physical properties of atoms and particles. However, the success of chemical thermodynamics provided a unifying quantitative science of physics and chemistry. Chemistry could be "reduced" to physics, at least in principle, but the question about life sciences being also reducible to physics remained open. At the end of the eighteenth century, Johann Friedrich Blumenbach's notion of "formative force" was taken up by Immanuel Kant in his *Critique of Judgment*, and it influenced nineteenth-century embryology as a way to conceive the epigenetic emergence of complex structures according to laws sui generis. That is probably one of the reasons why, especially in the nineteenth century and at the beginning of the twentieth, emergence was considered to be the effect of special vital forces acting in living beings and not reducible to physics and chemistry. In other words, emergentism was associated with vitalism.

In recent times, the attention of physicists and mathematicians has been drawn more and more to the physics of disordered and complex systems, for which classical thermodynamics was not sufficient and new mathematical tools of nonlinear systems dynamics were clearly needed (Katchalsky 1971, Haken 1975, Nicolis and Prigogine 1977, Eigen and Schuster 1979, Hopfield 1982, Atlan 1991b). This field of research has produced instances of emergent phenomena, in the sense previously limited to biology, in purely physical and chemical systems. Examples of such phenomena are structured vortices in hydrodynamics, spin glasses in solid state physics, and couplings of chemical reactions and diffusion in chemistry with the pioneering work of Turing (1952) on morphogenesis. Moreover, an essential property of biological organization is its "hierarchical" nature, that is, its organization on different levels of integration. Many problems of biology, as well as the richness of the behavior that seems specific to living systems, concern the nature of the articulation between levels (Atlan 1986, Salthe 1993). It is therefore natural to look for physical explanations of those behaviors in mechanisms of emergence similar to those produced in physical systems at a global level, by the coupling of simpler local constraints. However, in most cases, physical self-organization is viewed as the emergence of forms and structures, and little attention is paid to *functions*; indeed, the emergence of biological functions is still viewed as specific to biology. Emergence of a meaningful function, from the viewpoint of a mechanist self-organization paradigm, is the main topic of this article.

Early works on self-organization used the formalism of Shannon probabilistic information theory (von Foerster 1960, Atlan 1968). It was shown that the detrimental effects of noise in a communication channel at the elementary level could produce, under certain conditions, increased functional complexity at the more integrated level of a system containing the channel as one of its parts. This idea, developed later under the name of "complexity from noise" (Atlan 1972, 1974) was applied as a kind of principle for self-organization in several fields of biology and the human sciences (Dupuy 1994). The basic idea was that the effects of noise in a channel were different when appreciated at the output of the channel and on a system that *contains* the channel as one of its parts. While the effects of noise had to be counted with a minus sign in the first case, in the other they bore a plus sign and contributed to the creation of probabilistic information and complexity. More recently, because of the limitations inherent in probabilistic information theory concerning the question of the meaning of information, it was advantageous to switch to the more operational formalisms of boolean network and neural network computation (Kauffman 1970; Atlan 1983, 1987; Atlan, Fogelman-Soulie, et al. 1981; Atlan, Ben Ezra, et al. 1986).

With the help of these tools, I will ask the following questions: How can we understand intentional behavior as an emerging property of self-organizing systems? More specifically, how can new meanings emerge from old ones, or from meaningless, partly random structures? I will show that plausible mechanisms for functional self-organization can give at least partial answers. As mentioned earlier, the statement of these questions in operational terms and the attempted answers through computer simulations of self-organizing networks is at the same time emergentist and reductionist. It is *emergentist*, because we take seriously the possibility of nontrivial emergent properties in a multilevel system, in the sense not only that the whole is more than the sum of its parts but that global properties, both structural and functional, are qualitatively different and cannot be predicted in full detail *from* individual properties of the constituents. It is *reductionist*, since emergent properties are not merely described, even less used to *explain* the behavior of the system, but have to *be explained* causally, by nothing more than local interactions between the constituents that define, in mechanistic terms, the dynamics of the system. Indeed, from this viewpoint, such an approach is even more reductionist than traditional reductionism, since the aim of this research or, rather, this research program is something like a physical theory of intentionality.

2. Intentionality: Usual and Technical Senses

It is customary to distinguish the usual sense of intentionality as *the quest to attain a goal* from its technical sense in psychology and philosophy as *the source of meanings*. It is generally considered that the usual sense, the quest for a goal, is only a special case of intentionality in general, which includes desires, beliefs, knowledge, and all states of mind characterized by a directed activity or "aboutness" of consciousness, to which it gives meaning (see, e.g., Dennett 1978 and Searle 1983). However, this relationship between the usual and the technical senses operates in both directions. Not only is looking for a goal a particular case of intentional meaningful behavior but also, in the opposite direction, the meaning of something is defined by the specification of a goal. This is obvious for man-made machines and computers. The components of a machine have meaning only by virtue of what they are doing in the global operation *for which* the machine has been designed. In a computer program, the criterion for the meaning of the instructions is their role and effectiveness in solving the problem or accomplishing the task *for which* the program has been written. Thus, it is the goal, the task to be achieved, that determines the meanings of the components in a machine or the instructions in a program. In that case, the goal—and therefore the origin of meanings—is defined and imposed from outside by the programmer or by the designer. But is it possible to conceive of a machine where *the goal* to be reached, the task to be accomplished, would not be imposed from the outside but *produced by the machine itself*? This is an operational statement, probably reductive (but we shall return to this point), of the question: "Can a machine be intentional?" and more precisely, "Which kind of machines can exhibit some sort of intentionality?"

In other words, trying to answer the general question of intentionality in natural machines, we will have to proceed step by step, starting from a notion of meaning limited to the pragmatic sense of a biological function, appearing goal-directed, such that functionality and meaning are interpreted relative to the goal. Within this framework, we shall present instances of computer simulations aimed at answering the more operational and limited question: "How can a goal emerge as a source of meaning in a multilevel, self-organizing process, as a global property of a network of interacting automata?"

It is important to notice that this approach does not make use of functional analysis in any form of teleological explanation. Function and goal-directedness are not used to explain the behavior of the system. On the contrary, *they* have to be explained, and we want to use only causal explanations provided by the dynamics of automata networks to account for the emergence of function and goal. Of course, the automata in question are not themselves complicated machines endowed with autonomous faculties and behaviors simulating those of living organisms. They are not intelligent robots, whatever that may mean. Instead, they are physical or simulated physical units that obey relatively simple laws. It is their aggregation into networks that generates interesting global properties. In formal neural networks, each automaton is a unit that simulates in a very simplified form some of the physical and logical properties of neurons. In most of the computer simulations we will discuss further, we have used boolean networks, in which each automaton computes a binary boolean function. However, it was shown some fifty years ago by W. S. McCulloch and W. Pitts (1943) that the two formalisms of neural nets and boolean networks are equivalent. It is on the basis of these simulations that intentional behavior will be understood as a global property of complex dynamics, in networks made of idealized, neuronlike elements, locally determined by relatively simple, nonintentional physical constraints.

3. Physical and Phenomenological Reductionism

Philosophers, and especially phenomenologists, may find presumptuous our desire to develop a reductive physical theory of phenomena such as meaning and intentionality, traditionally considered to be the province of consciousness. This cannot be ignored, since Husserlian and post-Husserlian phenomenology constitutes a powerful attempt to understand what it means to understand and think, based on analysis of the flows of our consciousness as the origin of all meanings. However, even for Husserl, the ascendancy of vitalism made the idea of physical biology impossible, whereas this is precisely the type of biology that we practice today. That is why we have to consider a symmetrical relationship between *two kinds* of reductionism. On the one hand, the *phenomenological reduction*, Husserl's famous *epochē*, brackets the "naturalistic attitude," that is, puts aside the natural sciences, to be explained and founded later by transcendental phenomenology (Husserl 1947, 1950). On the other hand, the *physico-chemical, materialist reduction* brackets consciousness, to be explained later by the natural sciences. In other words, from a methodological perspective, the reductionism of the natural sciences is the mirror image of the reductionism of consciousness that phenomenology asserts as the first moment in its process. However, contrary to what Husserl believed, there is no fundamental asymmetry such that the intentionality of consciousness would be the unchallenged giver of meaning, including the meaning of the world of the natural sciences, as it would be disclosed by phenomenological analysis. In other words, contrary to the program of Husserl's *Cartesian Meditations* (1947), phenomenological analysis cannot serve as the absolute grounding of the natural sciences, because the natural sciences have developed historically with no need for an absolute grounding, according to local criteria for truth and effectiveness that free them from that need.

We must look to Spinoza for a philosophy that can support this symmetry between the physico-chemical reduction of consciousness and the phenomenological reduction of physical chemistry. We will come back to this later. In any case, the bracketing of consciousness is clearly the challenge made by the cognitive sciences, especially in recent years, in the wake of the new momentum provided by the physics and dynamics of complex and disordered systems. Thus, in the context of this work, I take my place intentionally—if I may say so—in the materialist context of physico-chemical biology, standing in what Husserl called the "naturalist attitude" and assuming that attitude, claiming that its undeniable limitations are no larger, deeper, or more radical that those of the phenomenological attitude.

4. Meaning in Biological Systems

Before going back to our question "Can a natural machine be intentional?" or in a more limited way, "Can a goal emerge as a source of meaning, in a self-organizing network?" *two remarks* must be made. One deals with the question of meaning itself: How do we judge that something is meaningful and, if possible, "how meaningful" it is? This will lead us to a schematic presentation of the concept of "sophistication" as a measure of meaningful complexity, derived from classical algorithmic complexity. The second remark comes from an analysis of different kinds of self-organization. Self-organizing systems can be distinguished as being more or less "truly" self-organizing, according to the origin of the meaning of what they are doing.

COMPUTER METAPHORS IN BIOLOGY: MEANINGFUL VERSUS NONMEANINGFUL
ALGORITHMIC COMPLEXITY

Classical programming theory does not ask about the meaning of programs, just as communication theory leaves aside the meaning of messages; both take meaning to be granted by the a priori position of a task to be executed or a content of a message to be transmitted. The theory of computation by artificial machines does not have to deal with understanding that meaning. As a result, we have the well-known paradox of classical algorithmic complexity theory: maximum complexity is achieved by an infinite, meaningless, random string because it cannot be compressed into a description shorter than itself. This is not a problem for a theory of computation of a priori meaningful programs, written by intentional people, but it is a problem for the assessment of meaningful complexity of *natural objects* not intentionally designed by humans.

In addition, we are encumbered by a handicap derived from biology, or more precisely, from its computer-science metaphors, such as those of the brain as a computer, even a parallel computer, and of embryonic development compared to the execution of a program written in DNA sequences. No doubt, the computer metaphor has some strengths. However, it has a major deficiency: it presupposes the solution to the problem it is supposed to resolve, namely, that of the origin of the meanings without which no computation can be meaningful and therefore represent intentional behavior. Such a meaningless computation is in itself no more than a syntactically correct expression, like an algebraic formula; it says nothing about the physical world. In the best of cases, it can do no more than simulate real behavior, and even then on the condition that an external intentional agent, a theorizing human being, projects meaning onto it. In other words, these computer metaphors also take meaning as granted, namely, in the form of biological functions interpreted *as if* natural machines were intentionally programmed.

EXPLICIT VERSUS IMPLICIT MEANING: SOPHISTICATION AS A MEASURE OF
MEANINGFUL COMPLEXITY

If we accept the need to depart from the classical theory of computer algorithmic complexity and return to the intuitive distinction between *program* and *data*, it becomes possible to define an operational measure of meaningful complexity for organized natural systems. Assuming that we can describe or reproduce a natural system by means of a computer algorithm, the classical algorithmic complexity (Kolmogorov 1965, Chaitin 1975) of the system is the length of its minimal description, typically composed of a program and the data it processes. However, from the viewpoint of computing complexity, data and program are interchangeable, since a program for a computer can always be treated as data for another one. Interchangeability of program and data is closely related to the fact that computing theory does not take into account the meaningful aspect of what is computed. Any minimal description written in binary language is a random sequence of os and 1s, since it is incompressible, that is, no order can describe it (by a correlation function or otherwise) in a shorter way. Nevertheless, this minimal description has a meaning: namely, what it describes as an object to be built or a task to be achieved. This indicates that, if meaning is implicit only, randomness and meaningfulness can go together; but if we want meaning to be dealt with explicitly, classical algorithmic complexity theory is insufficient.

In collaboration with Moshe Koppel, we have proposed a way to correct this flaw and capture the meaningful aspect of computing complexity by retaining the distinction between program part and data part in a minimal description (Koppel 1988; Atlan and Koppel 1990; Koppel and Atlan 1991a, 1991b; Atlan 1995b). We defined the "sophistication," or meaningful complexity, of the object or string generated by such a minimal description fed to a Turing machine as the minimal length of the program part only in that description. Thus, a random string R representing a nonstructured object has a near-zero sophistication, whatever its length, since its minimal description is "PRINT R," where the program part is reduced to the single instruction "PRINT," and the data part is the whole string R.

More generally, leaving aside the technical aspects of the matter, the program part in a description defines a structure characteristic of a whole class of objects or strings that share the same structure, whereas the data part specifies a given object within that class. As a simple example, let us consider a binary string S made of triplets, in which the first and third digits are the same and the second digit is their complementary one, such as: S = 010010101101101010101001001010101 . . . When this string is long enough, its true minimal description is made of a *program* that says "TRIPLE AND INVERT THE SECOND DIGIT" and of *data* in the form of a string that specifies which digit must be tripled and treated the way described by the program. In our example, the data string would be 0011101001 . . . Different data would generate a different string, but the structure defined by the program would be the same. Thus, one can see that the longer the program part, the more complex the structure—the more "sophisticated" it is, irrespective of the length of the data, which do not participate in the definition of the structure. Similar data would be processed by a shorter program, such as "DOUBLE," which would generate the less sophisticated string S′ S′ = 0000111111100110000011 . . . made of couples of identical digits.

We have applied these concepts to the critical analysis of the genetic-program metaphor, which views DNA sequences as analogous to computer programs. The origin of the metaphor may be traced back to the following observations and their implicit use in a fallacious chain of reasoning. (1) DNA is a quaternary sequence that can easily be reduced to a binary sequence. (2) Every deterministic sequential computer program can be reduced to a binary sequence. (3) Therefore, the genetic determinations produced by the structure of DNA function as a sequential program written in the DNA of the genes. Of course, (3) is wrong because it assumes that (2) necessarily implies its reciprocal, whereas if it is true that every program is reducible to a binary sequence, it is not the case that every binary sequence is a program. But if a binary string is not a program, what else can it be? It could, obviously, be a random string. However, this is difficult to accept in the case of DNA, in spite of our previous remark about possible implicit meanings appearing as random strings in minimal descriptions. But a binary sequence could also be *data* to be processed by a program. We have discussed elsewhere (Atlan and Koppel 1990, Koppel and Atlan 1991a) the respective advantages and disadvantages of the two alternative metaphors of DNA as program or DNA as data. In the latter case, the role of the program would be played by the whole cell's metabolic machinery, in the form of a network of biochemical reactions working like a parallel distributed program. Thus, the meaningful part of biological functions should not be restricted to DNA structures but should instead be seen in terms of the dynamics of cytoplasmic epigenetic regulations determining different patterns of activity of the genome in different cell types of an organism. According to this view, neither of the two metaphors, DNA as program and DNA as data, should be taken literally, since a more realistic view of the cell would be that of an evolving network of metabolic reactions, in which a given pattern of gene activity determines the structure of the metabolic network; the dynamics of this network can lead to a temporary steady state capable, in turn, of modifying the pattern of gene activity, and so on. Thus, DNA and metabolic regulatory activity can be viewed alternatively as data when their state of activity is constant and as program when their pattern of activity is changing according to the dynamics of their interactions.

The notion of sophistication as a measure of meaningful complexity helps clarify the concept of self-organization itself. I will argue that true self-organization can be defined as a property of systems with infinite sophistication. Before reaching this point, however, we have to analyze the kinds of functional self-organization that can be exhibited by automata networks and the conditions under which such self-organization can approximate a "true" emergence of meaning and intentionality.

In the following, we shall examine the behavior of networks we consider self-organizing in a "strong sense." In the example that we shall consider, meaningful functions of pattern recognition are achieved on the basis of self-generated criteria for recognition, which were not defined a priori, either explicitly, or implicitly—for example, through training sets of samples. Then, we shall analyze additional conditions for such networks to be self-organizing in a *stronger* sense: that is, conditions to reduce as much as possible generation of meaning from outside as interpretation by human observers. Finally, a "true" or "intentional" self-organization will appear to be achieved when the interpretation and meaning of the behavior of such a system are themselves properties emerging from an association of the dynamics of a self-organizing network with some memory devices.

By contrast to self-organization in a weak sense, in which the goal that defines meaning is imposed from outside, *strong* self-organization implies that even the task to be accomplished, the goal to be attained—that is, what defines the meaning of the structure and the operation of the machine—is a property that emerges from the evolution of the machine itself.[1] This is what we find in nonprogrammed natural systems, where we observe the emergence of structures and functions at a macroscopic level, based on general physico-chemical constraints at the microscopic level.

Such behavior can be simulated by computer programs, which are explicitly programmed for nothing in particular but can nevertheless execute something that has a meaning. A program of this sort, called "Soar" by its authors, A. Newell, J. Laird, and P. Rosenbloom (Newell 1990), is apparently an

---

1. [Atlan discusses the distinction between weak, strong, and intentional self-organizing networks in considerably greater detail in Atlan 1992b.—Eds.]

expert system programmed to resolve problems. Unlike a classic expert system, however, which halts when presented with a problem for which its knowledge base is inadequate, Soar never gives up. It offers a solution, even if inadequate and, taken to the limit, somewhat random, to any problem input to it, by searching through its "problem space." It applies extremely general rules, which are intended not to solve any particular problem but rather to move about in this space. Subsequently, it stores the problem and its "solution," thereby enriching its knowledge on the basis of unprogrammed "experiences." As Newell puts it, "Soar does not have to be programmed to do something. . . . Soar is goal-oriented, but not just because it has learned goals in its memory. It is goal-oriented because its goals arise out of its interaction with the environment. It builds its own goals whenever it can't simply proceed" (ibid. 228–29). Whether or not this performance is sophisticated enough, self-organization in a strong sense is characterized by the absence of a predefined goal and the emergence of what seems to be, *after the fact*, a functional behavior, that is, one with meaning.

5. Emergence of Classification and Recognition Procedures in Random Boolean Networks

Our own work on the emergence of classification procedures in random boolean networks (Atlan, Ben Ezra, et al. 1986; Atlan 1987a) provides a paradigmatic example of the simulation of functional self-organization in a strong sense. Random boolean networks had been studied for the first time by S. A. Kauffman to simulate cooperative properties of genes in eukaryotic cell genomes (Kauffman 1970, 1984). In collaboration with G. Weisbuch, F. Fogelman-Soulie, and J. Salomon, we studied some of their structural self-organizing properties (Atlan, Fogelman-Soulie, et al. 1981), namely, their usual evolution from homogeneous random initial conditions toward attractors exhibiting spatiotemporal macroscopic patterns. In this preliminary work, we found emerging stable *structures* made of subnets of elements stabilized in state 0 or 1, separated by subnets of oscillatory elements periodically cycling through relatively short limit cycles (see Figure 2.1 and item a in its legend). However, we have also shown that these networks provide interesting generic models of *functional* self-organization, where what "emerges" is not only a macroscopic structure derived from nonspecific microscopic constraints but also a meaningful function: when a random boolean network is stabilized in

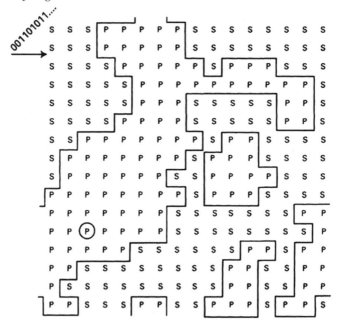

Figure 2.1.   The Schematic Representation of Functional Self-Organization in a Random Boolean Network (Atlan, Ben Ezra, et al. 1986, Atlan 1987a).

(a) *The emergence of macroscopic structures.* The elements of the matrix stand for boolean automata. They are connected in such a way that each of them receives two binary inputs from two of its neighbors and, after computation, in turn sends the same output to its two other neighbors. Each row and column is closed on itself. The computation is done in discrete steps, in which each element computes its output from its two inputs by means of one of the fourteen possible two-variable binary functions (leaving out the two constant functions). The boolean functions are randomly distributed on the elements of the network and do not change during the computation. The figure represents the states of the elements when the network has reached one of its attractors: S stands for stable, P for periodically oscillating with a relatively short cycle length.

(b) *The emergence of function.* The arrow indicates an element that serves as an input element for a recognition channel. After the network has reached its attractor, binary sequences are imposed on this element and perturb the state of the network for a period of time (at least two cycle lengths). It may happen that an oscillating element (circled P) becomes *stabilized* when *some* binary sequences are applied to the input element, and not when other binary sequences are applied to that element. The structure shared by all the binary sequences that stabilize circled P can be said to be "recognized." Thus, whether or not circled P is stabilized serves as an output for the recognition channel. Note that the functioning of that output element, and the structure defined by the class of sequences being recognized by this particular channel, are uncovered after the fact, since the network has not been programmed to achieve that goal. In other words, the behavior of the network as a pattern-recognition device and the criterion for recognition are emergent functional properties.

one of its attractors, it can function as a device performing classification and pattern recognition based on nonprogrammed, self-generated, criteria (Atlan, Ben Ezra, et al. 1986).

More specifically, input binary sequences are presented to some element after it has reached a stable constant state in the attractor configuration of the network. Since the sequences are imposed on the element as temporal successions of binary states, they obviously destabilize the previously stable element and some of its neighbors. However, some of these sequences produce complicated resonance phenomena with the cycles of other elements of the network, sometimes far apart, belonging to the periodically oscillating subnets in the attractor configuration. As a result, surprisingly, some of these oscillating elements are now stabilized and remain constant, while a perturbing sequence is imposed on a previously stable element of the network. Thus, the network functions as a device with an *input element*, to which binary sequences are presented, and an *output element*, separated by a complicated computing channel. (See Figure 2.1 and legend b.) The output element is normally periodically oscillating in the unperturbed attractor configuration, but it is stabilized *when the imposed binary sequences belong to a class defined by a given structure.* Each structure defines the criterion for the recognition of a sequence by a given couple of input and output elements in a given network. However, *the structure defining the class of sequences to be recognized has not been programmed*; it is a self-organizing property of the structure of the attractor, emerging from the microscopic structure of the network and from its initial conditions.

The interesting generic feature of this behavior is that, starting from random initial conditions, the dynamics of a given network, itself built randomly, defines the structure of a class of sequences that serves as a criterion for classification and recognition of sequences according to their belonging or not belonging to that class. In most cases, not only has this criterion not been explicitly programmed, but it is even impossible to predict its structure by just examining the microscopic structure of the network and even its macroscopic spatiotemporal structure after it has reached one of its attractors. Although the dynamics is determinist, in the sense that the microscopic structure of the network does not change after it has been set up at random, it is too complicated to allow for an a priori prediction of its outcome without actually running it and observing the results, as if we were observing a natural system and doing an experiment without knowing in advance the details of its outcome.

Thus, we have here a relatively simple model of emergence of function as a nonprogrammed criterion that gives meaning to a class of sequences being recognized. It is as if, having introduced two electrodes into the brain of a mammal (one to enter stimuli and the other to record a response), we were to find that we obtain a response only for a given class of stimuli. Then, we would assume that this classification had a functional meaning, and we would have to try to understand what it was, since the criterion for the differential responses would be produced by the internal structure and operation of the brain, and not be known to us beforehand.

6. The Role of Interpretation: Emergence of a Project

This model of the emergence of function can easily be criticized if one notices that the network cannot "know" or "understand" in any way that it is doing pattern recognition or giving meaning to a class of sequences. This is our interpretation as observers. In other words, only the eyes of an external human observer interpret the network's behavior as a meaningful function. Therefore, let us try to see what so far is missing for our network to behave at least *as if* it understood what it was doing by really creating meanings, not only in the eyes of the observer.

Generally speaking, if we want to build models of the true emergence of meaning, we need a theory of interpretation. Such models would simulate the third kind of functional self-organization, which I have called intentional; indeed, such a theory of interpretation would be the physical theory of intentionality—that is, of giving meaning—that we are striving for. That is why it is probably more efficient first to discuss limited models of possible mechanisms whereby specific interpretations may be produced and to try later to examine the conditions under which they can be generalized.

TRANSFORMATION OF A CAUSAL SEQUENCE OF STATES INTO A PROJECT-ORIENTED
PROCEDURE: THE APPARENT REVERSAL OF TIME

Since there is a need for an observer to interpret what the network is doing as a function, it seems that a necessary condition for the system to be the real origin of meaning is that it is somehow able to observe itself. The reticular formation in our brains seems to be necessary for such a function of self-awareness, although we do not know how. However, as will be discussed further, while this brain structure may be sufficient for self-observation in other mammals, this does not seem to be the case for human self-awareness, with its experience of temporality and its linguistic reflexive features.

Let us assume that such a module of self-observation is added to our network from section 5, in the form of a *memory* device storing the sequence of states that the network goes through until an attractor state is reached. Then a simple instance of intentionality could be produced, as an emergence of a *project*, and a procedure to achieve it. This is remarkable not only because humans seem to be capable of such a performance but also because some primates, such as chimpanzees, are able to transform objects into tools. As in the famous movie *2001*, some apes who originally made a sequence of movements with no apparent functional meaning discover some effects of these movements and repeat the sequence in a goal-seeking fashion, apparently with the intention of reproducing these effects. As with humans, their utilization of primitive tools oriented toward their future use cannot be understood other than as thanks to a capacity for projects that can determine behaviors by *interpreting the effects* of these behaviors, that is, by giving them a meaning that they did not have the first time they occurred.

A relatively simple machine capable of producing a project of this type might work in the following manner. As we have already observed in the behavior of our random boolean networks, a succession of states of the network produced by a purely causal dynamics of interactions among its elements may acquire a functional meaning for an outside observer. This is what happens when this succession of states leads to a final state in which the network does something that then appears to be a function determined by the series of preceding states. This series seems a posteriori—always to the eyes of an outside observer—to be that of the means used to reach this end. Suppose that this sequence of states is memorized as such. Then everything happens as if we had here a potential procedure that is memorized because the final state—with its effect—is stored as the outcome produced by this sequence. Later on, this final state may be recalled in memory, associatively, by a sequence of states different from the previous one, as is the case in associative memories, where the dynamics converge on the same attractor from different initial states (Hinton and Anderson 1981). But this triggers a recall from memory not only of this final state but of the entire procedure, since the two were stored together. This is how the network, being assumed to be able to observe itself, can see the effect of this final state as being the function of this state. We can thus imagine a mechanical and purely causal model of our capacity for projects, as already seems to be present in chimpanzees. This model allows us to understand the occurrence of the apparent *reversal of time* that is the main feature of the production of a *project*.

A project for the future is always the outcome of the reversal, in its representation, of an effect into a cause. What was the end of a causal sequence of moves and its effect the first time, in an action that preceded its representation and transformation into a procedure, is converted into a cause in the representation itself. This conversion is not particularly mysterious in the kind of model we are suggesting, because it is made possible by the flattening of time produced by all memorization. Once a temporal sequence of states is memorized, its past-future order is transformed into a symbolic one in which time disappears, since the entire sequence is present in memory at the same time—though this does not prevent the assignment of a temporal index to each event in the sequence, because the index is clearly symbolic rather than temporal.

A SATISFACTION FUNCTION AND ITS ORIGIN

We are now left with the following question: How is the system going to be aware of the meaning or desirability of a given state? Self-observation as mere memory is not enough. In addition to recording states that have been produced by chance, there must be a kind of satisfaction or optimization function that tells the system (when the final state is produced or recalled on a different occasion by a different route), that this state is desired or interesting, or more desired and more interesting than others. But if such an optimization function were imposed from outside, the system would be subject to the same criticism: namely, that the meaning originated with the programmer. Therefore, the satisfaction function must be produced by the system itself.

For this purpose, let us go back to our boolean networks and their recognition functions as a simplified model of the neural network of an ape and of its capacity to behave by chance in a certain way apparently meaningful to the eyes of a given human observer. In order to internalize the observer, we have assumed already that the network is associated with a memory device allowing for a stored observation of the states of the network. In order now for a satisfaction function to be generated by the system itself, made of the self-organizing network and its associated memory, we need the memory to record not only every incoming input sequence that has been recognized once and the pathway of causal sequences of states that has led to the attractor of the network as the final state capable of performing its specific function of recognition. We also need the memory to record *how frequently* input sequences of a given class have been presented to the network and recognized. This frequency distribution would determine the satisfaction function. Thus, the origin of this function would depend, on the one hand, upon the structure of the environment and the history of its encounters with the system, that is, the kinds of input sequences met by the system, and, on the other, on the structure of the system, which determines which features of the environment may or may not be recognized. Such a system would work as *a feature extraction* with no a priori definition of the feature, *neither explicit* (as in classical programming) *nor implicit* (as in adaptive learning from a training set).[2]

NONINTENTIONAL MECHANISMS OF INTENTIONAL BEHAVIOR

An interesting result of our mechanistic model of a project is that we can now identify the different parts of the classical syllogistic description of intentional end seeking as natural outcomes of different elements of the model. A classical way to relate intentionality to goal directedness or end-seeking is a description of a typical teleological behavior, making use of a means to produce an end in the form of the following syllogism:

1. Somebody B has the *desire* to be in state S.
2. B *knows* or *believes* that C is a cause of S.
3. Therefore, B produces C.

Obviously, this description presupposes intentionality, since it explains the production of C as a means of producing S by the existence of the intentional attitudes of desire, knowledge, and belief. However, the production of intentionality has itself to be explained by causal, nonintentional processes, and a whole chapter in the philosophy of mind is dedicated to that question. Usually, the existence of particular mental states, called intentional or directed states, or internal representations, is assumed to be the cause of intentional attitudes. Of course, the kind of relationship to be conceived between these mental states and brain states is at the core of the mind-body problem. In the context of naturalist models, the most popular and widely accepted relationship between mental and brain states is that of supervenience, which merely assumes the existence of some sort of asymmetrical relationship—several

---

2. For more on this model of intentional self-organization, see Atlan 2011 and Louzoun and Atlan 2007.

brain states may correspond to the same mental state but not the opposite—without spelling out the nature of this relationship.

Our model allows us to reverse the whole approach by not using intentional attitudes to explain intentional behavior but rather explaining both by different mechanisms that can be identified in the process of producing and achieving a project. In other words, a model of an intentional, purposeful, end-seeking process based on purely causal mechanisms may eliminate the need to resort to internal semantic representations, or intentional or directed states. In the above-mentioned syllogism, desire of S and knowledge or belief that C is a cause of S are identified with the following mechanisms that are parts of the suggested model for the emergence of a project:

1. The "desire of B to be in S" is simply the fact that a network B is driven to S by some stimulus acting upon its dynamics as specified by its own preexisting network structure.

2. The knowledge or belief that C is a cause of S is the storage in memory of a sequence of states, including C, that has produced S in the past starting from some stimulus (which might be C) and initial state.

3. The recall of that sequence of states by B is triggered by that of S because it was stored together with S when it occurred for the first time.

As mentioned earlier, the difference between state S produced by the second stimulus and interpreted as the desired state and state S produced by the initial, self-organizing causal sequence of states is due to the different pathways that have led to S in the two cases, starting from different initial states. This ability of a dynamic system to reach the same attractor state in different ways is a nontrivial but common property of complex physical systems. In general, the same attractor state can be reached not only by different routes within the network but also by large numbers of different networks made up of the same number of units, differently connected. This structural stability is an instance of the "underdetermination" of theories by facts; it also provides a possible natural foundation for intersubjectivity and mutual understanding.

7. *Extension of the Model: Consciousness as Memory*

A property of this model is that we can build on it and develop, at least in speculative and abstract terms, a conception of how a more elaborate intentional system could work in a way closer to our own cognitive system, which seems endowed with a practically infinite capacity for interpretation.

Human systems, both individual and social, occupy a curious intermediate position between natural systems, for which meaning has an internal origin and which can be observed from the outside only by interpretive projection, and artificial machines, whose external goal orientation and source of meaning are known because planned and observed directly by human beings themselves. Human systems are simultaneously machines, the external observers of these machines, and even sometimes, to some extent, their designers. Note that this in no way presumes that we must accept the reality of *free will* and must not presuppose the existence of causal determinations for intentions themselves. Rather, the intentionality that creates the project is recognized as a particular sort of efficient causality and, as such, a specific object of the sciences of man (and of related animals). The question of free will is set aside until we have detailed knowledge as to how specific intentions are causally determined—if that will ever be possible and if the underdetermination of theories by facts does not constitute an irreducible limitation to this knowledge. Accordingly, in the previous section, we have proposed a model of our capacity for new projects, a capacity that seems to be observable already in behaviors of the great apes. In this model, memorization of temporal causal sequences of states of a neuronal-like network was a necessary component to transform such a sequence into a project-oriented procedure. Then, differential memorization of different procedures depending on their relative frequencies could serve as an internal origin of a primitive interpretive activity in the form of a satisfaction function.

What is suggested by this model is that awareness or consciousness, although it may seem directed to the future, is merely memory of this process of constructing procedures and recollection of frequent states that happened to be the most frequent because of properties both of the network and of its environment. One step further on the way to humanlike intentionality would involve another level of memorization, namely, memorization of memorizing the procedures. This would allow us to extend our model to the experience we have of a *general* capacity for projects, independent of particular projects. In other words, it would produce a mechanical model of our consciousness as consciousness of intentionality, in the clearly limited and restricted sense that we have envisioned.

The idea that consciousness is basically memory of the past, not directed or oriented toward the future, does not mean that there is no novelty to be expected from the future. But this novelty is unconscious. Obviously, before it happens novelty is not known, not conscious. After the fact, conscious novelty is the result of indetermination and chance as they are transformed into structure and meaning by the dynamics of *memorized self-organization*. Contrary to the idealist view of consciousness directed toward external objects and giving them meaning, in our view (Atlan 1979), basically, consciousness is merely recording, storage, and retrieval of past events, whereas innovation and adaptation to newness, with the creation of new structures and functions, are produced by unconscious self-organizing processes.

In this view, what makes human intentionality different from that of animals is a qualitative jump accompanying the quantitative increase in memory storage capacities. This view is in line with primate studies of human language acquisition (Gardner and Gardner 1969, Linden 1974). What seems to be missing for young chimpanzees to learn human languages as well as children is not some capacity of high-level understanding allowing the formation of new sentences that are syntactically correct and meaningfully adapted to new situations in a dialogue; they obviously have that capacity, or at least are able to acquire it. What seems to be missing is merely enough memory to store and keep what has been learned and to retrieve it when necessary.

8. Infinite Sophistication as a Clue to True Self-Organization

Something is still missing from our model if it is to be relevant to human intentionality viewed as an endless capacity to give meaning. A last step is necessary to find a new organizational principle that would be unique to our semantics and intentionality in that it would explain our unique capacity for interpretation, for projecting and giving meaning to everything. What is missing is a capacity not only for memorizing procedures with their meaning, and then the process of memorization itself, but also for *modifying the meanings* of the procedures under almost any circumstances that modify the procedures themselves. Observation of oneself is not sufficient. In order to work like an apparently infinite source of interpretations, self-observation (or awareness) must be connected to one or several self-organizing devices able to produce novelty indefinitely, which it will indefinitely interpret with new meanings. Here also, a qualitative jump may result from a quantitative increase in memory and self-organization capacities. This is where we need to come back to our notion of sophistication.

Remembering that sophistication as a measure of meaningful complexity is the minimum length of a program part in a minimal description, let us consider the case of an infinite object (or the string that describes it) with infinite sophistication. This is what we expect to be the characteristic feature of a truly self-organizing system. Nothing prevents an infinite object from having a finite sophistication if it can be generated by a finite minimal program: obvious examples are infinite sets of numbers endowed with a given structure (such as prime numbers or solutions to given equations, etc.), which may be quite complicated, or rather "sophisticated" in that the set, with its structure, may need a long minimal program to be generated. However, the kind of objects we want to consider are assumed to have an infinite sophistication. A program of infinite length would be needed to generate them. It is clear that such programs—as well as infinite objects themselves—can be formally defined, but not written or implemented. In such an infinite program, only finite initial parts can be known at any given time, but, as soon as more of it is known, each time an additional part is uncovered, it appears to have a new structure when compared with the one that was known before, and so on ad infinitum. In other words, infinite sophistication would imply that the meaningful complexity of an object increases indefinitely as the object itself evolves. From this viewpoint, there is some similarity between the notion of sophistication and that of "logical depth" (Bennett 1988), measured by the time needed for an object to have evolved as it is.

Objects with infinite sophistication have the peculiar property of being neither recursive (i.e., computable) nor random. By definition, an object with infinite sophistication cannot be programmed by a computer algorithm, since any minimal program able to compute it should be infinite. Yet it is not random, since we have already seen that structureless infinite random objects have an almost zero sophistication (even though their classical Kolmogorov-Chaitin complexity, i.e., their minimal description, is infinite). Under the influence of the mathematical theory of computability, it is generally assumed that everything in nature is either computable or random. This assumption, known under the name of the physical Church-Turing thesis (Davis 1965), implies that true self-organizing systems could not exist in nature. It is based on considerations about the power of computation languages and their equivalence as far as the set of computable strings is concerned. Applied to the physical world, this thesis asserts that everything is computable, since everything obeys physical laws that are computable. The only exception might be random phenomena, if one admits that irreducible randomness exists in nature. However, in two provocative books, R. Penrose (1989, 1995) argues that the Church-Turing thesis cannot apply to all physical phenomena because of the limitations imposed by Gödel's theorem. He also stresses the distinction between computing and understanding, to show that human (and other?) brains capable of understanding might be instances of such physical systems, to which the Church-Turing thesis would not apply. This work is particularly interesting because it does not resort to a dualist ontology, in which a mind with no material substance would not be computable but nevertheless would be able to trigger actions and impose noncomputable properties on material bodies.

Now, it is clear that if something is programmable, it cannot be said to be truly self-organizing, since its organization is the outcome of a preexisting program, in a sense outside of, and independent from, its own existence. That is why the instances of self-organizing networks analyzed above cannot be considered to be truly self-organizing, even though their macroscopic structure and behavior cannot be predicted before being computed by a computer model simulation. We called them self-organizing in a weak or even a strong sense, but we left aside the discussion about what we called intentional self-organization. It is clear now that a truly self-organizing system should be neither programmable nor random. Infinite sophistication provides a formal definition for such systems.

Since no infinite program can be exhibited during our finite existence, we are left with the question of whether such nonprogrammable, nonrandom objects actually exist in nature. One can always argue that what appears random is produced by a hidden structure as yet unknown. This argument is supported by the fact, among others, that actual computer minimal descriptions translated into binary language appear to be random, in the sense that no correlation can be found in their sequence of binary signals and that they are not compressible, although they have been written with a given sophisticated structure in order to achieve a given task. Then, a computer model made of an algorithm including some kind of indetermination or randomness can approximate an object with infinite sophistication if it generates an infinite object, since the structure of that object cannot be completely predicted before it uncovers itself. Conversely, an object with infinite sophistication produced by nature would appear to us—through some finite parts of it accessible to our observation—partially structured but unpredictable due to what would look like noise or randomness.

This property of infinitely sophisticated objects being neither random nor programmable, not predictable but still structured and interpretable, is what we expect from meaningful things, to which new meanings are constantly assigned upon constant, unexpected reorganizations. Only if we had infinite knowledge would we be able to know whether this object is partially random or really has an infinite sophistication. However, knowing more and more about its structure and discovering it to be more and more sophisticated brings us nearly to the possibility of experiencing its infinite sophistication. If the sophistication of the sequences of states in our brains were large enough to approximate infinity, this would explain their apparent ability to create indefinitely new meanings by interpretation. Then, if such machines, producing behaviors of apparently infinite sophistication, could ever be made by Artificial Intelligence (AI), especially in the form of multilevel, partially random, architectures of automata networks, the question could be raised as to whether they would experience the same feelings of meaning and intentionality as we do.

Even then, we would have to leave open the question of real consciousness versus mere simulation of consciousness. We would face the question debated by the partisans of strong and weak AI (Searle 1990, Churchland and Churchland 1990, Dreyfus 1994): Does a computer program understand what it does, or does it produce a mere simulation of understanding, like a computer simulation of flying, breathing, digesting, and so on? For the proponents of strong AI, understanding is nothing but some sort of computation. Since a working computer is not simulating computation but actually performing it, nothing prevents a computer from understanding whether it performs the right computation. Clearly, this controversy is actually about our understanding of what understanding is and cannot be settled unless we agree on what understanding is. The same holds true for consciousness. If we hold that, for us, being conscious (as well as understanding) means computing in a particular fashion, then the computer truly computes and hence does more than simulate understanding and consciousness. We could hold that the computer, like a hardware-software Leibnizian monad, really understands and is conscious. If, by contrast, we hold—as I would be inclined to do—that our understanding and consciousness imply not only computation but also (for example) our digestion and breathing, that is, at least the metabolism of our neurons, then the computer merely simulates certain results of this metabolism, just as it simulates respiration without breathing.

9. Conclusions

Extending to infinity the analysis of self-organizing systems implies a kind of Spinozist approach, whereby our knowledge and understanding of nature are viewed as a part of an infinite understanding. This approach leads to some interesting conclusions about perception and action, and about the question of underdetermination.

PERCEPTION AND ACTION

Coming back to our association of memory and self-organization, we have seen that we can at least conceive how meaning and intentionality can be created by the more or less frequent repetition of events leading to self-observed behaviors of dynamic systems. In that sense, we can understand that pattern recognition and decision making are basically the same process, although the former seems to be a recollection of the past and the latter a projection onto the future. The differences are only a matter of emphasis on one aspect or the other in the unified process of creation of meaning by self-observed self-organization of behavior. Being more sensitive to the aspect of observation, we experience what we call perception or pattern recognition, and by accentuating the behavioral aspect, we experience what we call will and decision making. However, as Spinoza (pt. 2, prop. 49, corollary; Spinoza 2002a: 273) said three hundred years ago in his concise way: "Will and understanding are one and the same." This radical stand results from the psycho-physical monism and causalism of the author of the *Ethics*. Although it may be difficult to reconcile with our usual experiences, it provides the best theoretical framework for understanding the spectacular and apparently paradoxical data of B. Libet on the chronological order of action and decision making (Libet 1985, 1992). Conscious decisions to act seem to occur *after* the beginning of the action. In the context of our model, this observation is easily understood. It means that the slightly delayed observation and understanding of actions triggered by neuronal unconscious stimuli and already under way appear to us and are interpreted after the fact as a willed decision that causes the action. Similarly, as stated by M. Jeannerod (1992) in relation to the observation that consciousness takes time to appear, "the where in the brain determines the when in the mind." As mentioned earlier, our model of time reversal is a particular instance of a similar concept of memory storage producing a transformation in time and space. As such, it allows us to understand that discrepancies between apparent temporal courses of conscious events and their neural counterparts are possible and should not be puzzling.

In addition, in the context of our model of the creation of meaning by extracting features depending on contingent frequencies of repetitions, it is interesting to remember Spinoza's theory of Universals. Following nominalists such as William of Ockham, this author noticed how natural language makes use of general notions, called Universals, like the general notions of "man," "horse," "dog," and so on, whereas only individuals exist and are perceived in the real world. These notions are made by the superimposition and confusion of different images of the individual men, individual horses, and so on. It is interesting to quote Spinoza here, because, in view of our model, his wording is striking:

> We must observe that these notions are not formed by all persons in the same way, but that they vary in each case according to the thing by which the body is more frequently affected, and which the mind more easily imagines or recollects. For example, those who have more frequently looked with admiration upon the stature of men, by the name *man* will understand an animal of erect stature, while those who have been in the habit of fixing their thoughts on something else, will form another common image of men, describing man, for instance, as an animal capable of laughter, a biped without feathers, a rational animal and so on; each person forming universal images of things according to the temperament of his own body. (pt. 2, prop. 40, scholium 1; 2002a: 266–67)

We must add that this "temperament" is made by how the body, that is, the cognitive system, has been assembled and also how it has been more frequently affected.

UNDERDETERMINATION OF THEORIES BY FACTS AND MUTUAL UNDERSTANDING

Another property of dynamic systems like our automata networks or neural networks is related to the contingency of meaning suggested by our model, through what Wittgenstein calls the vagueness of natural language and the—apparently paradoxical—effect of this vagueness in facilitating mutual understanding. This property is a quantitative instantiation of what is known as underdetermination of theories by facts (Duhem 1914, Quine 1960), which analysis of network dynamics makes possible and conspicuous (Atlan 1989, 1991a, 1992b).

It can be shown that the number of possible different states for a network of N units is a power N, whereas the number of possible different network connection structures is a power N^2. The number of states is 2^N, 3^N, . . . or q^N, if each of the N units can be in 2, 3, . . . or q states. However, N^2 connections can exist between N units, and the number of connection structures, of different ways to connect the N units with one another, is at least 2^{N^2}. This number becomes 3^{N^2} if a connection can be positive or negative or nil, and in general p^{N^2} if each connection can be weighed with p different values: For example, the number of states of a binary network of 5 units is $2^5 = 32$. But the number of different ways of building a network by connecting 5 units with each other is at least 2^{25}, that is, approximately 10^7. As a result, the number of possible different networks, that is, of different discrete dynamic models of a system made of N observable units, is in general much larger than the number of possible observable states. The gap increases exponentially as N increases. This is typically a situation of underdetermination of theories by facts: if the observable facts are stable-state attractors of a network, it is clear that thousands and thousands of different network connection structures, that is, different dynamic models, will produce identical attractors. This is obviously an unpleasant situation for the theoretician: the difficulty is not to find a good predicting model or theory but to chose between thousands of different, nonredundant, equally good theories predicting the same observable facts.

If not only the stable states can be observed but also the transients of the dynamics, the underdetermination will be reduced, but not by very much, since for a deterministic given network, the number of different transients equals the number of initial states, that is, again 2^N (or 3^N, . . . or q^N, depending on the number of different states for each unit). The condition necessary to avoid underdetermination is that the number of states and the number of connection structures be of the same order of magnitude, that is, precisely $q^N \geq p^{N^2}$, or $q \geq p^N$. In other words, the number q of different observable states for each unit or variable in a model must be larger than the power N of the number of different possible values weighing each possible connection. This situation is achieved in general when the observable variables approximate continuous variables, that is, when the number of observable states for each unit approximates infinity. That is the case for most experimental systems built and observed under controlled conditions such that: (1) the number of independent variables is small, that is, N is small; (2) they can be observed and measured in an almost continuous fashion. Physical sciences have been established on the basis of such experimentally controlled systems. That is why they usually succeed in reducing this underdetermination of theories with enough observations to, ideally, falsify all the theories except one. However, that is not the case, in general, for natural systems made of large numbers of components observable under partially controlled conditions, such as biological systems observed *in vivo* in different stable states as responses to various stimuli.

This kind of underdetermination can be encountered already, for example, in small models of interaction between a few populations of lymphocytes involved in regulating immune responses to a given antigen (Atlan 1989). In view of the large numbers of neuronal units involved in the functioning of our brain compared with the amount of observable data on actual brain states, it is obvious that any modeling of complex brain functions is faced with a large degree of underdetermination.[3]

However, considering actual brain states rather than states of formal neural networks used to make models of them, the same underdetermination is viewed from a new perspective and acquires a completely different meaning. What is a flaw and a weakness for theorizing can be viewed now as an advantageous property of the structural stability and robustness of the dynamics of our actual brains. In view of the large variability in the detailed connections among the actual neural networks of our brains, this property allows us to reach identical or similar brain states through different pathways and in spite of different connection structures. This property may be one of the bases for our intersubjectivity, which enables us more or less to understand each other in spite of our different brain structures, produced by different genes and histories. In particular, we experience a similar convergence property in some situations of argumentation, when we realize that it is much easier to agree on a conclusion rather than on the way to reach it, for example, when we are arguing about possible behaviors to be justified by deductive reasoning from general principles. Contrary to usual beliefs, there is no need to agree on general principles in order to agree on the practical conclusions derived from the principles to be applied to a given particular situation. It was shown long ago (Winograd and Cowan 1963) that calculus, viewed as information processing, normally implies a loss of information because the *result* of a computation is common to many other computations and usually does not retain the

---

3. Notice that the underdetermination of theories that we mention here is somewhat different from the one previously discussed by Duhem (1914) and Quine (1960), mainly in the context of physical quantum theories (Friedman 1983). There, underdetermination was related to empirical equivalence between incompatible physical theories, sometimes, although not necessarily, associated with an intrinsic limitation, or "uncertainty principle," in the possibilities of absolute precision measurements. This kind of underdetermination is perhaps more akin to the ontological Kantian impossibility of knowing the thing in itself, whereas the one we are discussing here has more of an epistemological nature and is related to the operational complexity involved in dealing with relatively large numbers of interconnected elements.

memory of the particular one that produced it. Similarly, the same logical conclusion can be reached through several deductions following different routes from different initial assumptions. However, we may be dealing here with a loose analogy only, since there is no evidence that our actual ways of reasoning, of being convinced and convincing, are isomorphic with transient pathways in dynamics of neural networks. Again, metabolic and hormonal states, with their emotional counterparts, may interfere and modify the computation significantly.

MODELING THE MODELS? THE TRANSCENDENTALITY OF LOGIC AND ETHICS

According to the model presented and discussed in this essay, an intentional self-organizing system will extract from its environment classes of stimuli that have meaning for it and for sufficiently similar systems having a similar history, but not necessarily for us as observers, at least at first glance. We would have to try to learn to decipher this meaning by trial and error, and we may even need to use modeling, as for natural systems, where behavior is the result of an individual's history, made up of encounters with an environment partially unknown to us. However, although this kind of model can in principle help us to understand the formation of semantic contents and intentionality, it leaves completely untouched the question of the origin of logic, or reason, with its peculiar capability to distinguish logical truth from error. It might be the case that, for that matter, some sort of phenomenological logic would still be needed.

The same holds true for ethics, which not only determines cognitive judgments about what is good or evil but also produces norms for behavior to avoid evil and to pursue good. The normative aspect of moral judgments is superimposed on our cognitive activities and makes use of them but is different in nature. It seems to come from elsewhere and may have archaic roots, through many cultural transformations, in the physiological drives to avoid pain and to look for pleasure (Atlan 1991a, 1995c [Chapter 29, below]). In relation to our abstract cognitive capacities, it seems to be the case, as Wittgenstein (1961) told us, that, indeed, although in two different senses, logic and ethics are transcendental.

THREE

Noise as a Principle of Self-Organization

(1972/1979)

> Because they behaved with me at random, I will as well behave with them at random.
>
> —Leviticus 26:40–41

> I cook every chance in *my* pot. And only when it is quite cooked there do I welcome it as *my* food. And truly, many a chance came imperiously to me; but my will spoke to it even more imperiously, and immediately it lay imploringly on its knees—imploring shelter and love with me, and urging in wheedling tones, "Just see, O Zarathustra, how only a friend comes to a friend!"
>
> —FRIEDRICH NIETZSCHE, *Thus Spoke Zarathustra*, book 3

Natural and Artificial Machines

Since the beginning of cybernetics, generally recognized as originating with Norbert Wiener's 1948 *Cybernetics, or Control and Communication in the Animal and the Machine*, a sort of neo-mechanism has progressively imposed itself on biology. It consists in considering living organisms to be a particular type of machine—a natural machine, with reference to the artificial machines conceived and fabricated by men. However, it would be wrong to consider this attitude to be a continuation of the mechanism of the nineteenth and early twentieth centuries. Having consequences for the structure of biological knowledge as well as important methodological implications, this new mechanism has distinguished itself fundamentally from its predecessor, especially

---

NOTE: This text reproduces, with minor modifications, an article published in *Communications* 18 (1972): 21–36. [On the epigraph from Leviticus, translated by Atlan himself, see the final chapter of Atlan 1999b).—Eds.]

because of the radically new nature of the artificial machines it envisions. These new machines are observed and analyzed not only in their similarities to the old vision of the machine but also and above all in their differences from it.

These machines, even though they are produced by men, are no longer simple and transparent physical systems, where the observed differences from organisms are too extreme and evident to teach us anything at all. As W. R. Ashby said in 1962:

> until recently we have had no experience of systems of medium complexity; either they have been like the watch and the pendulum, and we have found their properties few and trivial, or they have been like the dog and the human being, and we have found their properties so rich and remarkable that we have thought them supernatural. Only in the last few years has the general purpose computer given us a system rich enough to be interesting yet simple enough to be understandable. . . . The computer is heaven-sent in this context, for it enables us to bridge the enormous conceptual gap from the simple and understandable to the complex and interesting. (Ashby 1962, 270–71)

Furthermore, the science of these artificial machines themselves is far from being closed: this neo-mechanism does not consist in a pure and simple veneer of mechanical schemas laid over living organisms but is, rather, a back-and-forth motion of interpenetration and cross-fertilization from this science to biological science and vice versa, with consequences for the evolution and progress of *both* sciences. Because cybernetic models are drawn from a science that is itself developing, where new concepts are still being discovered, their application to biology can be something more than the kind of reduction to an elementary mechanism common a century ago.

The Reliability of Organisms

Before we contemplate the problems of self-organization and self-reproduction, we should take note of the most important differences recognized between artificial and natural machines: namely, the ability of the latter to incorporate noise.[1] For a long time, the reliability of organisms (Cowan 1965)

---

1. "Noise" is taken here in a sense derived from the study of communication: it consists in parasitic random phenomena, which perturb the correct transmission of messages and which one normally tries to eliminate. As we will see, there are cases where, despite seeming paradoxical, a "beneficial" role for noise can be recognized.

was understood to be a performance incommensurable with that of computers. Reliability like that of the brain, capable of functioning continuously while its cells die off without being replaced (with unforeseen changes in the flow of the blood supply and fluctuations in volume and pressure)—not to mention the amputation of important parts, which nevertheless does not perturb except in a very limited fashion the performance of the whole—obviously has no equivalent in any artificial automata whatsoever. This fact already struck John von Neumann (1956, 1966), who sought to improve the reliability of computers and who could not conceive of such a marked difference in reactions to the environmental factors of random aggression that constitute "noise" except as a fundamental difference in the logic of the organization of the systems. Organisms, with their ability to "swallow" noise, could not be conceived as machines just a little more reliable than existing artificial machines but only as systems whose reliability had to be explained by qualitatively different principles of organization. Hence a whole field of research, inaugurated by von Neumann (1956) and continued by many others, notably S. Winograd and J. D. Cowan (Cowan 1965, Winograd and Cowan 1963), with the goal of discovering principles for constructing automata whose reliability would be greater than that of their components. This research has arrived at a definition of the necessary (and sufficient) conditions for the realization of such automata. The majority of these conditions (redundancy of components, redundancy of functions, complexity of components, delocalization of functions; Winograd and Cowan 1963, Atlan 1972) reach a kind of compromise between determinism and indeterminism in the construction of the automaton, as if, starting with a certain degree of complexity, a certain quantity of indetermination is necessary in order to permit a system to adapt itself to a certain level of noise. This result, of course, brings to mind an analogous result obtained by von Neumann in game theory (von Neumann and Morgenstern 1944)

The Principle of Order from Noise

A further step in this direction was taken during research into the formal logic of self-organizing systems: ascribing to organisms not just the property of resisting noise in an efficacious manner but the ability to utilize it, to the point where they transform noise into a factor of organization. Heinz von Foerster (1960) was the first, to our knowledge, to formulate the necessity of an "order from noise" principle in accounting for the most singular properties of living organisms qua self-organizing systems, notably, their adaptability. The "order from order" principle—implicit in the modern thermodynamic theories of living matter inaugurated by Erwin Schrödinger's 1945 essay *What Is Life?*—did not appear to von Foerster to be sufficient:[2] "Self-organizing systems do not only feed upon order, they will also find noise on the menu . . . it is favorable to have some noise in the system. If a system is going to freeze into a particular state, it is inadaptable and this final state may be altogether wrong. It will be incapable of adjusting itself to something that is a more appropriate situation" (von Foerster 1960: 43).

A series of works by Ashby (1958, 1962) lead in the same direction, even if the accent here is not explicitly placed on the idea of an organizational role for noise. This author first rigorously established a law for the regulation of systems, which he called "the law of requisite variety" (Ashby 1958). This law is important for understanding the minimal structural conditions necessary for the survival of a whole system exposed to an environment, which functions as a source of aggressions and random perturbations.

Let us assume a system exposed to a certain number of different possible perturbations. It has at its disposal a certain number of responses, and each sequence of perturbation and response puts the system in a certain state. Among all the possible states, only certain ones are "acceptable" from the viewpoint of the (at least apparent) goal of the system, which might merely be survival or might consist in carrying out some function. Regulation consists formally in choosing, from among the possible responses, those that will put the system in an acceptable state. Ashby's law establishes a relation between the variety of perturbations, the variety of responses, and the variety of acceptable states.[3] The greater the variety of perturbations and the smaller that of acceptable states, the greater the variety of available responses must be. In other words, a great variety in available responses is indispensable for

---

2. Since then, nonequilibrium thermodynamics has enriched itself with a principle of "order through fluctuations," which is very close to order from noise, at least in its logical implications, even if the formalism used is very different. See Atlan 1979: 104.

3. Variety is defined as the number of different elements in a set. The quantity of information is connected to this quantity inasmuch as it considers the logarithm of this number and it weights each element according to the probability of its appearance in a set of sets statistically homogeneous to the one being considered. The quantity of information as defined by Shannon is a more elaborate way (also more rich in applications) of expressing the variety of a message or of a system as defined by Ashby.

assuring the regulation of a system that aims to maintain itself in a very limited number of states while being submitted to a large variety of attacks. Or again, in an environment that is a source of diverse and unpredictable aggressions, variety in the structure and functions of a system is an indispensable factor in maintaining autonomy.

But one knows, by contrast, that one of the most efficacious methods for fighting noise, that is to say, for detecting and correcting errors in the transmission of a message, consists in introducing a certain redundancy, a repetition of the symbols constituting the message. We therefore already see how, in complex systems, the degree of organization can be reduced neither to its variety (i.e., its quantity of information) nor to its level of redundancy, but will rather consist in an optimal compromise between these two opposing properties.

The same author, who also investigated the logical signification of the concept of self-organization (Ashby 1962), arrived at the conclusion that self-organization in a *closed* system (i.e., one that does not interact with its environment) was a logical impossibility. The argument runs as follows. A machine can be formally defined in the most general possible way by a set E of internal states and a set I of inputs. The operation of the machine is the way in which the inputs and the internal state at one instant determine the internal state at the next instant. In set-theoretical terminology, the operation can therefore be described by a function f, which is the projection of the product set I x E onto E; the functional organization of the machine can thus be identified with f. One can speak of self-organization in the strict sense if the machine is capable of changing the function f by itself, without any intervention by the environment, in such a way that it will always better adapt to what it does. If this were possible, it would mean that f changes itself as a consequence of E alone. But this is absurd, since if one could define such a change in f as a consequence of E, this change itself would only be a part of another law of projection f', meaning that the organization of the machine is directed by another function f', which itself would stay constant. In order to have a projection f that truly changes, it is necessary to define a function $f(t)$ of time that determines these changes with respect to the exterior of E.

In other words, the only changes that can affect the organization itself—and that are not just changes in the state of a system that would form a part of a constant organization—must be produced from outside of the system. Now, this is possible in two different ways: either a precise program, injected into the system by a programmer, determines the successive changes in f, or these changes are determined by the exterior, but by random factors in which no law prefiguring an organization, no *pattern* permitting a program to be discerned, can be established. It is in the latter case that one can speak of self-organization, even if not in the strict sense.

This is therefore another method of suggesting a principle of order from noise in the logic of self-organizing systems.

One could find still others, notably in the analysis of the role of randomness in the structural organization (i.e., the connections) of complex neuronal networks, which we owe in particular to the work of R. L. Beurle (1962) and which Ashby (1967) also took up later.

Elsewhere (Atlan 1968, 1970, 1972), and with the aid of the formalism of information theory, we have tried to give to the principle of order from noise a more precise formulation,[4] whose principal steps we now propose succinctly to explain.

Notes on the Application of Information Theory to the Analysis of Systems

One of the principal theorems of this theory, due to Claude E. Shannon (Shannon and Weaver 1949), established that the quantity of information in a message transmitted in a noisy channel of communication must diminish by an amount equal to the ambiguity introduced by this noise between the entrance and the exit of the channel. Error-correcting codes, which introduce a certain redundancy into the message, can diminish this kind of ambiguity to the extent that, at the limit, the quantity of information transmitted will be equal to the quantity emitted, but in no case can it be greater. If, as many authors have proposed, one utilizes the quantity of information in a system, compared for this purpose to a message transmitted to an observer, as a measure of its complexity or, at least partially, of its degree of organization, this theorem seems to exclude any possibility of a positive, organizational role for noise. We have been able to show that this is not true (Atlan 1968, 1970), precisely because of the implicit postulates with whose aid information theory was applied to the analysis of organized systems (when this theory's field

---

4. We will see later that this formulation in fact constituted a return of the notion of order with respect to the definition that von Foerster had given. From a repetitive production of order, we have passed to a production of diversified order, of variety, that measures Shannon's H function.

of application, in its primitive form, seemed to be limited to problems in transmitting messages in channels of communication).

The total quantity of information in a message is a magnitude that measures, over a large number of messages written in the same language with the same alphabet, the average probability of the appearance of letters or symbols of the alphabet, multiplied by the number of letters or symbols in the message. The average quantity of information per letter is frequently designated the "quantity of information" or "entropy" of the message, because of the analogy between Shannon's formula, which expresses this quantity in terms of the probability of letters, and Ludwig Boltzmann's, which expresses the entropy of a physical system in terms of the probabilities of the different "states" in which the system can find itself. This analogy, the object of numerous works and discussions, is one of the reasons information theory soon overflowed the frame of problems in communication into the domain of the analysis of the complexity of systems. The quantity of information in a system composed of parts is thus defined by the probabilities that we can assign to each kind of its components, over a set of systems supposedly statistically homogeneous with it, or even from the set of combinations that it is possible to realize among its components, which constitutes the set of possible states of the system (Morowitz 1955). In all these cases, the quantity of information in a system measures the degree of improbability that the assemblage of the different components could have been the result of randomness. The more a system is composed of a large number of different elements, the more its quantity of information increases, since the improbability of it being constituted *such as it is* as a random assemblage of its constituent parts has increased. This is why this magnitude has been proposed as a measure of the complexity of a system (Atlan 1979: 79), inasmuch as it measures the degree of variety of the elements from which it is made. A theorem of the theory says that the quantity of information, expressed in units of "bits" of a message (written in any alphabet whatsoever), represents the average minimal number of binary symbols needed for each letter in order to translate the message from its original alphabet into binary. Transposed to the analysis of a system, this means that the more the quantity of information increases, the more the number of symbols needed to describe it in binary language (or some other language) increases; from which we get once again the idea that this quantity is a means of measuring the complexity of a system. Let us note immediately the qualifications it is advisable to make in accepting this conclusion, in particular due to the static and exclusively structural character of the complexity it treats, to the exclusion of functional and dynamic complexity, which are not tied to the assemblage of the elements of a system but to the functional interactions between these elements (Dancoff and Quastler 1953). The problem of the precise definition of the notion of complexity as a fundamental scientific concept (analogous to those of energy, entropy, etc.) is yet to be solved. However, as von Neumann noted while posing this problem and underlining its importance, "this concept clearly belongs to the subject of information" (1966: 78).

Ambiguity-Autonomy and Destructive Ambiguity

Be that as it may, the application of information theory in the analysis of systems implies a shift in the notion of information, from something transmitted in a channel of communication to something contained in an organized system. The formalism proper to information theory only applies to the first of these notions, and this shift, whose legitimacy has been contested, cannot justify itself except by comparing, at least implicitly, the structure of the system to a message transmitted in a channel that leaves the system and arrives at an observer. This does not necessarily imply the introduction of subjective or value-laden characteristics, which are by definition excluded from the domain of the theory if we consider this observer to be the usual ideal physicist, who confines interventions to operations of measurement but who nevertheless intervenes by these very operations.

Under these conditions, it is possible to show that the ambiguity introduced by the factors of noise in a channel of communication situated inside a system has a different signification (its algebraic sign is different) depending on whether one has in mind the information transmitted in the channel itself or the quantity of information contained in the whole system (in which the channel is one element among a large number of relations between numerous subsystems). It is only in the first case that the ambiguity is expressed by a quantity of information marked by a minus sign, in agreement with the theory of the noisy channel of which we have already spoken. In the second case, by contrast, the quantity of information that is measured no longer signifies a lost quantity of information at all, but rather an augmentation of variety in the entirety of the system or, as one says, a diminution of redundancy.

Indeed, suppose that there is a channel of communication between two subsystems A and B, in the interior of a system S (Figure 3.1). If the transmission from A to B effects itself without any error (zero ambiguity), B is an exact copy of A, and the quantity of information of the conjunction of A and B taken together is only equivalent to that of A. If the transmission takes place with a number of errors such that the ambiguity becomes equal to the quantity of information in A, this information is totally lost in the course of the transmission; in fact, there is no longer any sort of transmission of information from A to B. In other words, the structure of B is totally independent of that of A, such that the quantity of information of A and B taken together is equal to that of A added to that of B. However, inasmuch as the very existence of the system S depends on the channels of communication between these two subsystems, their total independence in fact signifies the destruction of S qua system. That is why, from the viewpoint of the quantity of information in the system, an optimum is realized when there exists a nonzero transmission of information between A and B, but with a certain quantity of error, producing an ambiguity that is also not equal to zero.

Under these conditions, when the quantity of information transmitted from A to B is equal to that of B minus the ambiguity, the quantity of information in the ensemble A and B is equal to that of B plus this ambiguity. In

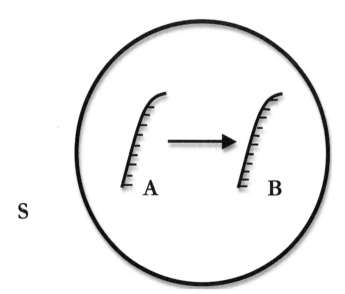

Figure 3.1. Channel Communication Between Subsystems

effect, the magnitude that measures the ambiguity is nothing but the quantity of information in B *inasmuch as B is independent of A*: it is therefore normal that this quantity be considered lost from the point of view of transmission from A to B, and to be a supplement from the viewpoint of the total quantity of information (i.e., the variety) in the whole of the system.

One thus sees how a positive "organizational" role for noise could be conceived within the frame of information theory without contradicting the theorem of the noisy channel: by diminishing the transmission of information in the channels inside the system, noise diminishes the redundancy of the system in general and by the same stroke augments the quantity of information the system contains. It is evident, however, that the functioning of a system is tied to the transmission of information in the channels from one subsystem to another and that alongside this "positive" role for noise as a factor of complexification one cannot ignore the classical role of noise as something destructive.

We therefore have to deal with two sorts of effects caused by the ambiguity produced by noise in the general organization of a system, which we have called destructive ambiguity and ambiguity-autonomy: the first counts as something negative, the second as positive. It seems to us that a necessary condition for the two to coexist is that the system be what von Neumann has called an "extremely highly complicated system" (1966).[5] Even if the concepts of complexity and complication have not yet been clearly and precisely defined (Atlan 1979: 74), the vague and intuitive idea that we have of it allows us to picture natural automata as systems of an extremely high degree of complexity, in that the number of their components can be extremely high (10 billion neurons in the human brain) and the relations among these components can be densely intertwined, with each one of the components in principle capable of being linked directly or indirectly to all the others. It is only in such systems that a positive role for noise, through the intermediary of ambiguity-autonomy, can coexist alongside its more destructive role. In fact, if one considers a system limited to a sole channel of communication between A and B, the autonomy of B with respect to A cannot signify anything but the malfunction and destruction of the system: in this case, a quantity of information in B inasmuch as B is independent of A has no meaning from the viewpoint view of the system, since the system is basically nothing

---

5. This is what Edgar Morin has called, in a manner that is probably more correct, a "hypercomplex system" (Morin 1973: 191).

other than this channel. For it to have any meaning, it is necessary to imagine A and B linked not only to one another over this channel, but each to a large number of other subsystems by a large number of other channels, on such a scale that even the total independence of B with respect to A does not translate into the dissolution of the system. Such a system, because of its numerous interconnections (and provided that its initial redundancy is sufficiently large), will still be able to function, and its total quantity of information will be consequently larger. This increase in information allows the system to realize even greater achievements, notably the possibility of adaptation to novel situations thanks to a larger variety of possible responses to the diverse random stimuli present in the environment.[6]

Elsewhere, expressing these ideas in a more quantitative fashion than we can do here, we were able to show (Atlan 1968, 1970) that, under certain simplifying hypotheses, a sufficient condition for ambiguity-autonomy to be able to compensate for the effects of destructive ambiguity is the existence of a change in the alphabet by an augmentation of the number of letters when one passes from one type of subsystem to another through a channel of communication between the two. This result can be understood as a possible explication of the change in alphabet actually observed in all living organisms when one goes from nucleic acids, written in a "language" of four symbols (the four nitrogenous bases), to proteins, written in a language of amino acids, with its twenty symbols.

Self-Organization Through Diminution of Redundancy

In a more general way, one can conceive of the evolution of organized systems—in other words, the phenomenon of self-organization—as a process of augmenting both structural and functional complexity, resulting from a succession of disorganizations, which are each time overcome by the reestablishment of the system at a higher level of variety and a weaker level of redundancy. This can be expressed rather simply with the aid of a precise definition of redundancy using the framework of information theory. If H represents the quantity of information in a message containing a certain degree of redundancy R, and if H_{max} represents the maximum quantity of information

---

6. See, above, Ashby's law of requisite variety.

that could contain the message if none of the symbols was redundant, the redundancy R is defined as:

$$R = \frac{H_{max} - H}{H_{max}} = 1 - \frac{H}{H_{max}}$$

such that the quantity of information H in the message can be written:

$$H = H_{max}(1 - R)$$

We define a nonprogrammed process of organization as a variation in H over time, affected by random factors in the environment. This variation is represented by the quantity $\frac{dH}{dt} = f(t)$ Differentiating the preceding equation, we obtain an expression for $f(t)$:

$$\frac{dH}{dt} = -H_{max}\frac{dR}{dt} + (1 - R)\frac{dH_{max}}{dt}$$

The rate of variation of the quantity of information $\frac{dH}{dt}$ is thus the sum of two terms that correspond, schematically, to the two opposing effects of noise or to the two sorts of ambiguity. The first term expresses the change in redundancy. If this is initially rather high, it will diminish under the effect of ambiguity-autonomy, such that $\frac{dR}{dt}$ will be negative and the first term $\left(-H_{max}\frac{dR}{dt}\right)$ will be positive, thus contributing to an increase in H, the quantity of information in the system. The second term expresses the variation of H_{max}, that is to say, at each instant the deviation with respect to the state of maximal complexity that can be attained by the system without regard for its degree of redundancy. One can show (Atlan 1972) that the process of disorganization with respect to a given (instantaneous) state of organization, when affected by destructive ambiguity, can be expressed by a decreasing function of H_{max} such that $\frac{dH_{max}}{dt}$ and the second term taken together are negative, contributing to a decrease in the quantity of information H in the system.

These two terms themselves depend on two functions of time $\frac{dR}{dt} = f_1(t)$

and $\frac{dH_{max}}{dt} = f_2(t)$, depending on parameters that express formally the nature of the organization under consideration. Thus, for certain values of these parameters, the curve of variation in the quantity of information H as a function of time will be such that H begins by increasing, attains a maximum at a time t_M (whose significance we will return to further on), then decreases. This sort of variation can be applied to a type of organization observed in organisms, where a phase of growth and maturation, with the possibility of adaptive learning, precedes a phase of senescence and death. The interesting point here is that these two phases, even while they happen in opposing directions from the viewpoint of the variation in H, are both the result of the responses of the organism at different states in its evolution to factors of random aggression from the environment: the factors responsible for the progressive disorganization of the system that later leads to its death are the same factors that previously "nourished" its development with progressive complexification. Of course, depending on the values of the parameters that determine f_1 and f_2, this effect may or may not be observed, and if it is, this would be in quantitatively very different fashions, depending on the "organization" of the system with which one is concerned. This is why we have proposed to use precisely this formalism in order to define quantitatively the concept of organization itself, such that the property of self-organization, that is to say, the apparently spontaneous increase in complexity (in fact provoked by random environmental factors), can be understood as a particular case.

Toward a Formal Theory of Organization

Thus the state of organization of a system is defined not just by its quantity of information H, which expresses only its structural characteristics, but also by its functional organization; this can be described by the rate of variation $\frac{dH}{dt}$ in the quantity of information in the system over time, itself the sum of two factors, f_1 and f_2, one connected to the rate of the decrease in redundancy, the other to the rate in the decrease of the maximal quantity of information. The different possible kinds of organization are characterized by different values of the parameters characteristic of these functions. One can show

(Atlan 1972, 1974) that two among them play a particularly important role: initial structural redundancy and reliability [*fiabilité*]. The first is only a structural characteristic, while the second expresses the efficacy of the organization's resistance to random changes and is thus a functional characteristic. There certainly exists a relationship between reliability and redundancy, in that the first depends upon the second, and it is through this that the necessary relation between structural and functional organization is established. All the same, one cannot be reduced to the other, and the distinction that we are thus brought to establish can be understood by reference, for instance, to the distinction that S. Winograd and J. D. Cowan (1963) introduced between "modular redundancy" and "functional redundancy" in their study of the reliability of automata: initial redundancy is the redundancy of modules, simple repetition of structural elements, while reliability is the redundancy of functions.

In order for a system to have self-organizing properties, initial redundancy must have a minimum value, because these properties consist in an augmentation of complexity through the destruction of redundancy. It is only under these conditions that the curve of variation in $H(t)$ will be able to have an ascending initial part. Then, reliability will also measure the duration of this ascending phase, such that the time t_M at which the maximum is attained will be later as the reliability is greater. Thus, if t_M is too short, the destruction of redundancy will be effected too quickly to be observed, the maximum will be obtained nearly instantly, and we observe a system in which the quantity of information apparently does nothing but decrease. Despite a sufficient initial redundancy, everything happens as if the system was not self-organizing. On the other hand, if the initial redundancy is insufficient but the reliability is large, then the system, which evidently cannot be self-organizing, nevertheless has great longevity: its reliability in this case has only the usual significance of resistance to errors, which can be expressed by the inverse of its rate of senescence.

Thus, tracking the values of these two parameters, one can distinguish different sorts of organized systems or, if one prefers, different "degrees of organization."

Principles of Self-Organization of Matter and Evolution by Selection

In a long essay on the possible chemical nature of processes of self-organization of matter, Manfred Eigen (1971) arrives, by a different method and a different formalism (rooted in chemical kinetics), at a logically very similar result. Taking his inspiration from the known mechanisms of DNA replication, protein synthesis, and enzymatic regulation, he analyzes the evolution [*devenir*] of "populations" of information-bearing macromolecules, both from the perspective of the total quantity of information in the system and from that of the different types of macromolecules synthesized. One of the problems thus studied is the conditions under which certain information-bearing macromolecules can be selected at the expense of others, in a system where the sole restriction is that the synthesis of these molecules be achieved by copying identical molecules. For the first time, the concept of directed selection—the foundation of theories of evolution—acquires a precise content, which can be expressed in terms of chemical kinetics, unlike the usual vicious circle one falls into when one describes natural selection as the survival of the fittest, where the "fittest" can only be defined by the fact of survival.

A value of selection is thus defined starting from the magnitudes of A_i, D_i, and Q_i, themselves defined for each type i of information-bearers, in the following fashion. A_i is a factor of amplification, which determines the speed of repetitive reproduction of the bearer i; multiplied by a constant K_0, it expresses the speed of the reaction of replication over the mold in which i is synthesized.

Q_i is a factor of quality between 0 and 1, which expresses the precision and the fidelity of its replications: if no error is produced in the copies, $Q_i = 1$; more generally, Q_i is the fraction of A_i comprising copies reproduced without error, while $(1-Q_i)$ is the fraction of copies that manifest errors, in other words the "mutants" produced by the erroneous replication of i. D_i is a factor of decomposition, which, multiplied by the same constant K_0, expresses the speed of the reactions by which the macromolecule i is destroyed.

Depending on the different initial conditions envisaged, the selective value of a species i is thus defined by either $A_iQ_i - D_i$, or by $\dfrac{A_iQ_i}{D_i} - 1$; that is to say, it can be expressed either absolutely or relatively, in terms of the excess of production of the macromolecular species envisaged over and above its destruction.

In both cases, the result reached by the analysis of the evolution of a population where i different species are present is the same: one of the necessary conditions for the total quantity of information in the population to grow over successive selections of certain species is that the factors Q_i are not 1, while remaining, however, much closer to 1 than to 0. This is simply the consequence of the fact that in the absence of error no novelty can appear.

And if, furthermore, we take into account that chemical reactions, on the molecular level, are stochastic phenomena, where the role of fluctuations is more important when we are concerned with small numbers of interacting molecules, then, in the absence of $Q_i < 1$, not only will the quantity of information not grow (due to the lack of innovation) but it will not even be able to maintain itself in a steady state and will diminish until all the species present have disappeared, without being replaced by new ones. This stems from the fact that there always exists a non-zero probability that, affected by fluctuations in the speeds of reactions, a species may, at any given moment in the evolution of the system, be destroyed more quickly than it is reproduced, to the point where there may no longer be a copy available for an ulterior reproduction, and thus that it may definitively disappear.

One of the most spectacular results at which Eigen arrives, applying his theory to systems constituted by a coupling of two subsystems with complementary properties (like collections of nucleic acids and collections of proteins), is a possible explication of the universality of the genetic code: it is the inevitable result of an evolution in which only this particular code could have been selected. He suggests, in the end, a number of "evolution experiments" with the aim of testing this theory.

Noise as Event

Thus, at least in principle, we see how the production of information as a result of random factors is nothing mysterious: it is nothing but the consequence of error production in a repetitive system, constituted in such a fashion as not to be destroyed almost immediately by a relatively small number of errors.

In fact, where the evolution of species is concerned, no mechanism is conceivable outside of those suggested by theories in which random events (chance mutations) are responsible for evolution toward greater complexity and greater richness of organization. Where the development and maturation of individuals is concerned, it is strongly possible that these mechanisms also play a non-negligible role, especially if one includes the phenomena of non-directed adaptive learning, where the individual adapts to a radically new situation for which it is difficult to call upon a preestablished program. In any case, applying the notion of a preestablished program to organisms is a debatable move, insofar as this concerns programs of "internal origin," created by the organisms themselves and modified in the course of their development. To the extent that the genome is produced from the outside (by its parents) one often compares it to a computer program, but this comparison seems to us to be totally abusive. If a cybernetic metaphor can be used to describe the role of the genome, that of data stored in memory [*mémoire*] seems to us more adequate than that of program, since the latter implies mechanisms of regulation that are not present in the genome itself. Otherwise, one cannot avoid the paradox of a program that needs the products of its own execution in order to be read and executed in the first place. By contrast, theories of self-organization permit us to understand the logical nature of systems where what fills the role of program modifies itself unceasingly, in ways that are not preestablished, as a result of "random" factors in the environment, which produce "errors" in the system.

But what are the errors? According to what we have seen, because of their positive effects, it does not seem correct to characterize them entirely as errors. The noise provoked in the system by random factors in the environment will no longer be truly noise from the moment it is used by the system as a factor of organization. This is to say that factors in the environment are not random. But, of course, they are. Or more exactly, it depends on the subsequent reaction of the system in relation to which, a posteriori, these factors are recognized as either random or as part of the organizing process. A priori, they are in effect random, if one defines randomness as the intersection of two independent chains of causality: the causes of their occurrence have nothing to do with the chain of phenomena that has constituted the prior history of the system until then. It is thus that their occurrence and their encounter with the system can constitute noise from the viewpoint of the exchanges of information in the system, where these encounters are susceptible to producing only errors. But from the moment the system is capable of responding to these "errors," not just so that it does not disappear, but rather so that the system uses them to modify itself in a way that benefits it or at least ensures its subsequent survival—in other words, from the moment the system is capable of integrating these errors into its own organization— then these errors lose, a posteriori, a little of their character of error. They retain this only from a viewpoint exterior to the system, in that the effects of the environment on the system do not themselves correspond to any preestablished program contained in the environment and destined to organize or disorganize the system.[7] On the contrary, from the interior perspective, insofar as organization consists precisely in a series of recaptured disorganizations, they do not appear as errors except at the instant of their occurrence

---

7. The thermodynamic mechanisms of order through fluctuation seem to put the accent on the internal character of organizational noise. This distinction is not a real one,

and in relation to a maintenance of the status quo (which would be as unfortunate as it is imaginary) of the organized system, which one pictures to oneself as soon a static description of it can be given. Indeed, after this instant, the errors are integrated, recuperated as factors of organization. The effects of noise then become events in the history of the system and its process of organization. They remain, however, effects of noise inasmuch as their occurrence was unforeseeable.

It thus might be sufficient to consider organization as an uninterrupted process of disorganization and reorganization, and not as a state, for order and disorder, the organized and the contingent, construction and destruction, life and death, to no longer be really distinct. However, this is not at all the case. The processes in which this unity of oppositions realizes itself (this realization not being a new state, a synthesis of thesis and antithesis, but a movement of the process itself—the "synthesis" being nothing besides this) cannot exist except inasmuch as errors are a priori true errors, inasmuch as order at a given moment is truly perturbed by disorder, inasmuch as destruction, while not total, is real, inasmuch as the irruption of the event is a real irruption (a catastrophe, a miracle, or both at once). In other words, the processes that appear to us as one of the foundations of the organization of living beings, as results of a sort of collaboration between what we are accustomed to calling life and death, cannot exist except to the degree that they are never really about collaboration but always about radical opposition and negation.

This is why immediate experience and common sense concerning these realities cannot be eliminated as illusions in favor of a unitary vision of a grand current of "life" that carries along both, even if this current is equally real. The simultaneous consciousness that we have of these two levels of reality is probably the condition of our freedom or of our sentiment of our freedom: it is permissible for us to accede, *without contradicting ourselves*, to processes that signify both our survival and our destruction. A true ethics, allowing us to use this freedom best, would be a law that would allow us at each instant to know how to intervene in this unceasing struggle between life and death, order and disorder, a struggle in which a definitive triumph of one over the other is always avoided—such a triumph would, in fact, be one of

---

since these "internal" fluctuations also result from the effect of the environment (temperature) on systems that cannot but be "open," therefore traversed by a flux coming from the exterior.

two ways to die completely, stopping the processes either in a definitively established and immutable order or, alternatively, in total disorder. It is remarkable that, while in the course of their processes our lives unite "life" and "death," thus realizing two sides of living beings, our cadavers also unite two sides of dead beings: rigidity and decomposition.

Translated by John Duda

FOUR

The Intuition of the Complex and Its Theorizations

(1991)

1. Complexity as Problem and as Explanation

For a long time, to qualify something as complex served to designate a difficulty, either in comprehension or in execution. At the same time, curiously,
this designation played an explanatory role—a purely verbal role, of
course—in relation to that which one could not otherwise explain: the declaration of complexity often enabled, and continues to enable, the justification
of a lack of theory and seemed to mitigate, although in an illusory manner,
an insufficiency of explanation. It is only recently that complexity, having
ceased to be a definitive declaration, has itself become a problem, an object
of study and of systematic research. This change in status is noteworthy and
constitutes an important fact in the recent history of the natural sciences,
first in biology, and more recently in physics.

In its early stages, molecular biology seemed to have explained everything.
It provided the experimental and theoretical tools that enabled the conception of an almost perfect continuity between the organization of inanimate

chemical objects—crystals and inorganic molecules—and that of living organisms. The greater complexity of the latter was invoked to explain the obvious differences that the simple observation of structures and functions could not possibly miss.

One can easily achieve an intuition of biological complexity by considering the immense number of possibilities engendered by the combination of a relatively small number of elements when each of these can be found in two or more distinct states. Two examples will serve quickly to convince us of this.

The molecules that are responsible for nearly all biological functions—enzymes—are proteins, that is to say, chains made up of smaller molecules, amino acids, linked end to end. The natural proteins use twenty or so amino acids, and the number of amino acids in a protein frequently exceeds a hundred or a thousand. Let us suppose, therefore, that every protein contains only one hundred amino acids—which is certainly a conservative estimate. This means that each position in a chain of one hundred elements is characterized by one of twenty possible states. In this case the number of different proteins possible would be 20^{100} (approximately 1 followed by 130 zeros). Some today think that, with time, and above all with the help of powerful computers, we should be able to explore all of these possibilities. Let us consider this.

The age of the universe is estimated at approximately 15 billion years, which is approximately 15 x 10^9 x 365 x 24 x 3600 = 4.7 x 10^{17} seconds. We see, therefore, that even if the fabrication of a particular protein chain of 100 amino acids took only one second, the age of the universe would have been grossly insufficient for the exploration of all possible hundred-element chains. Let us now consider the effect of the processing power of large computers, which considerably enhance the speed with which different combinations can be effected. Let us suppose that a combination of one hundred elements could be achieved not every second, but every thousandth of a second. Then, we determine the age of the universe in milliseconds by multiplying by 1,000, which yields 4.7 x 10^{20}—only three zeros have been added. Let us suppose that the speed of calculation could be further increased, such that the time necessary to process one possible combination becomes 10^{-50} seconds, which is a fraction of a second equal to 1 divided by 10 followed by 49 zeroes. The age of the universe measured in this unit of time is 4.7 x $10^{(17 + 50)}$ = 4.7 x 10^{67}. We are still far from the 20^{100} possible proteins. And yet the unit of time in question, 10^{-50} seconds, is already beyond everything physically conceivable: no physical phenomenon, of whatever scale, can be imagined as having such a short existence. This is to say that the task of exploring one by one all the possibilities for constructing proteins made up of even one hundred amino acids is a physical impossibility, given the age of the universe.

And yet among all these possibilities, only a tiny fraction seem to be compatible with the life of organisms as we observe them, because most modifications targeting the functional state of a protein, that is to say, its enzymatic activity, lead to the elimination of this activity. We must concede, therefore, that the constitution of organisms had to be accompanied by a process of selection that drastically reduced the number of possibilities retained. Biological evolution thus had to comprise mechanisms of selection set up a priori—for example, phenomena of amplificatory cooperation—that eliminated immense numbers of possibilities without exploring them.

These problems did not pose themselves in physics, at least not until very recently, before the development of a physics of heterogeneous and chaotic systems, about which we will hear more later in this conference [Colloque de Cérisy conference on Atlan's work, 1991b]. For a long time they appeared as specific to biology, because the physical and chemical sciences were traditionally occupied with much more *homogeneous* samples of material, where the properties observed are the consequences of averages calculated from immense numbers of molecules (on the scale of Avogadro's number, 10^{23}). Problems of combination were practically eliminated by the practices of averaging in statistical thermodynamics, while the diversity and heterogeneity of living organisms at the molecular level seemed incompatible with such practice.

Another example is cerebral states (or mental states, if we want to assimilate these to cerebral states in order to suggest a materialist, monist solution to the mind-body problem). The number of neurons in a human brain is estimated at 10 billion. Supposing that the number of possible states of activity for a neuron is limited to the strict minimum of two (active or inactive), we see that the number of possible activity states for a brain is on the order of $2^{10^{10}}$, or a number even more immensely great than those we have envisioned, because we could represent it approximately by 1 followed by several billion zeros. Certain elements of the nervous systems of much simpler animals, such as mollusks, have been studied precisely because of their relative simplicity: they have only about a hundred neurons, which have been mapped, and the determinate connections between these neurons have been identified. Even in the simplest cases, the minimum number of possible states is still about 2^{100} (on the order of 10^{30}), which we could compare to the age of the universe in milliseconds, for example, since this is about the unit of time characteristic of a change in state of neuronal activity; or we could compare it to the number of humans who have ever lived on earth, which is $3.5 \times 10^{12} = 3,500$ billion individuals since the appearance of *Homo sapiens* 35 thousand years ago, if we allow twenty years per generation and, very generously, a population of 2 billion for each of these generations. If every individual had lived one hundred years (that is 3×10^{10} milliseconds), the maximum number of cerebral states that all of humanity would have been able to explore, supposing one state per millisecond, would be roughly $3.5 \times 10^{12} \times 3 \times 10^{10}$, that is. about 10^{23}, which is to say, still very few compared to the number of states possible for a mollusk brain.

All of this should not discourage research, however—quite the contrary—but should only indicate that methods better adapted to this type of problem need to be imagined. And some are already beginning to be used: breakdown into levels of integration that appear as pertinent over a course of study thanks to the emergence of collective behavior, which need not address the details of elementary behavior; parallel calculation; the use of randomness and probability under different forms (in the choice of initial states, in the mode of calculation). These methods already enable us, in a certain measure and more or less following the laws of interaction between elements, to avoid the difficulty of the combinatory explosion.

The laws of interaction between elements are evidently the source of the a priori selection mechanisms I evoked above. We can see here the habitual effect of calculation, which generally reduces the diversity of results in relation to that of input data. (An operation produces the same result from a large number of possible input combinations: for example, 11 can be obtained by adding $6 + 5$, $7 + 4$, etc.) But there is more, because these laws, though they act locally, have global effects that often appear—precisely because of the large number of possible conditions—significant and unexpected, such that the question of the move from the local to the global has become one of the central problems in the study of the effects of complexity.

But this work has only really begun to develop in the past few years. In the meantime, we understand that exponential growth, which is at the beginning of the combinatory explosion that leads very quickly to immensely large numbers, has always played a key role in the mathematical approach to the study of the complex. We will return to this later on, in the definition of algorithmic complexity, formalized by the way in which time of calculation varies with the size of the problem to be resolved.

At the same time, we are able to consider, in a qualitative and intuitive manner, this combinatory explosion as an explanation for imperfectly understood collective behavior. Whence comes the idea that, since living organisms can be reduced to physicochemical objects by excluding "vital fluids" and the "work of Life," which continued until very recently to be invoked, it had to be their incomparably greater complexity that accounted for their properties specifically as living beings. Such was the creed of a time that marked the end of an era: that in which vitalist conceptions had not yet crumbled before mechanicity and chemical and physical biology. But, at the same time, this creed was the beginning of a new era, one that John von Neumann heralded in the fifties and we continue to inhabit: if it is true that the complexity of living organisms is the origin of their specificity, then, since this conviction is inscribed in a realm of analytical thought continuous with the science of physical chemistry and nineteenth-century thermodynamics, the questioning process of this science cannot rest with this explanation. Inevitably, in this mechanistic and causal conceptual context, a question arises: What is complexity made of?

We know that the concomitant development of the information sciences began to provide the means both to pose the question and to attempt to respond to it. This is how von Neumann, first a quantum physicist, and later an inventor of the first electronic computers, could foresee in his posthumous work, *Theory of Self-Reproducing Automata* (1966), that complexity (or complication, the difference between the two not being very clear in his text) would be the privileged object of the sciences of the twentieth century, as were energy and entropy in the nineteenth. The current multiplication of colloquia and publications on this subject seem to have proven him right.

The clarification of the concepts of energy and entropy was a long process, during which fairly vague initial intuitions served as a foundation for precise quantitative definitions, which were, however, only operational in certain domains. Their generalization was possible only later, when different methods of their use could be compared and unified thanks to a logico-mathematical formalization that was developed alongside these schemes, that of statistical thermodynamics.

Von Neumann himself inaugurated the process regarding complexity: even before complexity was defined in a univocal manner, he made it play a precise role in a generative process, which enabled a certain intuition about the possible role of a quantification of complexity: the existence of a *threshold* of complexity served to establish a difference in kind between classes of mechanisms. Following his earlier work on automata whose reliability would be greater than that of their component parts, he posed as a principle that, for certain mechanisms, exceeding a threshold of complexity enables them to produce effects more complicated than their own structure.

Far from seeing this mode of thinking as circular, we must rather envision a spiraling process of clarification in which a slight shift that appears tautological—the definition of the complex by the complex—is effected by an *operation*, here, the postulate of a threshold in a generative process in which we can speak of complexity without yet needing to know what complexity *is*. To demonstrate something by "just doing it" can also be an apt method of conceptual clarification.

This is why the organizers of this colloquium were truly inspired in bringing together specialists from such different disciplines, who, in their various fields, have had to take into account the problem of the complex. Further, it was important to establish conditions of exchange between those who have attacked, from various angles, the problems of formalization (mathematicians, physicists, computer scientists) and those (biologists, psychologists, sociologists, and philosophers) for whom the complexity of the materials of their discipline is an experience that demands above all a phenomenological description.

Since we are not yet at the stage where existing formalizations—which are, moreover, multiple and nonunified—would encompass and exhaust these descriptions, it was important to maintain this diversity. In this way, I hope, it will be possible for us to learn the most from one another.

For my part, I will confine myself to presenting a few reflections on the role of the viewpoint of observation in our appreciation of the complexity of things. In particular, I want to try to analyze several questions posed by the differences and the possible meeting points between the complexity of natural objects and that of the objects we make ourselves.

2. The Reliability of Automatons and the Threshold of Complexity

The goal of von Neumann's work on the reliability of automatons was to simulate the properties of biological systems, particularly the brain. The functioning of the brain does not seem much disturbed by the fact that neurons die in great numbers every day without being replaced or that its physicochemical conditions change constantly—without even mentioning accidental destructions of greater or lesser impact—in such manner that, if modifications of the same order occurred in more familiar machines, they would very quickly cease to function. This sort of observation led him to research on machines more reliable than their component parts. The lower the likelihood of breakdown, the greater the reliability of the machine. Usually, one considers this probability to be the sum of the probabilities of breakdown for each of the components, which implies the intuitive notion that the greater the number of components, the greater the risk of breakdown. It is this notion, which the observation of the brain seems to contradict, that motivated von Neumann to research a principle of machine construction in which the reliability of the machine would be greater (that is to say, the probability that the machine would break down would be less) than that of its parts. Here, he pointed to the role of a threshold in the types of coding used to connect, in a redundant manner, components (Atlan 1972). And this led him to suggest the role of a threshold effect in the processes of complexification that I spoke about at the beginning. In these early works, and in those of Winograd and Cowan (1963), who completed them, Shannon's probability theory of information provided an adequate formal framework.

Since then, much work has been carried out in a different direction, incorporating a different formalization: one that concerns the complexity of algorithms. Once universally programmable computers—to whose production von Neumann himself contributed to such an extent that sometimes they are called von Neumann machines—became functional and widely used, it became imperative to be able to compare programming algorithms so as to know a priori, as far as possible, what type of program would be preferable to another for resolving particular classes of problems.

In this research on the complexity of algorithms, which has led to a theory at present already very elaborate (Aho et al. 1974), the concern is no longer to consider natural systems, such as living organisms or machines intended to simulate them. Rather, the essential problem here is that of the efficiency of a programmable machine in accomplishing certain tasks. The context is completely different in that an objective is posed at the beginning: one wants to resolve a certain problem, or to construct an object, or to have it constructed by a programmed machine; a question arises, therefore, concerning the complexity of the task to be accomplished, and more exactly, concerning the program one should write to attain the objective.

Taking into account the quantity of work accomplished in this domain, it would be interesting, from the viewpoint of a biologist, to see how these two sorts of problems—the simulation of natural systems in which a particular aim is not specified and the efficiency of programs in achieving certain given objectives—might overlap. One can hope that it might be possible to transpose some of the results from the theory of algorithms to problems of theoretical biology. The work that I will present later with M. Koppel (Koppel and Atlan 1991a) is an attempt in this direction.

But first, it is important to understand how we are concerned with two approaches to complexity that differ from each other, on the one hand, and that also differ from the intuition we derive about complexity from common language, on the other.

3. Algorithmic Complexity and Natural Complexity

The complexity of a problem is measured by the difficulty involved in resolving it, assuming one has an automatic procedure—that is to say, a computer program—to perform this resolution. If such a program does not yet exist, the question is to know the magnitude of the difficulty of writing and running such a program. This question has been theorized in a precise manner thanks to a certain number of hypotheses, which correspond to the practical conditions of programming. These conditions are obviously limited from the perspective of a reflection on "thought" or "intelligence" in general. They are, however, sufficiently generalizable to cover the entire field of logico-mathematical thought, that is to say, the part of our intellectual production that results from our activities of logical reflection and computation and that we seek to have performed by programmed machines.

How does one measure the difficulty of a problem to be resolved—whether it is a question of a mathematical problem, or of an object to be produced, or, in a general manner, of a task to be accomplished? Different formulas have been proposed, which can be shown to be reducible to each other, clearly enabling the generalization of which I have just been speaking. Perhaps the most intuitive is the minimum time necessary for a machine to run a program that will lead to the solution or, perhaps, the minimum capacity of memory the computer one uses for this task must possess. Of course, these magnitudes depend upon the type of machine envisioned (the speed of its components, their architecture, etc.). But there exists a way of normalizing all this by imagining a sort of universal ideal computer, called a Turing machine, that would be capable of resolving every problem that any kind of computer is capable of resolving. Thus, the complexity of a problem to be resolved is measured by the necessary calculation time for a Turing machine to resolve the problem. This time is directly proportional—multiplied by a factor determined by the speed of calculation—to the minimum number of instructions in a program for a Turing machine (that is to say, its minimum length) capable of resolving the problem.

One can allow oneself, in these sorts of equivalences, to ignore important technical factors like the speed of calculation, because what is important for the measure of difficulty of a task to be accomplished is not so much the absolute value of the time of calculation as *the manner in which this time varies with the size of the problem to be resolved.* This is why the unit of measure of the calculation time does not matter much here and can be reduced to the time necessary for the execution of a Turing machine program. The *size* of the problem consists in the number of variables and parameters to be taken into consideration, or the number of input data to be incorporated, such as, for example, the dimension of a matrix if the problem is posed in the form of matrices, or the number of connections of a circuit, and so on. It is evident that the calculation time will necessarily increase with the size of a problem, but the question pertinent to its complexity is: *How* does it increase? Slowly or rapidly? In proportion to the size of the problem? Or in a slower manner, for example, in proportion to the logarithm of this size? Or, on the contrary, more rapidly, like a power of N, if N measures the size of the problem? Or yet more rapidly, in an explosive manner, like an exponential function of N?

One can easily conceive that the solution of a problem is proportionally more difficult—its programming "algorithm" is proportionally more complex, in the sense I have been describing—when the necessary computation time increases very quickly with its size. And things happen as if there were a difference in kind—a threshold of difficulty—between the calculation times or the lengths of Turing machine programs that don't increase more quickly than by powers of N and those that increase in an exponential manner. The latter define classes of problems that are practically insoluble, unless the amount of data to be taken into account is very small, which obviously diminishes the interest of an automatic calculation.

I don't want to go into the details of this theory at the moment. On can see fairly well how the complexity of an algorithm capable of solving a problem can be measured by the length of a minimal program for a universal computer, that is to say, ultimately, a number of bits, if one considers this program to be written in binary language. (We will see later on [Koppel and Atlan 1991a] that the distinction between the program and the data in these measurements of complexity is not a simple question. This distinction is not generally necessary when the goal is only to measure the complexity of a calculation as a function of the size of the problem, as I have just described. By contrast, the distinction seems crucial once we want to consider the *meaning* of what a program does.)

We will certainly see in the rest of this colloquium enough examples to show the practical interest of these measurements.

But it is important to bring out that which is implicit in this theory, in the form of presuppositions that obviously circumscribe its field of application.

First of all, the objective is always to resolve well-posed problems. In other words, a goal to be attained needs to be well defined, and we then seek to find the path to arrive there most efficiently. The choice of a criterion of efficiency can even itself be a parameter and thus enter into the very definition of the goal to be attained.

In all cases, a specific aim is imposed from the exterior, and it determines the value of an automatic procedure; we judge that value, the efficiency of one program compared to another, in relation to a goal fixed a priori. By contrast, in the case of a natural system, one not programmed by a human being with an definite objective in view, we don't know a priori what its aim is; if it has one, this can only be internal to the system itself or imposed by a metasystem of which the system in question would be a component. In each case, the aim of the natural system is itself an object of research. And for any research scheme that does not accept animistic explanations, the aim in question can only be mechanical, like that of every physical process oriented by evolutionary laws toward dynamic attractors that correspond sometimes, but not always, to minimums of energy or, in a general manner, to the extreme values of thermodynamic-state functions.

In other words, in the case of natural systems, our experience of complexity cannot rely upon the definition of an evident objective, known a priori. Rather, insofar as a natural system is not known in its details to the point that we would be capable of describing it by an algorithm capable of producing it with all its structural and functional properties, the partial knowledge we can have of this system can enable us to measure our ignorance. And we can use this measure of ignorance, which concerns structure as well as the eventual relationship between structure and function, use to measure complexity. We know that this is exactly what we do when we use the Shannon entropy formula to measure the complexity of objects. But this is possible only on the condition that these objects are elements of statistically homogeneous classes and that their components are distributed following the characteristic probabilities of each of these classes.[1] Under these conditions, the complexity of an object measured by its Shannon "entropy" is proportionally larger as the probability that it was randomly assembled is smaller. In fact, this value measures, as we know, a quantity of missing information, an a priori uncertainty—thus our ignorance—regarding specific determinations responsible for its structure and its functions. Even if this measurement can be made in bits, because the inverse of a probability can always be measured by a number (not necessarily integral) of binary states, we see that its significance is quite different from that of the complexity of an algorithm. I have suggested designating the latter by the term *complication*, corresponding to situations in which one can concede that everything is known but that the description of this knowledge may be long, thereby distinguishing this from complexity as a measure of our ignorance, sometimes—but not always—rendered possible by the type of observations we can make, which enable us at a minimum to circumscribe a natural system, to name it, and sometimes to experiment upon it.

In the most general case, even when the conditions for the application of a probabilistic measurement of uncertainty are not met, the complexity of a natural object still appears to us as a lack of knowledge of determination and of an order whose existence we yet sense, "an apparent disorder about which we have reason to think that there exists a subjacent order" (Atlan 1979: 77–78). These reasons have to do with our observation of a *function* that emerges from this apparent disorder, as if it were directed by a specific aim whose mechanisms we have yet to discover.

One can demonstrate the existence of a gradation of knowledge (or the lack thereof) in these different experiences of complexity. Probabilistic uncertainty is obviously only one particular case of uncertainty, already attenuated by our knowledge of a distribution of probability over a statistically homogeneous whole. As for complication, it implies—necessarily in the case of a finite object, and sometimes in the case of an infinite sequence—an uncertainty reduced to zero, as can be shown by choosing, as the level of

---

1. A classic example is the letters of an alphabet, which make up sentences, which are elements of the class that constitutes a language (see Atlan 1972).

probabilistic description of the components, the level of the system as a whole (Atlan 1979).

We see therefore how these two notions are profoundly different.

4. Loops and Infinite Searches

One senses as well, however, that some common ground must exist, particularly when one leaves, for one reason or other, the domain of the finite.[2] One example concerns cases in which we have to do with a problem that is definite but we do not know whether or not it can be solved. It is possible, for example, that we may know how to solve it for certain particular cases but not for the general case, and that we may not even know whether a general solution exists. Thus we can only measure algorithmic complexity in the particular cases in which a program brings us to a solution and not in the general case. In other words, we clearly have a goal to attain, but we don't know if it is attainable. This, however, does not keep us from trying, that is to say, from writing programs having a certain complication that we can only measure a posteriori. For we do not know a priori, for some cases of the same problem, whether the program will reach a solution after a finite period of calculation or whether it may seem likely to run indefinitely. It is as if the goal assigned to the program (the general solution of the problem) were unattainable, which does not stop us, once again, from calculating—in fixing a limit time, for example—and sometimes finding interesting partial results along the way. Thus, a posteriori, it is always possible to assign as objectives the results obtained.

For another thing, there are several ways in which a program can fail to stop, which do not have the same significance. The program could be engaged either in an infinite loop or in a search on an infinite amount of given data. This should also be considered.

Even an infinite loop, which is generally considered to be a sign that a program has failed, can be seen from another point of view and acquire the significance of a goal in itself, like a limit cycle, that is to say, an oscillating attractor in a dynamic of a discrete automaton.

---

2. A theorem establishes implicitly that the algorithmic complexity of a string approaches its Shannon entropy as the length of the string approaches infinity (Zvonkin and Levin 1970).

5. Levels and Metalevels: Complexity for Understanding

To conclude this section on the relationship between (algorithmic) complication, (natural) complexity, and the degree of explanatory ignorance, I want to cite a revealing passage from a classic manual on the conception and analysis of computer algorithms (Aho et al. 1974). We see here how the traps of natural language enable us to catch in the act, if I can put it that way, in an auto-referential manner relative to formal theories of complexity themselves, the role of our difficulty in comprehension, in its most general, most intuitive, and least formalized sense, in our appreciation of that which is complex.

The formalization of the complication of tasks to be accomplished does not provide an account of the complexity that results from our difficulties in comprehension. To convince us of this, it will suffice to cite the conclusion of the chapter of a manual that establishes the theorems of equivalence—which I mentioned earlier—between measurements of algorithmic complexity in different practical conditions of calculation. The calculation time effectively depends not only on computer speed but also on the programming language used. Two models of elementary language are studied, known respectively as RAM (Random Access Machine or Memory) and RASP (Random Access Stored Program machines),[3] in relation to which theorems of equivalence are demonstrated. These theorems say that these two models of language are equivalent in the sense that they are polynomially related.[4] This means that every algorithm whose complexity is a polynomial function of size

---

3. These two models differ in that the first (RAM) does not store programming instructions in memory, so that one cannot modify in the course of the program instructions given previously. Rather, it uses indirect address: that is to say, a register of memory identified by its address i can have a content $c(i)$, which serves as an address for another register; the content $c(c(i))$ is thus indirectly addressed by the intermediary of the content of the first. In the second language model (RASP), by contrast, programming instructions are stored in memory, and one can therefore manipulate them by treating them as operands as the program is running. In this case, one does not use indirect address, because it is not necessary.

4. These two theorems are formulated as follows. Theorem 1.1: "If the instruction costs are either uniform or logarithmic for each RAM program of temporal complexity T(N), there exists a constant K such that there exists an equivalent RASP program of temporal complexity KT(N)." Theorem 1.2: "If the instruction costs are either uniform or logarithmic for each RASP program of temporal complexity T(N), there exists a constant K such that there exists an equivalent RAM program of temporal complexity at most equal to KT(N)."

N of the problem to be solved in one of these languages will also be so in the other (as well as, incidentally, for a Turing machine).

The authors summarize these results as follows: "It follows from Theorems 1.1 and 1.2 that as far as time complexity [measured by calculation time] (and also space complexity [measured by the quantity of memory necessary]) is concerned, the RAM and RASP models are equivalent within a constant factor, i.e., their order-of-magnitude complexities are the same for the same algorithm." Then, in a manner surprising and revelatory for our inquiry, this sentence follows in conclusion: "Of the two models, in this text we shall generally use the RAM model, because it is somewhat simpler" (Aho et al. 1974: 16–19).

We must ask ourselves, evidently, in what way one is simpler than the other—less complex—when we have just demonstrated their equivalent complexities. This can obviously be the case only in a different sense, indicated by the "somewhat," and in a different context, that of the metalanguage that the text of the manual itself employs. The authors of this manual, in their pedagogical enterprise, which cannot do without natural language, declare themselves to prefer one of the two models because, although they are formally equally complex, it appears simpler to them. Obviously, this lesser complexity or greater simplicity concerns the possibilities of exposition and comprehension (of the formal theory of the complexity of algorithms), and this complexity is different from that which is defined in the body of the theory, thus different from that which I suggest calling complication.

After having shown that two models have the same complexity, one cannot say that one is simpler than the other unless one now employs a different notion of complexity and simplicity. The authors can only do this because this different notion appears in the expository metalanguage of the models of language.

One could imagine that we are, in spite of everything, dealing with the same notion of complexity, and yet that this would not be contradictory, precisely because this notion would be used at two different levels of language. But in this case we would have to be able to formalize the metalanguage—that of the manual—in an artificial language that would enable an algorithmic transition from one level to the other. That is to say, we would have to be able to write a computer program that would render comprehensible—without the aid of natural language—the theory of the complexity of programs.

Might this perhaps be possible one day? Then, the complication of algorithms and the complexity of understanding would perhaps be one and the same notion.

6. Complexity, Chance, and Meaning

Up to this point, we have encountered three different notions of complexity. Two are formalized and designate, respectively, the difficulty a programmed machine encounters in accomplishing a task and the probabilistic uncertainty (or missing information) of an observed structure. The third designates a nonquantitative intuition about our difficulty in understanding an explanation, an idea. It is clear that this difficulty in comprehension cannot be specified, insofar as we do not know exactly what kind of operation we perform when we understand or do not understand something. In other words, this complexity essentially concerns the meaning of that which we are trying to analyze and understand, whether this is something observable in nature, or an abstract problem to solve, or an artifact manufactured by a procedure that can be automated. And it is the difficulty of formalizing the meaning of words, sentences, and things that we encounter here as the difficulty of formalizing this sort of complexity. Yet the two other sorts of complexity, algorithmic and probabilistic, are only formalized to the extent that they are not explicitly concerned with questions of meaning. The theory of probabilistic information, as well as the theory of programming algorithms, attains its quantificational objectives without having to worry about the questions of knowing how we understand or how meaning is created. In fact, it takes for granted that meaning exists, that of the information transmitted in a channel of communication, as well as that in the instructions of a computer program.

But the mathematical theories that deal with these questions have no need, in order to be operational, to give an explicit account of meaning. In information theory, the fact that messages and structures have a meaning is evident but remains implicit, because we are concerned here only with problems of coding and the efficiency of transmission, without needing to envision the effective meaning of the messages to be coded and transmitted. The same formalization is used to measure the probabilistic complexity of a structure, in supposing that there exists a channel of communication between the structure observed and the operation of observation. The measurement is the same for every page written in a given language, whatever its meaning, because it depends only upon the number of signs contained in the page and the distribution of the frequencies of usage of the different signs of an alphabet, a characteristic of the language used. But the fact that we are dealing with a spoken or written language implies that the signs carry meaning, as well as that we can verify this by observing their effects on receptors (auditors or readers) who understand the language. In the same manner, a structure about which we know only the distribution of the frequencies of usage of its components in a class of equivalent structures will not appear different from a heap of randomly piled pieces, except that we observe in it one or more meanings in the form of functions that this structure accomplishes. But these meanings, once again, are only implicit in the theory, which limits itself to measuring frequencies or to calculating probabilities without explicitly taking these meanings into account.

Similarly, it is obvious that a computer program written to solve a problem or accomplish a task has a meaning—precisely to solve this problem or accomplish this task. Yet the theory of algorithmic complexity is founded on logical considerations about the length of calculation and decidability that don't need to address this meaning explicitly: the definition of the aim of the program, that is to say, of the particular problem to be solved or the task to be accomplished, is not necessary for the theory to function. This aim is always evident, and it is not necessary to inquire into its origin when we analyze the logical and operational properties of the program.

The fact that meaning—of information transmitted or of the program to run—plays no explicit role in these theories appears in a spectacular manner in the paradoxical relationship that we discover in these theories between complexity and randomness.

According to the most widely accepted definition (Kolmogorov-Chaitin), the algorithmic complexity of a problem to be solved or a task to be accomplished is the minimal length of the program and the input necessary to arrive at the solution or accomplishment.

In general, a structured sequence of elements, provided that it is long enough, can be produced by a program that can achieve this structure while being shorter than the sequence itself. This is obvious in all the cases where a program of finite length is capable of producing a sequence of infinite length. From this follows the paradox that an infinite random sequence—without structure—represents the most complex task to achieve, because the only way to produce it is to duplicate it and thus there exists no program shorter than this sequence that is capable of producing it. But this is only a paradox in relation to our global intuition of the complex, in which it seems meaningfulness must occupy a central place in opposition to the statistical disorder of the random. This does not impede the functionality of the theory in its domain of application—from which, by definition, such random sequences are excluded—because it is obvious, and implicitly admitted, that a program written to accomplish a given task cannot be a random sequence of signs.[5]

We find the same problem—with the same solution—in the definition of the complexity of a natural object given by the Shannon entropy function. As with every measure of entropy, we are concerned here as well with a magnitude whose maximum corresponds to statistical homogeneity, to the disorder of the random. The only difference between our experience of disorder in nature and that of complexity so estimated derives from our observation that the latter accomplishes one or more functions. At a minimum, in the absence of direct observation—as in the case of a disorderly anthill, in which the functions of spatial organization, division of labor, and storage of reserves can still be observed—we take as a working hypothesis, on the basis of more or less convincing arguments, that the apparent disorder is responsible for the accomplishment of one or more functions. In other words, we attribute a meaning to that which we designate as complex rather than simple disorder, even while, formally, the magnitude for measuring them—entropy—is the same.

This is also true of a programmed machine, in which the meaning of a component or of a program instruction concerns its function, its role in the functional organization of the whole of the machine. But the origin of these functions and therefore of these meanings is very different in the two cases. In a machine, because the task to be accomplished is specified from the outside, this task is the source of meaning: a component or an instruction will appear absurd and meaningless if it contributes nothing to the accomplishment of this task, whereas its meaning will emerge through its more or less evident and efficient contribution to this accomplishment. On the contrary, in a natural organization, such as an organism or an ecosystem, we cannot

---

5. Let us note, however, that a minimal program presents itself as a random series according to this definition because, being minimal, it cannot be produced by a program shorter than itself. The only reason we consider it to be something other than a random sequence is its meaning, that is to say, we know by some other means that it is capable of producing a certain function when it is read and interpreted by a Turing machine (Koppel and Atlan 1991a).

suppose that a goal has been established a priori by a consciousness, either external or internal. With such a goal-establishing consciousness, this would come down to pointing to God as an explanatory principle in the form either of a Creator external to nature or of a cosmic consciousness encompassing everything; either way, the explanatory value is null, because no one can know what the intention of God might have been, if there was intention in the organization of a living species, an organism, or an ecosystem—which clearly contrasts with the situation of a machine fabricated by humans.

In the two cases of natural probabilistic complexity and algorithmic complexity, we encounter the same paradox: a formal identity between maximum complexity and random complexity (that is to say, disorder with a maximum statistical homogeneity). And in these two cases, the solution to the paradox consists in ignoring it by supposing that some meaning exists a priori, which eliminates in one stroke the hypothesis of randomness.

Only very recently have attempts been made to resolve this paradox through work on algorithmic complexity incorporating a definition of complexity that would be meaningful and would show itself to be null, as one would expect, and not maximum, in the case of a random sequence. The work that I will present later in collaboration with Koppel (1991a) follows directly in this line.

As for natural complexity, I will very briefly recall some attempts to take into account in one way or another the role of the creation of meaning in the process of self-organization.

In living organisms, the meaning of information is expressed in an observable physiological function, about which we have no means of knowing whether it was conceived and realized with a precise intention that could be defined in a noncircular manner (other than having sight in order to see, breathing in order to breathe, etc.). This is why various formalizations were used to try to model the mechanisms of self-organization by which nonintentional systems, without aims imposed from outside, are able to organize themselves such that the meaning of information becomes an emergent property of a dynamic. I will content myself with citing two series of work in which I have tried to pose the problem of the creation of meaning in biological systems. The first concerns what I have called the principle of complexity from noise.

The probabilistic theory of information is extended here in order to establish the necessary conditions for self-organization with increasing complexity, that it to say, the creation of information. The question of the creation of signification is treated here in the negative, taking advantage of the fact that the quantification of information with which we are concerned does not explicitly take into account its meaning. Self-organization functions like a sequence of disorganizations recollected in reorganizations. Complexity from noise, in this formulation, is the expression of an affirmation by a double negation: the destruction by noise (first negation) of information from which meaning is absent (second negation) can be equivalent, on the condition that we change our level of integration and observation, to the creation of a new complexity. This implies the creation of meaning, without which destruction by noise could not be accompanied by reorganization.

In a second, more recent attempt, in collaboration with Françoise Fogelman and Gérard Weisbuch (Atlan, Ben Ezra, et al. 1986), we used simulations of automata networks to try to pinpoint, this time in a positive manner, the emergence of functional significations in automata networks with self-organizing properties. It is well known that many automata networks present properties of structural self-organization, in that their dynamic causes them to evolve from homogeneous initial conditions toward attractors in which macroscopic spatiotemporal structures can be observed. This is a very active field of research at the moment, in which the question is to understand the passage from the local to the global, the determination of macroscopic properties emerging (from the network) from effects of cooperation between microscopic properties (individual automatons). Moreover, these networks can yield examples of functional self-organization, such as the simulation of a machine, constructed in part by chance, capable of classing and recognizing forms on the basis of criteria that are themselves auto-generated: the recognizable forms and the criteria of recognition are observed a posteriori, because they are the products of self-organizing processes and were not explicitly programmed. It is only a posteriori that the mechanism of the system of recognition can (sometimes) be grasped. As in the observation and analysis of a living natural system, one not programmed by humans, the criterion for meaning must be deciphered after the fact, sometimes with difficulty, because it is the global product, not explicitly programmed, of a great number of local interactions and the meaning that it (the criterion) thus creates in its classificatory activity can seem a priori strange from the viewpoint of a rational being possessed of a sort of intentionality that tends to plan things out in advance.

We can see here a way to simulate the way in which nature, in opposition to the work of an architect, seems to work by bricolage, as François Jacob has said. Let us note, though, that prescientific techniques also appear as bricolage (Lévi-Strauss). Their efficacy thus may perhaps have the same origin as that of the bricolage of nature. Actually, the recent realization that our activities of scientific discovery themselves must often be attributed to the "cunning reason" of Mētis (Detienne and Vernant 1974, Elkana 1981), which also invokes this bricolage, thus seems to make our science and our fabrication of artifacts revolve in the gyre of nature. After all, we should have expected this. Only the God of Descartes neither tricks nor wants to deceive us. It is because physicists thought for a considerable time that their task was to discover his activity as Architect and to imitate it that their activity has seemed to reveal only that of Thēmis, that of the classical architect and geometrician, which measures and plans, eliminating the unforeseen and ignorant of the games of the living.[6]

---

6. Calculation done by threshold automata networks on the model of neural networks once imagined by McCulloch and Pitts (see Atlan 1972) has been much developed recently in the realm of neoconnectionist theories. But it also enables us to simulate immunoregulatory phenomena and, in general, phenomena of biological regulation that imply several coupled processes. Here it is not necessary to reference billions or even hundreds of elements in interaction; small networks, on the order of ten or fewer automata, are very useful while affording a richness of behavior as great, relative to their size, as that of large neural networks. In particular, one can already see in them a large *underdetermination of theories by facts*, and the small size of these networks enables us to analyze clearly the origin of this and even to quantify it. This underdetermination of theories thus appears as probably the most spectacular incident of *natural complexity*. The analysis of these smaller networks demonstrates that this property is not specific, as had been thought, to human phenomena—psychological, linguistic or otherwise—and thus does not result from a relation to the theorization of facts issuing from mysterious aspects of the mind-body relation. Much more simply, this property derives from the rapid augmentation of the number of possible theories or models, different from each other but equivalent in their power to predict facts limited by the conditions of observation, as soon as these facts are determined by the coupled effects of several factors or processes. Each theory is represented by a given structure of connections in a network whose automata represent the different factors or processes in play. In general, observed facts are represented only by the stable states of a network, its attractors. Its much more numerous transient states often cannot be observed in the absence of an ad hoc experimental setup. *And yet numerous different structures of connections can generate the same attractors*, and the number of such structures increases very quickly with the number of interconnected automata. One can easily see (Atlan 1989: 247–53) that if N is this number, the number of possible structures is 2^{N^2}; if each connection has only two possible values, o if it does not exist, 1 if it does exist; and, in a general manner, p^{N^2} if each connection can be weighted and the strength of each connection can have p values. The number of different possible attractors as stable states of a network of

7. *Natural Languages and Formal Languages*

The analysis of language offers a privileged domain for approaching the relations between natural complexity and artificial complexity. Formal languages can be reduced to programming languages and analyzed in terms of algorithmic complexity. As for natural languages, their analysis poses some of the questions most pertinent to the complexity of natural organizations: the origin of meaning and sense here is usually analyzed in a structural manner, in the form of the relations between syntax and semantics. But the acquisition of language as a dynamic phenomenon is a particular case of developmental biology with clearly evident genetic and epigenetic components. It is difficult, however, to separate these two aspects and to analyze them independently.

N automata cannot be larger than the number of possible states, that is 2^N if we are dealing with binary automata and q^N if each automaton can occupy q states. Thus, in a general manner, for a given number of automata, the number of network structures is much larger than the number of states and, a fortiori, the number of attractors. One can conceive of cases in which q would be much greater than p (more precisely, if $q > p^N$), such that the number of possible states would be greater than the number of possible structures of the network. This implies a very great precision in observation, such that the number of discriminable facts also becomes very great. These cases, therefore, approach those in which the observable states can be described by systems of differential equations with continuous solutions (that is to say, in which the number of possible states is a priori infinite). Under these conditions, obviously, the underdetermination of theories by facts can be reduced and even disappear. But these conditions are much more rarely found in the observation of natural systems than in the observation of artificial systems assembled in laboratories such that the variables are separated into pertinent observed variables and nonpertinent (controlled) variables, and in which the greatest precision is attained in the observation of pertinent variables. It is for this reason that modeling systems by differential equations is often not very useful when the systems in question are natural systems comprising several coupled processes, because this modeling would demand, in order to make use of all its predictive force, empirical facts about the observable states and interactions that are, in general, not accessible. Modeling by discrete automata networks enables us not to renounce every kind of quantitative modeling as such, but the price we pay is that the number of possible solutions is often much smaller than the number of models realizable with a given number of automata. In other words, the dynamic of automaton networks constitutes a degenerated form of the dynamic of systems of differential equations, and the underdetermination of theories by facts is a consequence of this degeneration. Insofar as this underdetermination derives from the lack of knowledge of empirical data that would be necessary to increase the degree of determination, it expresses nicely the complexity of imperfectly known natural systems.

We can also imagine the phenomenon in an abstract manner, like a particular case of biological self-organization with increasing complexity. Then, the first stages of development appear as states of nondifferentiation with a high redundancy, in the sense that different signs and symbols (auditory or visual) here are at first equivalent, because their meanings are vague, overlapping, and ambiguous. Only later does language become more differentiated and diversified, as a result of a reduction of redundancy such that ultimately each sign can become effective in a specific manner. In this conception, the evolution of the semantic component of natural language is seen not as an acquisition of polysemy and metaphorical significations departing from literal, well-defined, nonambiguous significations but precisely as the opposite. A nonambiguous semantic content should not be taken for granted at the beginning of the process, such as an innate property of syntactical structures, but rather should be seen as the end of the process, evolving from a structure that is nonspecific, polysemic, and, to a certain extent, presyntactical toward a strictly defined syntax, accompanied by totally univocal meanings, as in artificial logico-mathematical languages. The last thus appear as a final acquisition, resulting from the progressive reduction of an initial redundancy with the creation of one-to-one correspondences between symbols and their effects, that is to say, their meanings.

This reduction of ambiguity can appear to contrast the process of complexity from noise, in which complexification emerges from increasing ambiguity. But in fact this is only the consequence of a change in the level of observation. The meanings of language with which we are now concerned represent characterizations of each part of the system, that is to say, of each sign or symbol, seen from the exterior by an observer of language considered as a self-organizing system. The ambiguity between the parts increases, in the sense that each of these parts becomes independent from the others in a "semantic" space; the communications between them become looser as the ambiguity in these communication channels increases: the part acquires a specific value or an effect on the receiver who understands—whence the word—its meaning in a manner independent from that of the others. By contrast, at the beginning of the process, in the undifferentiated state, rich in polysemy and apparently metaphorical significations, several different signs are equivalent, carriers of common significations poorly defined to our eyes as exterior observers, which implies that their interdependence is great, that is to say, that ambiguity in communications between them is minimal.

It is interesting to analyze what is produced when the process evolves to its end and runs up against a limit form, which constitutes an artificial logico-mathematical language. Then, there plainly exists a correspondence without ambiguity between each symbol and its meaning, but at the same time, insofar as it is a formal language, we can see here a structure in which symbols carry *no* meaning. In other words, when language reaches the stage where its meanings are without ambiguity, it can be formalized and exist without meaning. This would seem to indicate that the source of the meaning of words and things in a nonformal language is precisely their polysemic nature, more than an initial semantic structural content without ambiguity.

Thus, to apply the principle of complexity from noise to the acquisition of natural language brings us close to the analyses of some post-Chomskian psycholinguists (Shanon 1991). The logical and chronological anteriority of phenomena here is inverse in relation to what linguistic theories constructed around formal grammars suggest: the complexity and the specificity achieved, in language, by literal meaning emerge at the end of a process, the results of a progressive restriction of initial nondifferentiated, apparently metaphorical meaning. (This does not necessarily imply that truly metaphorical meaning is produced starting with the earliest stages of the acquisition of language, before the comprehension of literal meaning has been acquired by the child. This nondifferentiated meaning appears "metaphorical" to us by analogy with the experience of metaphors that we have as adults, which evidently depends upon our capacity to recognize literal meaning and to distinguish it from metaphor. Consequently, it is possible that a final stage of development consists in the acquisition—or the recall—of metaphorical meaning in adults, following the acquisition of a minimal capacity to create and to understand literal meaning. We would find again here, on another scale, a mechanism of recharge in redundancy similar to that which we have suggested [Atlan 1978] as a possible function of the paradox of sleep and dreaming. Cerebral electrical activity here seems more diffuse, more synchronous, extended throughout the entire cortex, thus indicating a diminution of activities of inhibition. Simultaneously, the effects of differentiation that produce identification and ambiguity-free definitions, characteristics of learning in the state of waking, seem to disappear; [metaphorical] associations sometimes full of sense thus appear in dreams, although they are normally prohibited and absurd while awake.)

By contrast to these processes, grammars have taken the form of determinist procedures written in formal languages. As such, they appear, in large part, not to be pertinent to the problem of the acquisition of natural language. Not only do the natural practitioners of a language seem not to have any sort of innate, rational grammar, but such grammars must be imposed upon them by directed instruction in school, just like mathematics and logic. The distinction established by Quine between "guiding" and "fitting" rules (Quine 1972) is particularly enlightening here. The rules discovered by grammar have as a goal giving an adequate account of the real production of well-formed sentences by natural speakers. Even if their success were total, they would not be identical to the unconscious rules that direct this production. Quine compares this situation to that of an electron, which is not taken to know (even unconsciously) the rules deducted by Schrödinger's equation, which give an adequate account of its behavior. It is true that the temptation remains to think that the electron—and the natural speaker—"obey" these rules and are directed by them. To yield to this temptation is more justified in the case of the electron, since it itself is nothing other than a solution to Schrödinger's equation, in the sense that it does not manifest itself outside of the context of experimental and theoretical physics. Can we say, by the same token, that the language of a natural speaker is nothing other than a grammar in the sense that it does not manifest itself outside of the context of linguistic theories? In his article, which concludes with a plea against absolutism, Quine demonstrates that the goal pursued by logical analysis is different from that of a grammar, even though both are operational. In logical analysis, the expressions of ordinary language are paraphrased in technical symbols that offer a gain in a certain kind of clarity and efficacy in view of a specific operation to be carried out (logical analysis in order to eliminate paradoxes, or the algorithmic production of well-formed sentences). The tools used in this paraphrase, such as logical symbols and quantifiers, or algorithmic decision trees, are not more implicit in ordinary language than other specific technicalities, such as the fourth dimension in physics to take account of time, or binary language for the programming of computers. It is true that those who work in the natural sciences are tempted to believe that the specific tools of technical languages exist implicitly in the natural phenomena that they succeed in explaining or controlling. It seems even less justified to believe that logical or linguistic structures exist implicitly in ordinary language seen as a natural phenomenon.

In fact, not only do the rules not "direct" the phenomenon, but even the "adequation" of their explanatory power is not perfect. The criteria of well-formed sentences in a language are a grammarian's criteria, which are not only insufficient to account for meaning but often contradict the criteria of a natural speaker that enable him or her to attribute a meaning to a sentence or not. This speaker must learn after the fact—at school—the grammarian's criteria as something artificial, just like logic and mathematics. In the course of this training, the speaker is often brought to eliminate sentences that do have a sense for him but that are wrongly formed and thus incorrect or to include well-formed sentences that were a priori without meaning for him. The speaker thus learns progressively to use the two kinds of criteria for meaning together, in the same way that we learn to apply logic to existence by combining logical decision-making criteria with spontaneous criteria that are not necessarily logical to begin with. Thus, the existence of an innate competence to form an infinite number of sentences that make sense does not necessarily imply that this competence is assured by the logical and rational structure of a grammar. The most obvious characteristic of logic is the use of negation, with the principle of noncontradiction that is derived from it, as well as the usage—which is also derived from it—of a demarcation by social consensus between that which has sense and that which does not. Generative grammars, like other grammars, are nothing but systematizations of this socially directed process of apprenticeship in language, which proceed by establishing written rules for these demarcations. Wittgenstein's remark about classical grammars remains pertinent: "Grammar does not tell us how language must be constructed in order to fulfill its purpose, in order to have such-and-such an effect on human beings. It only describes and in no way explains the use of signs" (Wittgenstein 2001: 117e, §496). What differentiates generative grammars from classical grammars is the level of the logic of their rules: whereas classical grammars are content to establish rules of demarcation, generative grammars seek rules of production. As such, they are very useful for the construction of artificial languages in a logical manner by determinist algorithmic procedures, whereas traditional classical grammars are not much help for this task. This is not, however, the case in the production of natural languages. Rather, it is possible that the use of probabilistic algorithms and associative memories capable of integrating noise in one way or another would be more effective, because it seems that a certain degree of indetermination and randomness is necessary to produce the effects of self-organization.[7]

---

7. In Shanon and Atlan 1990, we can find an illustration of this that uses the von Foerster theorem presented later in Koppel, Atlan, and Dupuy 1991.

8. The Simple

We have been so habituated—by school—to seeing the world through the bars of our scientific theories and their analytical exposés that we naturally think in terms of simple objects (elementary particles, atoms, molecules, words, phrases, "atomic propositions," percepts, individuals, stars, planets . . .). The complexity of nature appears to us as the fact of the combination of fewer or greater numbers of such objects in temporal and spatial relationships with each other.

But we must see that this representation itself involves something artificial. Simple objects do not exist in nature. They are the result of our activity of breaking down observed phenomena and reducing them to more elementary phenomena that we can manipulate—concretely or conceptually—more easily.

In other words, the experience of the complex is primary; or more exactly, it is the use of our capacities to describe the world and to analyze it with the aid of language that simultaneously creates the experiences of the complex and of the simple. As Wittgenstein says: "It seems that the idea of the simple is already to be found contained in that of the complex and in the idea of analysis, and in such a way that we come to this idea quite apart from any examples of simple objects, or of propositions which mention them, and we realize the existence of the simple object—*a priori*—as a logical necessity. So it looks as if the existence of the simple objects were related to that of the complex ones as the sense of ~p is to the sense of p: the *simple* object is *prejudged* in the complex" (Wittgenstein 1979: 60e).

In other words, the notions of the simple and the complex are not intrinsic properties of things, but depend critically upon the logical and empirical conditions in which we come to know them. Our use of negation, that is, our capacity to designate things that do not exist but are possible, is indissociable from our perception of the simple and the complex. The conditions of observation themselves are determining in that, as Wittgenstein says a little later: "To perceive a complex means to perceive that its constituents are related to one another in such and such a way" (Wittgenstein 1961:111, 5.5423). This perception is relative to a focalizing of interest, whose image is suggested to us by the possible perception, according to the accompanying perspectival diagram (Figure 4.1), of two different cubes arranged so that, observing the same figure and focusing first on the *a* angles, then on the *b* angles, "we really see two different facts" (ibid.).

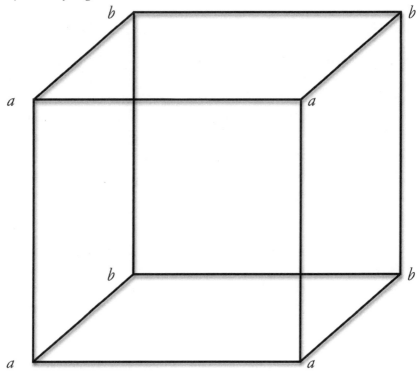

Fig. 4.1. Cubes Arranged along A and B Angles

In the end, we come back to the effects of meaning that I have already evoked in our intuitive perception of the complex and the simple. In fact, we are here concerned with one of the possible ways to distinguish between sense and meaning. Following Wittgenstein's terminology, simple objects would be vehicles of signification, those of the names that designate them. Only complex objects would be vehicles of sense, that of the propositions in which the *relations* between nouns are expressed: "A name means an object. The object is its meaning" (Wittgenstein 1961: 23, 3.203); "only facts can express a sense, a set of names cannot" (Wittgenstein 1961: 21, 3.142). Given the difficulty of taking into account all these aspects (logical, empirical, epistemological), it is not surprising that we are still far from knowing how to formalize this distinction and that, in the best of cases, one can hope only to retain an intuition of the English *meaning*, in which the effects of sense and signification are confused.[8]

8. This distinction is not completely absent in English, where *to make sense* is more natural and less "scientific" than *faire sens*.

As H. Simon had already seen, breaking down organisms into different *levels* of integration enables us to simplify considerably both the conception of artificial organization and the analysis of natural organizations. Because of this, one can effectively neglect the phenomena pertinent to an elementary level (e.g., elementary particles or molecules) if one is considering a level separated from it by at least two others (e.g., respectively, the cellular level or the level of a population of organisms). This is equally true when the scales of observed phenomena differ greatly not only in space but also in time. But these simplifications are obtained at the price of an assimilation of natural organizations that we observe to the artificial organizations we conceive and construct. It is this assimilation whose program is exposed by H. Simon in *The Science of the Artificial* (1969). But we must recognize that it has its limits. It is not, therefore, surprising that we would later be confronted with situations in which certain habitual breakdowns, which had until that point seemed natural, are contradicted by new observations. Whence the experience of particular complexity suggested by situations of "entangled hierarchies" (Hofstadter 1979), in which an element pertinent to a certain level of integration finds itself suddenly implicated in a phenomenon observed on an entirely different scale of time and space, thus playing at the same time the roles of the part and of the whole. Some effector proteins, such as hormones, play the role of regulatory proteins relative to gene expression in the cells that synthesize them; we are surprised by this, because we have every reason to think that it is the expression of genes that determines the synthesis of this type of protein and not the other way round.

In the same manner, we are amazed to find neurotransmitters in the digestive system and digestive hormones in the brain. But we are the ones who have carved up the organism into nervous system and digestive system, while the organism, it would seem, is not always aware of this.

Thus, after we have simplified an ensemble of phenomena—an organism, for example—by dividing them into different physiological systems and levels of integration, complexity reappears each time we ask ourselves the question: How is the organism determined, for itself, by the different properties that we discover in it or project onto it in our analysis? Does a breakdown into parts and levels and their relationships exist for the organism? And the aims of its behavior, its rational strategies at different levels—do these exist for it? Complexity comes from the fact that we have indices (of the efficacy of observations and of predictive mastery) that enable us to think that all this exists; thanks to this, we create a feeling of legitimate simplification. But it is denied each time we run up against that which we do not know, that which surprises us in relation to our simplified knowledge and that brings further indices that all this may have no intrinsic reason to exist. For, as Bateson would have us note (Bateson 1980), it is not a given that nature is rational. We habitually suppose, he tells us, that nature does not obey false syllogisms, which sometimes characterize delirious—or poetic—thought, such as "Grass dies; men die; men are grass" and that it obeys only *sound* syllogisms, such as "Men are mortal; Socrates is a man; Socrates is mortal." But this implies that nature, well before the emergence of language and grammar in the human species, would have already made the grammatical distinction between a subject and a verb.

It seems likely, therefore, that the complex reappears, in its specificity, on the occasion of a type of combat that we constantly wage against nature in order to master it. It is our enterprise of the rational simplification of nature that causes us to discover the complex in things when this enterprise collides with its own limits, encountering the defenses that things oppose to it.

For nature, despite our wishes (to simplify our existence?), does not work the way we do when we construct machines. We remember the episode in the film *2001: A Space Odyssey* in which the discovery in a deserted space of a regular parallelepipedic block is proof of an anterior activity of fabricating artifacts, proof that beings possessing thought like ours produced this object. For nature alone, which produces the desert and nonhuman life, according to all evidence does not produce a stone shaped in a parallelepiped.

Contrary to our way of proceeding, nature works with imperfect symmetries. The right and left sides of the human face, the symmetries of orders greater than two, are never perfect in nature. Things happen as if imperfections and approximations were characteristic of what we observe in nature in relation to an ideal of symmetry that is also an ideal of simplicity: we have inherited this ancient form of thought, in which the ideal form is that of the sphere, which offers a maximum of symmetry and, therefore, of simplicity.

The question now is to know whether, and to understand how, on the contrary, the perfect symmetries we conceive are not, rather, imperfections, flawed by their abusive simplification in relation to what nature does in its complexity. But in this work of understanding, should we renounce—and can we renounce—our ideal of simplicity?

Translated by Cara Weber

PART TWO

Organisms, Finalisms, Programs, Machines

FIVE

The Genetic Program

(1999)

From Genetic Reductionism to the Emergence of a New Paradigm

Already in 1961, an article published by Ernst Mayr brought together the principal elements of the frame of thought issuing from the great discoveries of molecular biology. It led very quickly to a new form of genetic reductionism. In this article, entitled "Cause and Effect in Biology," Mayr highlights two distinct approaches that, according to him, characterize the types of questions posed in biology.

The first approach is that of functional biology, which seeks to understand the mechanisms of biological functions. In this frame, the questions posed are of the type "How?" and the method is a reductionist physicochemical one, which attempts to work backward from observed phenomena to their physicochemical mechanisms. However, Mayr cites at length a text by Claude

---

NOTE: [This is the first part of *La fin du "tout génétique"*? It is continued in Chapter 24, "Does Life Exist?"—Eds.]

Bernard, who insists that the biologist is in need not just of physicochemical reduction but also of the existence of what he calls a "plan of organization." For Bernard, this concept of the plan of organization was essential; it consti- tuted a kind of complement to physics and chemistry, indispensable for understanding physiology. Evidently, this idea had, and always will have, a lingering whiff of vitalism, but Mayr shows how molecular biology, thanks to the discovery of the genetic code, permits a reformulation of Bernard's idea that frees it from its vitalist context. The idea of the plan of organization implies a purpose or, taken to the limit, even a form of intentionality on the part of life or the living, something unacceptable to the mechanistic scientific consciousness. Yet immediate observation, notably of embryonic develop- ment, very strongly suggests the existence of just such a finality. According to Mayr, then, this problem is resolved by the foundational discoveries of molecular biology, namely, of the structure of DNA and above all of the genetic code. These discoveries allow one to treat the mechanisms of protein synthesis as mechanisms of transmission of information and to characterize the important macromolecules (DNA and proteins) as molecular carriers of information. The discovery of the genetic code, key to the correspondence between the structure of DNA and that of proteins, for the first time allows an efficacious introduction into biology of the notion of information, in effect understood as genetic information. From this insight, in an extraordinarily hasty move, Mayr in the same paragraph passes from the genetic code to the notion of the genetic "program." For him, it would seem, a genetic program is inscribed in the nucleotide sequence of DNA and furnishes a mechanistic, nonvitalist explanation of the directed development of organisms. This devel- opment only *appears* to be finalist: it is goal-oriented [*finalisée*], but in a mech- anistic fashion. In this connection, Mayr invents some expressions that would become widespread, notably that of "mechanistic purposefulness." Every- thing happens as if a purpose were involved, but this is all just a matter of appearances, since the realization is mechanical. This realization is effected via the intermediary of a program, which functions in the manner of a com- puter program, that is to say, without intention or with an intention mediated and mechanized by the machine. Likewise, the development or functioning of a living organism can be explained by the simple unwinding [*déroulement*] of a mechanical program, without needing to invoke a spiritual (or intellec- tual) intention. Mayr also invents or, more precisely, recovers a word invented earlier by Pittendrigh: teleonomy. The intent here is to replace the classical Aristotelian idea of teleology (development in view of a goal) with that of teleonomy—a distinction that would mark the fact that the goal is not intentional. Pittendrigh spoke of the "non-purposeful end-seeking process" in order to define teleonomy in opposition to the purposeful, end-seeking process of teleology.

The second approach that characterizes the problematics of biology according to Mayr is evolutionary biology, which does not ask how living beings function but rather why they are what they are and why they function the way they do, this question being posed not in the finalist sense of "For what purpose?" but in the sense of "How did this happen, how did things get this way?"—"How come?" instead of "Why?" This approach is thus biology's historical component, and of course Mayr here takes up the classical neo-Darwinian theory of mutation and natural selection. Evolution produces organisms that are seemingly teleological [*finalisés*], but in reality we know that they are not, since they function through the intermediary of programs.

Here we have, then, a brief summary of the foundations of a well-established and widely taught paradigm, which has developed remarkably and which today continues to confirm and reinforce itself in a spectacular manner. However, we are also witnessing, and have been for some time, a progressive weakening of this paradigm. How are we to explain this paradox?

In spite of the conceptual weaknesses of this frame of thought, to which we will return, over the past thirty years its exploitation has proved to be extremely fecund and has permitted an extraordinary development of experimental results. This exploitation has culminated in the Human Genome Project. Certain people have claimed that, since everything is written in the genetic program, once we have deciphered this program, we will be able to understand the totality of the nature of an organism, whether we're talking about *Escherichia coli*, *Homo sapiens*, or any other living being. Out of this has emerged the idea of sequencing the human genome to discover within it, as in the source code of a computer program, the logic of the nature of human organization.

This maximal exploitation of the paradigm has shone light on its weaknesses, and the idea that "everything is genetic" has started to be seriously shaken. This idea, according to which the totality of, or what is essential in, the development and functioning of living organisms is determined by a genetic program, is gradually being replaced by a more complex model, which rests on notions of interaction, of reciprocal effects between the genetic (to which no one denies a central role) and the epigenetic, whose importance is being progressively discovered. Does this shift announce a change of paradigm? Recall that, according to Thomas Kuhn's famous *The Structure of Scientific Revolutions*, we understand by paradigm a kind of meta-theory, a frame of thought, inside of which a consensus is united in defining the pertinent questions that will orient the direction of experimentation and define "normal" science, until a change intervenes. Such a change is more than a new theory; it is a total change of perspective.

In March 1997, an article appeared in *Nature Biotechnology*, signed by Richard Strohman, a Berkeley professor of cellular and molecular biology. In this article, which made quite a splash, Strohman analyzes, as indicated by the subtitle ("The Coming Kuhnian Revolution in Biology"), a paradigm shift that he claims is underway in the domain of biology. For Strohman, the components of the emerging paradigm are "epigenesis and complexity" (the title of his article), which he associates with a third component, self-organization. The principle thesis of his article can be summarized: "the Watson-Crick era, which began as a narrowly defined and proper theory and paradigm of the gene, has mistakenly evolved into a theory and a paradigm of life: that is, into a revived and thoroughly molecular form of genetic determinism" (Strohman 1997: 194). In this regard, Strohman highlights a certain number of errors, which he calls epistemological or theoretical, that have been propagated in the course of the last thirty years and that we are only beginning to recognize, thanks principally to discoveries made possible by the dominant paradigm itself, which has therefore contributed, as is often the case, to hastening its own revision.

Strohman insists on two principal errors. The first consists in the fact that the discoveries from which the era of Watson and Crick developed were limited to mechanisms of DNA replication, to the genetic code and to the mechanisms of protein synthesis. This was not so bad at the time, since these discoveries addressed the most difficult questions then confronting biology. I recall that, when I was a student, we were taught that just about everything was understood except the synthesis of proteins. And this was essentially the case . . . except, of course, for the material support of genes, which was absolutely unknown. The impressive cascade of discoveries upon which molecular biology was founded provided in a few years the resolution of a whole series of major problems concerning the material basis of genes, the mechanism of protein synthesis with all its intermediary steps, and so on. The theoretical or epistemological error here consists in claiming to explain the whole set of cellular functions on the basis of this new body of knowledge. This abusive extension has lead to some erroneous approaches. It is at the origin of the "everything is genetic [*tout génétique*]" mindset that reigns today, persisting even though everyone knows that it is necessary to correct this picture with more complex representations incorporating epigenetic mechanisms of functioning and development.

Many "dogmas" were established on the basis of these initial discoveries. They contributed to fashioning the dominant paradigm, even though, for some of us, their validity and their generality were largely refuted by the facts. The first of these ideas, explicitly denominated "the central dogma of molecular biology," corresponds to the schema suggested by the study of *Escherichia coli*: "one gene—one enzyme—one function (or character)." We have known for a long time that this is not true in the case of eukaryotic cells: many genes can contribute to the expression of a character, and one gene can contribute to the expression of many characters.

Another important idea, which today one knows to be inexact except in the exceptional cases on the basis of which it was established, concerns the three-dimensional structure of proteins, which helps to determine their function and, notably, their enzymatic activity. For a long time, it was believed that a single, stable, folding conformation of a protein in space was determined in a univocal fashion by the unidimensional structure of the protein, that is to say, by the linear sequence of amino acids, themselves coded by the sequence of DNA nucleotides.

Finally, a third dogma, a little more subtle, is today still accepted without discussion: the confusion between coding and programming, which Mayr declared explicitly and to which we will return. All of this leads to a schema that represents a unidirectional flow of information, from DNA to RNA and then to the proteins responsible for biological functions.

To this image, one can oppose a schematic representation of epigenetic phenomena (Figure 5.1). In this flow of information, there exist a certain number of loops: certain proteins determine the state of activity (not the structure) of the DNA; between proteins and functions, one is obliged to include networks in which function is not the result of a single protein but of the interactions of multiple proteins. Finally, these epigenetic networks themselves have a retroactive effect on the state of activity of the DNA and RNA.

More generally, a new paradigm is being established, renewing interest in molecular bearers of information that are not reducible to structures of DNA alone. This is accompanied by a displacement of the center of interest, from the "everything is genetic" model—where everything could finally be reduced to the source, that is to say, to DNA—toward an analysis a little

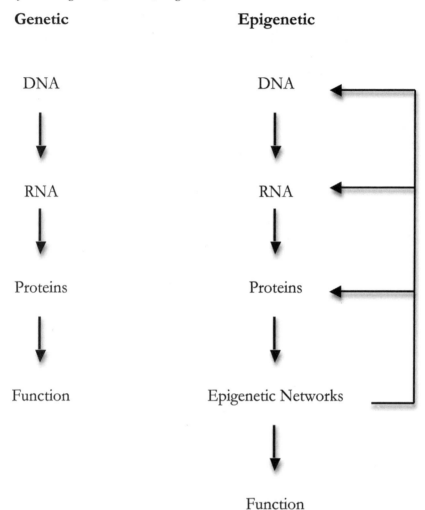

Figure 5.1. Schematic Representation of Genetic and Epigenetic Phenomena (Strohman, 1997).

more complicated, one that is obliged to establish, at each step, the feedback loops involved and to study the structure of these networks of interactions. Finally, in order to represent this state of things, the quantitative or semi-quantitative formalisms of dynamic systems and of networks of automata are becoming more adequate than the relatively simple image of the computer program.

The unexpected discoveries that resulted in these new directions were made possible by the techniques of molecular genetics, such as the fabrication of transgenic organisms and the targeted inactivation (knockout) of specific genes, which today have proved to be indispensable in the study of epigenetics. At the origin of many of these surprises, one finds, on the one hand, the first results of the analysis of the genomes of model laboratory organisms and, on the other hand, the discovery of several spectacular facts, generally considered "impossible" in the frame of the dominant paradigm.

In the analysis of genomes, we can already extract three properties of DNA that sequencing has established: pleiotropy, redundancy, and complexity. Pleiotropy is the fact that the same gene has different functions in different organisms and at different stages of development: for example, the mutations of the BRCA-1 gene, which lead to a predisposition toward breast cancer in women, are linked to anomalies in embryonic development among mice (Kinzler and Vogelstein 1996). As for redundancy, this consists above all in the existence, already described in the 1970s, of repetitive DNA, which for the most part codes for nothing and whose function, if it has one, remains unknown. But today we also know of redundancy in coding genes as well, with repetition at the level of the gene and of fractions of the genome, even at the level of the entire genome itself: the genome of mammals could thus have been produced by duplications, with variations, of a basic genome similar to that of flies and worms (Miklos and Rubin 1996).

Furthermore, one sometimes observes unexpected phenomena of functional redundancy: the permanent inactivation of a gene coding for a protein, which seems to play an essential role in the regulation of a function or a metabolic pathway, might not result in visible disorder in cellular functioning—as if the deficit thereby produced is compensated for by activating other genes or by epigenetic phenomena. This is particularly evident in the case of proteins involved in what we call "signal transduction" from membrane receptors, the study of which constitutes a chapter of cellular biology that has grown considerably in recent years (Miklos and Rubin 1996; Lichenstein and Atlan 1990; Bray, Bourret, and Simon 1993; Falke et al. 1997; Artavanis-Tsakonas, Matsuno, and Fortini 1995; Rutherford and Lindquist 1998). Finally, in a general way, the relation between the structure of a genome and the state of expression of the genes has made apparent a functional complexity much larger than had been suspected (Miklos and Rubin 1996).

Among the new discoveries and realizations of recent years, which were all considered impossibilities within the frame of the dominant paradigm until they came to light, I would like to highlight the discovery of prions and of "prion-like" phenomena. These phenomena constitute new examples of epigenetic heredity determined by the three-dimensional structures of proteins and are in no way reducible to the structure of DNA or RNA (Coustou et al. 1997, Wickner 1994). I would also mention the cloning of organisms through nuclear transfer, which makes apparent the role of maternal cytoplasmic factors in the "reprogramming" of the genome of an already differentiated cell (Kono 1997). We will return to all of this.

The second theoretical error pertains to Mayr's second approach to biology. It concerns the theory of evolution and the fact that thanks to—or because of—the massive utilization of the notion of the genetic program, for several decades developmental biology—that is, embryology—was excluded from the synthetic theory of evolution. In this domain as well, the study of molecular mechanisms and of the genetics of development has recently helped to modify profoundly received ideas about the continuous accumulation of small genetic variations as the mechanism of evolution. The synergistic effects of different mutations, which modify the course of development by interfering with the mechanisms of stabilization and fidelity in the transduction of intercellular communication signals within embryonic tissues undergoing differentiation (Artavanis-Tsakonas, Matsuno, and Fortini 1995; Rutherford and Lindquist 1998), have provided a molecular substrate to the old intuitions of Richard Goldschmidt regarding "hopeful monsters," which had always been rejected out of hand by adherents of the formerly dominant paradigm. In the same vein, paleontological theories that postulate macroevolution by discontinuous jumps (Eldredge 1995), punctuating long periods of stability, have received obvious support from the discovery of these mechanisms, through which one can conceive of the possibility that variations might affect the totality of processes of differentiation and thus produce new developmental plans.

At present these theories remain very controversial, however, due to the long tradition of excluding developmental biology from the "modern synthesis." Until the discovery of developmental genes, this tradition of exclusion continued to be reinforced by the dominant paradigm of molecular biology, since on the theoretical plane, even if one does not know the mechanisms, embryonic development was supposed to have been understood, since it was assumed to be reducible to the execution of a program.

The Dangers of a Metaphor

Let us start by analyzing certain aspects of this notion of the genetic program, which has been imprinted on the brain of each and every one of us. First off, on the strictly logical level, where did this idea—that one can find, written in the structure of DNA, a program more or less analogous to a computer program—come from? What does this purported analogy consist in? My first observation is that this idea, which has enjoyed enormous success, is not even discussed in the article by Mayr that I mentioned earlier: it has imposed itself as something obviously self-evident.

Now this idea turns out to be a kind of sophism, constructed in the following manner:

> First proposition: DNA is a quaternary sequence, easily reducible, like all quaternary sequences, to a binary sequence (in other words, one can consider DNA as a sequence of zeros and ones);
>
> Second proposition: All sequential computer programs, that is to say, all classical computer programs, are reducible to binary sequences (since this is, properly speaking, what machine language is);
>
> Conclusion: Genetic determinations function in the manner of a computer program written inside the genes in DNA.

This deduction is obviously false. The error in reasoning consists in believing that the second proposition is a reciprocal one, that is, that all binary sequences are programs. All programs can be reduced to binary sequences, but not all binary sequences are necessarily programs. It is therefore not evident that DNA, although it is reducible to a binary sequence, is therefore analogous to a computer program or that it has a logical structure akin to such a program.

It is clear, on the contrary, that no one to date has discovered in DNA any such structure. The coding sequences of DNA are translated into sequences of amino acids by the intermediary of the genetic code, but that's the extent of it. One must not confuse an encoding with a program. DNA presents only a very few syntactic elements and, so far as we know, not even the smallest semantic element that would permit one to view it as structured (even formally) like a language. If one admits, therefore, that DNA is not necessarily a computer program, then what is it? There are two other possibilities. The first is that it is purely and simply a random sequence. Even though the relationships between random sequences and computer programs are not so simple, as we will see, I will not examine this hypothesis in detail. It appears difficult to admit that DNA is only a random sequence, since modifications of this sequence have important consequences for the development or the functioning of the organisms concerned. The other possibility is that DNA is a collection of data more than it is a program. It is this possibility that I would now like to explore. Before I do so, I would like quickly to recall that this type of analysis has been, for a long time already, the basis of a different approach to the same problems.

Complexity from Noise

It is possible to refuse to take the computer metaphor seriously, to reject the metaphor of the program, while conserving the essential concept, that of genetic information. From the moment one is dealing with information-carrying molecules, one can, in effect, appeal to the probabilistic theory of information. To Mayr's first question, concerning the origin of the apparently teleological orientation of an organism's development, one could thus respond not with the metaphor of the computer program but with the simultaneously logical and physicochemical study of the mechanisms of *self-organization*. It is within this frame that, for quite a few years, we have attempted, with physicists, chemists, and cyberneticians, to explore the necessary conditions (or even the necessary and sufficient conditions) for the appearance of processes of self-organization in matter. In physics and physicochemistry, we can cite the work of Ilya Prigogine and his school, of Aharon Katzir-Katchalsky, of Manfred Eigen, of Herman Haken, and so on, which has followed in the wake of the colloquia of the 1960s on self-organizing systems. It was during one of these colloquia that Heinz von Foerster proposed his principle of "order from noise," which Prigogine and Nicolis adapted to thermodynamics and rebaptized "order through fluctuations." In the meantime, my own work contributed to generalizing the implications of this principle and making its formalization more precise, leading me to propose the formulation "complexity from noise."

The mechanism can be summarized as follows. In, for example, the transmission of information between the nucleotide sequences of DNA and the amino acid sequences of proteins, one knows that there always exist errors, equivalent to what one calls "noise" in a channel of communication. One easily conceives that these errors can produce a negative effect that, in the formalization of information developed by Shannon,[1] translates into reduction in the quantity of information transmitted (Figure 5.2). The effect of

---

1. For an introduction to Shannon's probabilistic theory of information, see Atlan 1972.

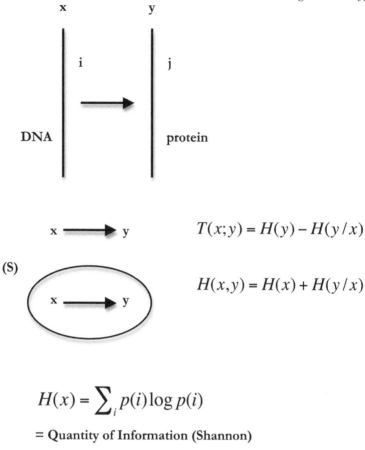

Figure 5.2. Self-Organization: Complexity from Noise

noise is a diminution of the information carried by the protein, compared to what this quantity would have been had the transmission been exact, in other words, if the protein corresponded rigorously to the DNA. But if, instead of considering the transmission of information from its origin to its destination, one now envisages the total quantity of information contained in the system

(S) of which this channel of communication is but a part, one can show rather easily that the quantity of information produced by noise is added to the system, rather than subtracted from it. The following makes this intuitively clear: the errors that result in a protein that is not an identical reproduction of a DNA sequence introduce a new variability, representing increased diversity compared to what would be present in the absence of errors. This diversity can obviously be the source of malfunctions and bring on negative effects, but in certain cases, it can on the contrary be the source of an augmentation of complexity and eventually of an augmentation of functional complexity, with a sum-total positive effect for the system.

This mechanism functions in a certain number of circumstances. It is, for example, at the center of the classical theory of evolution, since mutations—which are, precisely, errors in DNA replication—are considered to be the source of the progressive augmentation of diversity and complexity in living beings. Another domain in which this phenomenon is quite spectacular is the maturation of the immune system. The extraordinary diversity of the immune system's repertoire is in fact produced by somatic mutations and the aleatory rearrangement of chromosomes, which take place in the course of the maturation of lymphocytes considerably more often than in other tissues. There one has a totally characteristic example of complexity from noise, since the complexity of the immune system, at least on the level of the great diversity of its repertoire, results directly from a mechanism of this type. One finds the same mechanism at work in other processes of differentiation where "developmental noise" contributes to creating diversity and specificity through the diminution of initial redundancy.

Program, Data, and Meaning

In order to return to the nature of genetic determination and to the computer metaphors of "programs" and "data" as applied to DNA, I must make a small theoretical detour. We have seen that a binary sequence does not necessarily represent a program. The distinction between programs and data is intuitive for anyone who has used a calculator. There is a program—for instance, multiplication—and this program takes some data, which is then treated by the program, that is to say, in this case multiplied. But it so happens that computer scientists have developed a powerful theory of the complexity of algorithms, in other words, of the complexity of what can be done in a mechanical fashion by a computer. And this theory tells us that there is no fundamental difference between program and data: a computer program can be treated as data, by another program and possibly by another computer, and, vice versa, data can be incorporated as a part of a program. For this reason, programmers consider this distinction to be artificial and superficial. All of this is perfectly exact if one envisages the working conditions of real computers and of programs conceived by programmers for well-specified tasks. In this framework, the important question of meaning is always bracketed, outside of the theory, since meaning is always defined initially as something external to the program: it's a goal to be attained, an objective assigned to the program, a task to accomplish, or a problem to resolve. This is not at all the same situation in the case of natural systems such as organisms.

In artificial, intentionally constructed systems, meaning is always exogenous and the theory of programming doesn't bother with it, since it considers it to be already there, always implicit. The algorithmic complexity of a task to be accomplished, defined by the classical theory of programming, is the minimal length of a description written in a standardized computational language necessary for a normalized computer—called a Turing machine—to accomplish this task. This translates into a splendid paradox: algorithmic complexity is at a maximum when the task under consideration consists in reproducing a random sequence of data. There in fact does not exist, by definition, any way of reducing the description of a random sequence: the only way to reproduce it is to copy it verbatim. The incompressible character of such a sequence is therefore responsible for its large complexity.

Another aspect of this apparently paradoxical relationship between noise (randomness) and meaning can be found in the random character of all computer programs once translated into machine language, that is to say, into a series of 0's and 1's. In programming language, the program obviously has a meaning, namely, that of the task it is meant to accomplish, for which it was written, and in relation to which each of its instructions makes sense, inasmuch as it contributes to the realization of this task. But this meaning disappears, even in the eyes of the programmer himself, when the program is translated into binary. Moreover, the binary sequence thus produced is in general incompressible: there does not exist an algorithmic description shorter than the sequence itself that can reproduce it. And yet this is the standard definition of a random sequence. Put in other terms, meaning, as it is easily understood by all programmers familiar with higher-level programming language, disappears when it is translated into the different lower-level programming languages, until machine language is reached. This phenomenon is only a consequence of the exogenous character of the meaning. It is not an intrinsic property of the program. Imposed from the outside, it is preserved during the procedures of encoding and decoding by which it is translated into machine language. This, in addition, is what explains the fact that the machine has obviously no need to "understand" this meaning in order to execute the program, whatever way we ourselves may have of understanding what it means to understand a meaning.

Finally, recall that complexity from noise as a principle of self-organization also plays on the same paradox, but taken this time in the opposite direction: chance can be the source of new meaning in the eyes of those who are not familiar with the details of the organization of a machine. This is what has allowed me to articulate this principle within the frame of Shannon's probabilistic theory of information, which measures information while ignoring its meaning. The creation by noise of functional complexity—in other words, meaningful complexity—functions within this formal frame in the manner of a double negation (Atlan 1979: 88 [see Chapter 3, above]).

Thus, inasmuch as meaning is known but for this very reason remains implicit, it can remain outside of the theory. But the situation is different when one is concerned with natural objects. The meaning that they produce is not defined from outside. They produce their own objectives, or rather, what appear to us as their objectives when we describe their activities in terms of functions. And when we use computational metaphors to do so, we must differentiate between the part that is the program and the part that is the data in order to be able to analyze explicitly the meaning of what they do.

More precisely, in the description of a natural object, one must distinguish between, on the one hand, a "program"—which yields the possible meaning of its structure or its function, therefore making explicit an at least apparent goal, and defines a class of objects sharing the same structure and the same goal—and, on the other hand, "data," which specifies a particular object to which this function will be applied. A very simple example will show how this distinction functions. Suppose that one would like to realize an object described by the sequence 0011110000001 10011. One notices immediately that this sequence is produced by the doubling of each of the ones and zeros of the sequence 01 1000101. In order to produce the first sequence, it is therefore sufficient to write a program that doubles each bit in a sequence and to furnish this program with the second sequence as a particular piece of data upon which to operate. The program will, moreover, allow the production of a great many different sequences, which will share the same structure and will therefore have in common membership in the class of sequences constituted by doubles.

When one is concerned with natural systems, one must modify the definition of the algorithmic complexity of an object (of a sequence, a task to accomplish, etc.) by separating the program from the data. One must use a definition like the one I created with my colleague Moshe Koppel (Atlan and Koppel 1990, Koppel and Atlan 1991a, 1991b) of what we called "meaningful complexity," in opposition to classical algorithmic complexity, in which the meaning is not taken into account precisely because it is implicit. We called this notion of complexity "sophistication": this is the minimal length of the "program" portion of the minimal description. This definition corrects the paradox raised earlier, since a long random sequence, which is maximally complex in the terms of the classical theory of algorithmic complexity, has a sophistication of nearly zero. In fact, in the realization of a sequence of random data, the data portion is the sequence itself, and the program portion is extremely short, since it basically can be reduced to the single instruction "copy the data."

DNA: Program or Data?

THE TERMS OF THE ALTERNATIVE

Let's return now to DNA, in order to pose the two terms of the alternative. Note first that this concerns not just the way in which DNA functions but also the way the rest of the cell does. Whether we consider DNA to be a program or data will have different implications for the functions we attribute to the cellular machinery. If DNA is a program, the biochemical network of the cellular metabolism will play a role in the interpretation of the program, since an "interpreter" is always necessary in order to read and execute a program. On the contrary, if DNA plays the role of data, the cellular machinery will play the role of program, since data must be treated by a program. In fact, the cellular machinery as a whole can be seen as a network of biochemical reactions and transports that function like a distributed program in a parallel computer capable of handling data. As such, the cell is considered to be a "state machine,"[2] of which the principal properties are the following:

---

2. [A "state machine" is a kind of model used in computer science to describe abstract automata as systems of well-defined transitions between a set of "states."—Trans.]

1. The state of a cell is the set of concentrations of its constituents inside its microcompartments;

2. A network of biochemical reactions and transports moves the state of a cell from one state to another over time;

3. Protein activity—dependent on the three-dimensional structure of proteins—is at once a determinant and an effect of the state of the cell;

4. Memory is coded not just in static structures, but also in stable dynamic states, normal or pathological, which are transmitted in cellular division.

Let us say that reality is situated somewhere between these two poles: it is obviously not a matter of replacing the metaphor of the genetic program with that of genetic data, to be taken in its turn totally at face value. These two metaphors have, nevertheless, a larger or smaller degree of truth depending on the circumstances. Let us therefore analyze the respective advantages and inconveniences of the two terms of the alternative. In the first case, classically, DNA is a program that is interpreted and executed by the cellular machinery. In the second case, DNA is data, and the cellular machinery plays the role of a distributed program. In both cases, one must pose the question of the *domain* covered and the question of the *range* or the resulting product (Figure 5.3). If DNA is a program, the domain that can be processed is the set of DNA sequences that a cell will be able to read and execute. The theory of algorithms establishes that the set of programs susceptible to being read by an interpreter constitutes an undefined domain, since it is impossible in advance to know if a program can be executed or not: it's necessary to try to run it. The reading of a program involves an extremely close relationship between the interpreter and the program, and it is therefore a priori unlikely that any program whatsoever can be read and interpreted. Today one knows that practically any strand of DNA, even if it comes from an entirely different species, can be read by any cell whatsoever, on the obvious condition that it is presented with adequate material support. Concerning the *range*, this is the set of all the structures specified by the different programs, since we have seen that the program portion defines the structure of a class. If we admit that each sequence of DNA is a different program, a very large number of programs are interpreted and executed, and a large number of different class structures must be produced.

What happens if we now admit that DNA plays the role of data? With the program now being realized by the biochemical network, the *domain* that can

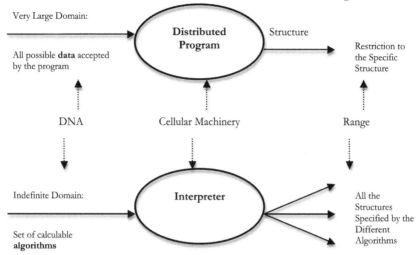

Figure 5.3. Summary of Functions Attributable to DNA If One Considers It to Be Data Processed by the Cellular Machinery:
- All the data can be processed (if it is presented in an adequate format): all programs cannot be interpreted;
- ⇔All the coding sequences of DNA can be expressed (if they are presented in an adequate format), even if they come from different organisms;
- Different data specify variations within a common general structure;
- ⇔Different genomes specify different ways of carrying out the same general biological functions by the biochemical network.

"A central core of metabolic pathways and cellular processes is very largely conserved among the organisms which have served as models for the analysis of genomes" (Miklos and Rubin 1996).

be processed is the set of data accepted by the program. This domain is extremely large, since we know that a program is capable of accepting any data whatsoever, provided again that this data is presented with adequate material support. This option corresponds better to reality, which shows that practically any piece of DNA can be processed by a cell. Cellular machinery plays the role of a "distributed" program, a notion that was difficult or even nearly impossible to imagine thirty years ago, because computer science had not yet reached it. Today, we know, on the contrary, that it is possible to construct machines with the structure of networks.

Each element in a network is relatively simple, but the structure constituted by these elements taken together, along with their various interconnections, is capable of operating in a way that, under certain conditions, can reproduce the operation of a classical computer program. In certain cases, such a network is even capable of performing better than a classical program. Consider, for instance, the properties of learning, homeostasis, memorization, and self-organization, which are more difficult to model using classical computer programs. This is why computer scientists are currently very excited by what is called parallel computing, which purports to reproduce the operation of a neural network. (In reality, such computing techniques are ultra-simplified compared to real neural networks.) The algorithmic calculations made by "neural nets" also simulate the operation of networks whose structures are those of distributed programs. As we have seen, nothing prevents us from considering the ensemble of biochemical reactions and transports within a cell as constituting such a network. The computational properties and the robustness of relatively simple biochemical networks are beginning to be studied (Artavanis-Tsakonas, Matsuno, and Fortini 1995; Rutherford and Lindquist 1998; Bray 1995; Barkai and Leibler 1997).

If one admits that DNA is data processed by such a distributed program, what will the *range* be? Since it is defined by the structure produced by the program, the range will obviously be the same whatever the DNA provides to the system. This is, in fact, the case: the structure of cells and of the major biological functions is practically the same for all organisms, the specific differences being produced by DNA, just as in a class where all the elements share the same structure, the specific differences between elements coming from the different data furnished to the program. All together, this term of the alternative (DNA as data) therefore seems a bit better than the other (DNA as program) in terms of its correspondence to observed reality.

It is obviously not a question, for all that, of purely and simply replacing the classical model of program-DNA prevalent today with a model of data-DNA: no one would dream of denying that the structure of genes determines that of the network of biochemical reactions. The metaphor of data-DNA, if interesting because it calls into question the idea that everything is genetic and introduces the necessity of taking into account the interactions between genetics and epigenetics, cannot found by itself a satisfying alternative model.

What one observes is in fact the result, due to loops in the system, of the superposition of different dynamics corresponding to different scales of time. Thus, the state of activity of the genes (the data) is modified by the state of the biochemical network (the operation of the program), but, in return, everything happens as if this modified DNA state then goes on to play the role of a program by modifying, in turn, the structure of the biochemical network . . . this is why the model that seems closest to reality doesn't rest on just one of these metaphors but on a kind of balancing act between the two, on different time scales, constituting what, with Katchalsky, we have called an "evolving network," that is to say, a network whose structure changes as it goes about its operations. Over short durations (a few minutes to a few hours), the cellular machinery treats the data produced by the active genes and attains a stable functional state. But over longer durations (hours, days, or more) certain stable states of the network modify the differential state of activity of the genes. This results in a new structure of the network of reactions, which produces a new stable state, and so on.

The notion of epigenetic inheritance designates the transmission, during cellular division, not just of the structure of the genes but also of their state of activity. The different parts of the same genome can present different states of activity at the same time, and these states can evolve over the course of time. This differential state of activity of the genome is determined by epigenetic mechanisms. Some examples of this transmission of states of activity are rather trivial. When an already-differentiated cell divides, it in general produces similar differentiated cells: a skin cell will divide into skin cells, and a liver cell into liver cells, even though the two types of cells, in the same organism, have exactly the same genome. What is transmitted in the moment of cellular division is therefore not just the DNA sequence but also its state of activity. This is, obviously, an extremely important phenomenon, to which, however, hardly any attention is paid. At a given moment, a certain number of genes are active and others inactive, and it is interesting to ask how this state of differential activity, which doesn't depend only on the genome, is transmitted.

This phenomenon of epigenetic inheritance has been known for a very long time, since one observes it in the transmission of the differentiated state in somatic cell lines. But the question arises whether this phenomenon is equally observable when it is no longer a matter of somatic cells but of germ cells. Classically, the response has been negative—this is the foundation of the principal distinction between *soma* and *germen*, which is accompanied by the more or less explicit hypothesis that the germline cannot do anything but reproduce the structure of the genes, without being determined by the cellular environment of these genes.

In fact, a certain number of cases of transmission of a state of differential activity of the genome at the level of the division of germ cells have been brought to light. R. Holliday, one of the researchers who has worked on this question, speaks of "epimutations" (Holliday 1987). One observes, in fact, transmissions of states of activity of the genome that have all the effects of a mutation without being one, since there is no change at the level of the structure of nucleotide sequences. The most studied case is that of the transmission of a state of DNA methylation. Methyl groups ($-CH_3$) can come to be linked at certain sites in DNA and in doing so can modify the state of activity of this DNA. The number of methyl groups associated with a gene can inhibit or facilitate its expression. During cellular division, the two daughter cells receive DNA, which finds itself in the same state of methylation; therefore they inherit a character determined by the state of activity of this DNA. But the transmission of this methylation state is not as simple as one might think, since it is not sufficient for the radical $-CH_3$ merely to attach itself to a location in order to remain there. Like all chemical reactions, methylation results from a dynamic equilibrium, and its maintenance requires the presence of the enzyme that catalyzes the reaction, methylase. In order for the state of methylation to be transmitted to the daughter cells, it is therefore necessary that, at the moment of cell division, methylase is also transmitted in a concentration sufficient to maintain the state of methylation.

Another case where this type of mechanism is beginning to be studied is histone acetylation. Histones are proteins associated with DNA in the composition of chromosomes, and their acetylation sometimes determines the state of activity or inactivity of the genes with which they are associated. More generally, modifications in the chromatin architecture in which DNA sequences are integrated seem to play a determinant role in the regulation of genetic expression (Pennisi 1997; Ait-Si-Ali et al. 1996; Fryer and Archer 1998). Another example of epigenetic heritability is that of "genomic imprinting," where the activity of certain genes depends on their parental origin: even though their structure is the same, certain genes are expressed or not depending on whether or not they come from the mother or the father, with this state being conserved across the division of germ cells, then the fusion of gametes. The state of activity transmitted in this moment therefore depends on the cellular environment of the genes in question. Finally, the prion-like phenomena observed in yeast and fungi is another example of epigenetic heredity: a certain state of protein folding is produced and amplified in an autocatalytic fashion, and this state is transmitted during cellular divisions (Coustou et al. 1997, Wickner 1994).

In a more general fashion, such a phenomenon of transmission can be observed each time there is a positive feedback loop. An autocatalytic reaction of a protein can produce a change of state, such as phosphorylation, modification of the three-dimensional structure, or some other characteristic on which the biological function of this protein depends and which spreads to the greater part of the molecules of this protein in a given milieu. It could also be a matter of a more or less complicated feedback loop between a gene, the product of this gene, and another region of the genome that controls the expression of the gene in question. Such feedback loops are very widespread, with some being very complicated, namely, in cases where the retroactive effect is not direct but involves many products, many other genes, and so on. In the case of a very simple loop, it is easy enough to show that such a system can present two stable states (Figure 5.4). In one state, the gene is inactive and the concentration of its product is zero. Consequently, there is no stimulation of the zone that controls the gene, which remains inactive, and the concentration of the product remains at zero. In the other stable state, on the contrary, the gene is active, and the concentration of the product attains a maximum value. The fact that the system finds itself in one or the other of these stable states depends on the initial concentration of the product. If this initial concentration is even just a little bit below a threshold value, the system necessarily evolves toward its state of inactivity. At the time of cellular division, it is therefore sufficient for the product to be divided between the two daughter cells in a more or less equal fashion, such that the concentration remains the same, in order for the daughter cells to remain in the same stable state. By contrast, if for one reason or another a change in the concentration of the product occurs and the threshold is crossed, the system will topple from one state to the other. This extremely simple schema, which holds in a great number of circumstances, allows us to understand how what is transmitted at the time of cellular division is not just the structure of a gene but also its state of activity, which depends on the feedback loops involving the gene's products.

EMBRYONIC DEVELOPMENT AND EPIGENESIS

The problems posed by the mechanisms of embryonic development also provide an occasion to analyze the illegitimate extension, to take up Strohman's terminology, of the notion of the genetic program. The development of the embryo visibly obeys a rigorous determinism, since starting from an egg one

1. Autocatalytic reaction of a product

Examples: phosphorylation of protein, changes in conformation…

2. Self-regulation of a gene by its own product

(after E. Jablonka, M. Lachmann, and M. Lamb, 1992, *Journal of Theoretical Biology*, 158, pp. 245-268)

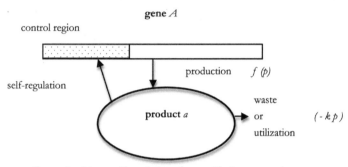

System implying a self-regulation of gene *A* by its own product *a*. One has $dp/dt = f(p) - kp$, with p = concentration of *a*.

Two Stable States

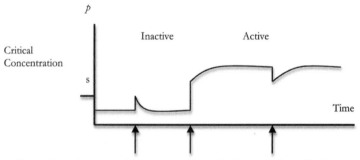

Type of behavior expected from such a system. At the time indicated by the arrows, a stimulus which modifies the concentration of the product *a* has been applied. *s* is the threshold concentration above which $f(p) > ks$.

Figure 5.4. Epigenetic Memory

obtains an adult, and this determinism in fact resembles the execution of a program, which, at the start of the process, would be contained in some part of the egg. For more than thirty years, the answer to the questions of what the nature of this developmental program is and where it is localized has been, obviously, that this program is nothing but the genetic program carried by DNA. This response has begun to be put back into question today, following a number of discoveries. This second abusive extension has had important effects—because of it, embryonic development has virtually disappeared from the center of interest in evolutionary biology and, more generally, molecular biology, inasmuch as it lead to the acceptance of molecular biology as the solution to all problems of embryonic development.

In the frame of the classical response ("the developmental program of the egg is nothing other than the genetic program"), one claims that the genes contained in the cellular nucleus actively control not just the structure and function of adult organisms but also embryonic development. Now the study of certain phenomena of embryonic differentiation, as well as recent experiments (which we have not yet finished discussing) concerning the cloning of organisms from already differentiated cells, have increasingly brought to light the importance of the activity of the cytoplasm (Kono 1997). Rather than seeing in the structure of genes a centralized system for controlling the organism, it seems more adequate to consider, on the contrary, that the organism controls the activity of the genes, in this case, the genes of development. A very instructive example is the evolution of the egg of *Drosophila* during the first two hours following fertilization. What happens with *Drosophila* might or might not be literally extendable to other species, but the mechanisms discovered in this fly present some very interesting properties (St. Johnston and Nüsselein-Volhard 1992, Dubnau and Struhl 1996, Rivera-Pomer et al. 1996).

It is in *Drosophila* that developmental genes, genes in which a mutation produces global catastrophes at the level of embryonic development, were discovered. The discovery of these developmental genes, a little over a dozen years ago, was hailed as an illustration of the notion of the genetic program: we had laid our hands on genes that resembled "master genes," that is to say, control genes that are responsible for the activity of many other genes and that we could consider as playing the role of instructions in a program. But the way in which developmental genes function makes it necessary to question this notion. A single observation brings us to suspect this right away: one of the principal characteristics of these developmental genes is that they have been conserved in the course of evolution, which is to say that their structure is practically the same in species as different as *Drosophila*, mice, and humans. This should have made us suspect something: since the same developmental genes direct the development of organisms as different as those of a fly and mouse, their activity should depend heavily on something else, which can only be their genomic and cellular environment.

In *Drosophila* (Figure 5.5), the first moments of embryonic development present diverse particularities. On the one hand, the genome of the egg is

mRNA

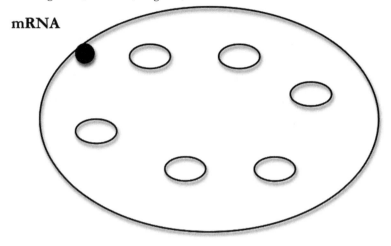

Bicoid Protein   **Caudal Protein**

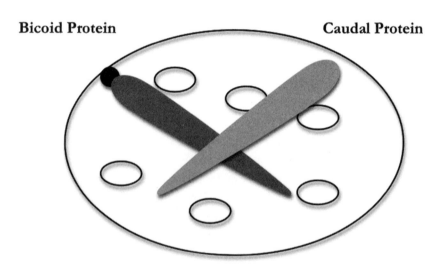

Figure 5.5. *Drosophila* Embryo (first two hours)

not expressed in the course of the two hours following fertilization, which is to say that no protein is synthesized from this new genome. On the other hand, during these two hours, the genome undergoes unceasing self-replication, producing many nuclei, which result in a multinucleated cell. Simultaneously, a number of other phenomena that will be the basis for the first differentiation of the embryo take place, notably, the definition of the poles that will become, respectively, the head, the tail, the back, and the abdomen.

This differentiation-polarization stems from an extremely simple phenomenon, determined by a piece of messenger RNA of maternal origin that finds itself attached to some part of the cell membrane. This RNA, which was there before fertilization, remains silent—that is to say, it is not translated—until fertilization. After penetration by the spermatozoon, a mechanical action that is probably accompanied by a modification in the structure of the membrane, this messenger RNA begins to be translated, setting off the synthesis of a "bicoid" protein, which then diffuses throughout the entire cell. One meets here with a system at once simple and complicated, studied for a long time and very precisely by my teacher Aharon Katchalsky, who, following Turing, described it as a "chemico-diffusional system." In such a system, which associates chemical reactions (taking place at a certain speed) with a process of diffusion of a product of these reactions (which happens at another speed), the relation between these two speeds can determine the state of the system. It can, for example, in certain cases be at the origin of a structured stationary state, that is to say, of a nonhomogeneous spatial distribution of the products of the reaction, which remains stable over time. It is therefore an extremely simple example of chemical organization, where the appearance of a spatial heterogeneity in the concentration of a certain product results simply from the coupling between the speeds of reaction and diffusion. This is exactly what we observe in the egg of *Drosophila*: the bicoid protein is synthesized at a point in the cell with a certain speed, and it simultaneously diffuses. A relatively stable gradient in the concentration of the protein is therefore realized, the concentration dropping off as one moves farther away from the place of synthesis. This gradient, along with others, will ultimately determine the polarization of the embryo, the head corresponding to the place on the membrane where the messenger RNA is attached.

Other pieces of messenger RNA of maternal origin, present and equally silent, then start to be translated, their activity depending, this time, on the concentration of the bicoid protein. One example is the "caudal" protein, whose speed of synthesis is inversely related to the concentration of the bicoid protein: this leads to a second gradient in protein concentration, opposed to the first. Other proteins of this type are progressively being discovered. Now, it happens that all these proteins are products of developmental genes. In other words, the developmental genes do not act directly; moreover, the genes present in the nucleus of the fertilized cell are not the ones that act. The developmental genes exert their influence through the intermediary of the products of messenger RNA, which was synthesized in the maternal cell before fertilization and whose expression is modulated by changes in the state of the membrane and in the cytoplasm of the egg. As we have seen, the constitution of concentration gradients of a variety of proteins thus synthesized will form the basis for the orientation of the embryo. During these two hours, when this structure of gradients is establishing itself, the nuclei are doing nothing but dividing. When their genomes finally begin to express themselves, they each do so in a different environment, since they are exposed to different concentrations of these cytoplasmic proteins. And these inequalities in concentration will be the basis for the differences in the activity of the genome in all these nuclei. It is thus that, little by little, the great regions of the future embryo are constituted, starting from the nuclei, whose states of genetic activity are different and which transmit these differences following the mechanism outlined above.

The example of the initial development of *Drosophila* is spectacular because of the temporal separation between the processes. In other organisms, which do not manifest this phenomenon so distinctly, one can conceive that the two types of process are interlaced over time. One understands well, in any case, how the differential state of activity of the genome in different cells of the embryo, which is the origin of this embryo's differentiation, is itself produced bit by bit by the differentiation between different cytoplasmic states. To the question of knowing what the nature of the developmental program is, it is therefore impossible to respond that this program is contained in the genome. This program is contained in the local interactions between cytoplasmic phenomena and cellular nuclei. It is therefore, properly speaking, neither localized, except in the totality of the space of embryo, nor fixed in time, since it produces itself bit by bit as development unfolds. In other words, if one is going to hold onto the metaphor of the program—a position that, in my opinion, should be relinquished—one must imagine a program contained in the totality of the space occupied by the embryo (and then by the organism that will result from it), whose structure modifies itself progressively as it realizes itself. In order to understand the nature of such a "program," delocalized in space and time, which tends to conflate itself with the very process of development, one can refer, by analogy, to models of self-organizing networks.

There exist extremely simple computational models, which their authors call "games of life." One plays these games on a grid, in the birth and death of small points, on the basis of a few elementary rules directing the interactions between these points: for example, a point is "born" (from the division of one of its neighbors) when it is surrounded by three points; it "dies" when it is no longer sufficiently surrounded, and so on. One thus sees, on the grid, the appearance and disappearance of forms that depend on the laws of interaction one has chosen. In these games of life, the question of the nature of the developmental program is very simple: it is contained in the rules of the game, which are the same across the board but which produce spatially heterogeneous results depending on the initial state of the system and its (in part aleatory) history. Now, if one poses to oneself the question of the localization of the program, the only possible response is: in the totality of the grid, since, in order to be able to play, it is necessary to possess not just the rules, but the total state of the grid.

The transposition of these questions and their answers to an organism is direct in at least one case: that of an egg enclosed within a shell. Let us consider, therefore, a closed egg, a chicken's egg, for example. We can admit, as we did earlier, that the nature of the developmental program is, quite simply, the set of physicochemical reactions that can occur between the different molecules present. These molecules obviously include the DNA of the egg, but they also include the proteins, the lipids, the sugars, the ions and the small molecules that participate in these reactions. The program is localized in the totality of the egg, including its nutritive material, which generally is considered to play no role in development. In reality, like the grid in the game of life, the nutritive material transforms itself, progressively, into a baby bird. One therefore sees here how the developmental program is contained in the totality of the space occupied in the end by the embryo, and how its nature is the set of biochemical reactions that lead to the development. Among these reactions, certain ones transmit a genetic memory, different for each species, and others comprise the biochemical network, which is more or less the same for all organisms. The "rules" of the game are the physico-chemical laws that direct these reactions, and the initial conditions are the structures of the egg's genome and of its biochemical network, based upon which the universal laws of physics and chemistry are canalized in a different manner in each organism.

Translated by John Duda

Underdetermination of Theories by Facts

(1991)

Sometimes we find ourselves in the following situation: we are looking for an explanatory theory that is both quantitative and predictive, to account for a certain, necessarily limited number of facts we have observed. In certain cases, several different theories exist that are not redundant in relation to one another and predict with the same exactitude the facts that have been observed. No empirical means, then, exist to decide among these theories, even though their meanings and the possible implications of their generalizations are completely different. This is what we understand by "underdetermination of theories by facts."

This phenomenon is often pointed out in the context of the human sciences, especially in psychology and linguistics. In that context, it has been possible to believe that the underdetermination of theories by facts was due to uncertainties related to the mind-body problem, with its metaphysical connotations and all the different possible solutions that have traditionally been suggested.

In fact, we already encounter this underdetermination in systems that are much simpler.

A method of calculation that has recently gained currency in the cognitive sciences, calculation by means of neural networks and, more generally, calculation by means of networks of automata, brings this phenomenon to light in a spectacular fashion without it being necessary to occupy oneself with the enormous networks meant to simulate our brain's billions of neurons. In fact, this method of calculation can very usefully be applied to studying the behavior of small networks of no more than ten units. These small networks of automata serve to model biological processes that are relatively simple when studied in isolation—such as a biochemical reaction, or a transport of matter through a membrane, or the activation and multiplication of a cell population—but whose cooperative functioning quickly becomes inextricable once the number of coupled phenomena exceeds just three or four.

Let us take a very simple example to illustrate this, the example of a network made up of 5 interconnected units in which each unit can at any moment be in one of 2 possible states. The total number of states in which the entire network can be is thus 2^5, that is, 32. Such a network can be used to study the collective behavior of five binary processes that function in connection with one another. (A network of this type has been used to represent, in a very simplifying manner, five different cell populations, which can each be in either a state of rest or a state of active multiplication, while at each moment their active or nonactive character depends on the stimulations or inhibitions each of the populations receives from itself and the four others.) The 32 possible states of the network thus serve to represent the observable states in which the system to be studied can be seen as a whole. In particular, certain states among these appear as stable states toward which, starting from different initial states, the network evolves spontaneously and in which it stabilizes. For this reason, these stable states are called the "attractors" of the network, and the study of the collective behavior of the system in question consists in determining the attractor, or attractors, of the network. The interest of this method of calculation lies in the fact that the nature of these attractors can be calculated on the basis of the *structures of the connections* between the five units of the network. Each structure of connections corresponds to a set of hypotheses about the nature of the interactions between these units, that is, about the way in which they function together in a connected manner. Put otherwise, we can say that every structure of connections represents a theory about the structure of the real system to be studied, and the attractors that these connections allow us to calculate represent the observable states that the theory in question allows us to predict. The ideal of the work of theorization would evidently consist in a single theory, that is to say, a single structure of connections, allowing us to predict the states that are effectively observed in reality. Unfortunately, it is easy to see that, apart from totally particular favorable circumstances, we are generally very far away from this ideal. In fact, the total number of possible structures of connection—that is, of different theories—is much larger than the number of attractors. The number of possible connections linking 5 units one to the other is 25; in each structure possible for the network as a whole, each of these 25 connections can be either activating, inhibiting, or null. The result is 3^{25}, that is, about 10^{12} different ways of constituting a possible structure of connections for the network. This number is much larger than the 32 possible different states of the network, so that, most of the time, *many different structures of connections produce the same attractor or attractors.* Put otherwise, many different theories predict the same observable facts.

This is exactly what we call *underdetermination of theories by facts.*

It is often said, generally, that an indetermination of theories is due to a flaw in the work of theorization, which has failed to respect the economical rule of Ockham's razor and accepted superfluous properties into the theory. Put otherwise, the *indetermination of theories is often confused with their useless redundancy.* True, adding to a theory superfluous characteristics that do not change its predictive value entails a multiplication of different theories that predict the same facts. Put otherwise, it is clear that *a useless redundancy in the theorization creates an underdetermination* of theories by facts, and a good work of theorization, employing Ockham's razor, owes it to itself to avoid this flaw.

But *the inverse is not* true, and our network example shows this. *An underdetermination of theories by facts does not necessarily imply a useless redundancy in the theory itself:* several structures of different networks, with the same number of automata, can have the same attractors and correctly predict the same facts, even though none of them is more or less redundant than the other.

To be sure, we must never be satisfied with this state of affairs and always try to reduce as much as possible this nonredundant indetermination. There are several ways to achieve this.

One consists in looking for new empirical data on the network such that certain theories that produced correct predictions can nonetheless be eliminated because their hypotheses now appear unfounded.

Another way of reducing indetermination consists in augmenting the number of specific predictions (evidently, by the empirical means of verifying them). This allows us to define and to look for supplementary pertinent facts, for example, not just the attractors but also the way of reaching these attractors, that is, the path running through the space of states, and its speed. This way, the set of observable behaviors to predict grows, and certain theories that previously were accepted are now eliminated because they no longer predict the totality of pertinent facts. Nonetheless, there is *no guarantee* in general that enough empirical data can be available to allow us to *completely reduce indetermination to a single theory*. When we consider the *immense numbers* of possible structures of connection we have already encountered, the *probability of success* (in this reduction) decreases very quickly when the number of heterogeneous linked phenomena to be taken into account by the model increases—that is, when the *minimum number of automata* necessary in the model increases.[1]

This underdetermination of theories by facts is an operational way of appreciating the *complexity* of a natural system to be modeled. And the analysis of the problem in terms of networks of automata shows that complexity *is not* a property of the *all or nothing* type, in the sense that natural phenomena are *not* reducible either to a single complete description, whether analytical or numerical, or to mysteries where nothing can be rigorously understood. On the contrary, most biological systems, when *they are studied* in vivo *and not in perfectly controlled conditions* in vitro, are situated somewhere between these two extremes, and that is what makes them complex, that is, not completely and indubitably understood, but not untreatable or unanalyzable, either.

This phenomenon is a very general one. We encounter—and have studied—it in immunology. It can also be encountered in cellular biology, where the concern is to understand the functioning of a single cell produced by the coupling of a certain number of biochemical reactions and transports of matter, where the conditions of observation *in vivo* limit the possibilities of measuring important variables and parameters (such as concentrations, speeds of reactions and of transport, chemical affinities or reaction constants, etc.), such that there, too, a more precise representation by systems of differential equations can hardly be used. It should, of course, not come as a surprise when this underdetermination of theories is encountered in much more complex systems, such as the nervous system of mammals or, evidently, our own.

---

1. See pp. 91–92, above, and p. 437n4, below.

But a small number of elements in interaction or of coupled processes suffices to produce an underdetermination. It is not necessary to reach the billions of neurons of a mammal brain or to occupy oneself with psychology and the classical neuro-philosophical mind–body problem.

Rather, it is possible that the study of the immune system or of other biological systems of the same kind can shed some light on the approach to the mind–body problem by depriving it of its specificity: the underdetermination of theories by facts in psychology is not due to some sort of metaphysical dualism of substance or even to an epistemological dualism, but only to the complexity of the system studied, that is, to the fact that it cannot be completely reduced by means of a sufficient corpus of empirical data.

This not necessarily an argument in favor of a metaphysical monism of any kind. It would, rather, lend itself to support a kind of relativism, or "multiplism," if we may say so, which would allow us to accept that each field of research presents some degree of underdetermination of theories or, in other terms, to accept that a complete and indubitable comprehension may be out of reach even if all constituents, taken in isolation, are totally known.

In a general way, for each field of research, we must therefore weigh differently the possibilities of certain knowledge that we do have according to the type of objects and phenomena we engage with: the more complex and singular a phenomenon is, the more every theory susceptible to rendering account of it is underdetermined, thus uncertain.

Translated by Nils F. Schott

SEVEN

Internal Purposes, Vitalism, and Complex Systems

(1991)

Internal Purposes and Vitalism: The Legacy of Critical Philosophy

It is important to realize to what extent our way of thinking has been influenced by the idea that nature strives toward one or several goals, that our understanding of nature depends on the discovery of these goals, just as if one or several hidden intentions existed in nature. In consequence, the study of the temporal processes of biological and historical development would seem to amount, if not to unveiling these intentions, at least to making sure that they exist. The achievement at the end of the process, reaching the goal at the end of the journey, would seem to determine the chain of events leading up to it. The existence of final causes acting in the manner of an intentional causality—just as, when we have a project, we try to carry out that project through a temporal succession of stages that are all determined by the project, which will itself be realized only at its end—has from ancient times been associated with the observation of living beings, whether that relates to their behaviors, which are similar to ours to the extent that they seem to be finalized in projects that define strategies (of hunting, protection, seduction, mastery) or to their development and their adaptation to an environment, where the final, accomplished form seems to condition previous stages. That is why it seemed natural to think that one or several souls had to direct the movements and transformations of living beings, which for that reason were called "animate," whereas "inanimate" was always synonymous with the nonliving.

Accordingly, recognition of the existence of natural purposes always amounted to an animism, in the proper sense of the term. The great change brought about by the scientific revolution of the seventeenth century was to confine animism to grasping and studying living beings, after it was expelled from the physical world, most notably from celestial bodies. Those were the good times of scientific vitalism, which backed up Kant's deism and also inspired the great atheistic and romantic philosophies of life, like those of Schopenhauer and Nietzsche.

Yet at the same time modern scientific inquiry was establishing itself on the basis of the principle of sufficient reason, that is, on the search for mechanical causalities, in which causes are to be sought only in what comes before effects, and not in what comes after them.

The achievements of Laplace's physics and Lavoisier's chemistry, which were based on that principle, made it possible for them to replace Aristotle's physics of entelechies and the alchemy of the Renaissance. That is also why final causes, which were confined to the living world, have become a problem, *the* problem, of biology. More precisely, they have remained a splinter, a recurring issue continually repressed and reopened at the heart of biology, like the contradiction, which has indeed proved fruitful, between a vitalist inspiration irreducible to the physicochemical and the institution of biology as an experimental science whose models are those of physical chemistry and mechanics (by contrast, in ancient times physics was inspired by observation of the living, the animate). That is how the question of purposiveness and intention, which we introspectively experience in our capacity to make plans, has haunted biology for about three centuries, ever since it has tried to establish itself as the science of the animate by borrowing its models from the sciences of the inanimate.

Today, this question seems to have been eliminated by physicochemical and molecular biology, which might appear to have succeeded in removing souls from the living.

But the question has been revived with renewed vigor ever since the human psyche, in its turn, has become an object of science; in particular, since people started wondering about the structures of an unconscious psyche. No questioning that concerns the unconscious can avoid the question of its purposes and intentionalities, for the dynamics of the unconscious can be grasped either using the model of the machine or using that of the subject whose experience we have.

Of course, the model is always corrected according to the specificity of the unconscious, but it is always there as a reference. "Desiring machine," "talking machine," "machine that creates meaning"—all refer to the unconscious as a specific machine, but as a machine nonetheless. By contrast, we can also consider the unconscious to be the double of a suffering, talking, sensing, and thinking subject, that is, the double of a patient, in the literal and the figurative meaning of the term, with her strategy, her tricks, her discourses. In that case, everything happens as if she were ascribed the properties of a conscious subject, hidden behind consciousness but functioning as its double.

The stakes here can be formulated with regard to what appears as most obviously finalized in the unconscious but also has to do with consciousness: namely, desire.

By definition oriented toward the future, desire is desire of something that is supposed to occur later, eventually, even if it is sometimes a repetition of the past (as is the case with conscious desire, actually).

And that desire—is it the desire of a subject, of somebody, even if that somebody does not know it? Or is it the expression of an optimizing mechanism of the kind suggested by metaphors, be they energetic or otherwise? In other words, are mechanistic metaphors of desire (which may be energetic, but also economic or, more recently, informational) to be taken literally, or only as objectifying metaphors that coexist with, if one might say so, the "subjective" experience of the subject?

It is remarkable that Spinoza, the founding father of an uncompromising rationalism, already raised this issue at the dawn of the scientific revolution. His *conatus*, which is neither a specifically human and conscious desire nor a principle of static homeostasis, is in fact an optimizing principle that directs, in a mechanical manner, the undivided affections of mind and body—which we usually call our feelings—just as it directs anything whose existence is at once perceived in extension and conceived in thought.

From this point of view, Spinoza, at the very birth of modern rationalism, pushes to their limits the conclusions that will be drawn from it three centuries later: the only purposiveness that is rationally acceptable, not only in biology but also in psychology, is an optimized mechanism, not an intention originating in a free will. This gives even more philosophical value to his ethics for those of us who cannot refer unreservedly to Kant's ethics any more, considering its unacceptable vitalist foundations. Today, biological purposiveness has indeed become mechanical. Accordingly, to the extent that we conceive the unconscious as rooted in the biological, can we still conceive that it has intentions, or even makes plans, in the way that consciousness seems to?

In other words, do there exist in nature intentions and plans other than those we experience consciously, independently of the status of reality or illusion that we readily grant to such experience? A quick flashback, back to the time when biological purposiveness was not mechanical but literally animistic, may help us to think about this question.

Until the Renaissance, the immediate experience of living beings—both the observation of animals and the introspective experience of human lives—served as the essential benchmark, or obligatory point of reference, for any consistent, rational worldview, whether in philosophy or in religion. Notions of a world soul, of a divine life, and of a macrocosm structured like the human microcosm were as obvious, for any enlightened consciousness, as the existence of molecules is for us today. It is striking that this primacy of the experience of life over that of the inanimate world can also be found under other skies—in Hindu, Greek, and Jewish thought, then later in that of the Christian and Muslim middle ages. All the stars, being endowed with movement, were perceived as animate beings, endowed with a soul, an intellect, and maybe a will, which would serve as an intermediary for the Creator's will or the great plans of the soul of the universe.

These ideas have reached us, either barely transformed, as in contemporary popular or pseudo-scientific astrology, or through great thinkers with mystical bents, such as Sri Aurobindo, Teilhard de Chardin, Henri Corbin, and Abraham Itzak Hacohen Kook, to whom one might perhaps add Arthur Schopenhauer and Carl Gustav Jung.

But in the meantime a scientific revolution has occurred, that of the seventeenth and eighteenth centuries, and these ideas of a cosmic life can be found today only in the writings of those who draw inspiration from mystical traditions. Contrary to what many admirers of bygone times believe, this intellectual revolution is irreversible: it originated in the fact that the stars had lost their souls, and it is quite unlikely that, one fine day, they will recover them. But it is interesting to analyze and trace the consequences of this fact over the three centuries during which our sensibility and knowledge have evolved—above all, because of a remarkable looping effect, with regard to the consequences for our theoretical and practical knowledge of living beings (i.e., for biology itself) and for the representations of ourselves that, little by little, under the influence of biology and the human sciences, we have come to have.

Since the scientific revolution, birth, life, and death, the voluntary movements of men and animals, and their involuntary movements of attraction and repulsion by way of sympathy and antipathy no longer serve as a compulsory model and reference point in analyzing and understanding natural phenomena. They were replaced first by the movements of the stars, now established as being repetitive and mechanical, and soon thereafter by machines, which efficiently applied the laws discovered by the sciences of the inanimate: physics and chemistry. The compulsory model or reference that replaced life was the machine: first the clock, then the steam engine, and today the computer.

For almost a century and a half, vitalist theories kept alive a fruitful tension between this mechanical representation of nature and the idea that the observation of living organisms nonetheless provided a privileged access to forms of intentional purposiveness. These operated on the model of intentional purpose characteristic of the artist and the craftsman. The effectiveness of the adaptation of means to ends sometimes went together with aesthetic properties, and our perception, just as with a work of art, was that of "purposiveness without purpose." This perception could cause in us the feeling of the sublime, in which, in a certain way, aesthetics, morality, and science met. It is not difficult here to recognize as the main representative of this attitude—and its most rigorous and cautious analyst—Immanuel Kant, the founder of critical philosophy and of modern humanistic rationalism.

To be sure, Kant distinguished carefully among formal ends (i.e., purposes that can be mathematized, which today we would call "physical" in reference to *extremum* principles in physics), external ends (e.g., the ways in which rivers or snowy mountains serve the end-seeking of those who use them as routes for communication or as reservoirs of water), and internal ends, characteristic of natural organization. Kant recognized the existence of the last only in living beings, which he held to be the only organized beings. And it is with respect to them that reason could not help issuing teleological judgments, that is, judgments recognizing the presence of final causes in them. In nature only living beings imply the existence of purposes, and even organizing intentions, modeled after our arts and crafts. By contrast, other forms of end-seeking merely accompany mechanical causality, which is the only causality to have pride of place in the sciences, for it meets, or has to meet, the needs of our reason in accounting for natural phenomena. In no way do they imply that we recognize natural ends, even less a final purpose in nature. Teleological judgment operates as a "propaedeutic to moral theology" by recognizing man and his freedom as the final end of nature. To be sure, the domains of reason are clearly distinguished; only scientific reason, which works with mechanical causalities, is "determining," whereas teleological reason is merely "reflecting" and subjective. Any union of the two is excluded because of the limitations of our understanding, and it is clear that, even in man, one can speak of a purpose of nature only as "noumenal"—from the "supersensible" viewpoint of his freedom.

But all this magnificent construction rests on the particular status granted to living beings, the only ones that are organized beings, the only phenomena with respect to which our reason, according to Kant, cannot and could not help thinking teleologically.

The argument goes as follows: (1) some internal purposiveness must be located at the foundation of our knowledge of many things in nature—essentially, organized beings; (2) We, men, with our limited understanding, can conceive and understand purposiveness only through the representation of the world as the product of an intelligent cause, of a God. Of course, this attitude is related to our faculty of judgment, as "we men," and not to our faculty of knowing and demonstrating "from pure objective principles," for our understanding is limited. That is why the previous proposition cannot be transformed into a dogmatic proposition that would claim to be objectively valid: "There is a God."

But, as such, it is enough "from every human point of view and for every use to which we can put our reason, whether speculative or practical," for it "is founded on an absolutely necessary maxim of our power of judgment" (Kant 2007: 227, §75). And this alleged necessity is no other than that of putting intentional purposiveness "at the basis of our knowledge" of living beings (ibid.).

Let us now see how this necessity is expressed:

> It is, I mean, quite certain that we can never get a sufficient knowledge of organized beings and their inner possibility, much less get an explanation of them, by looking merely to mechanical principles of nature. Indeed, so certain is it, that we may confidently assert that it is absurd for human beings even to entertain any thought of so doing or to hope that maybe another Newton may some day arise,

*to make intelligible to us even the genesis of but a blade of grass from natural laws that no
design has ordered.* (Ibid., 227–28, my emphasis)

However, Kant continues:

> are we to think that a source of the possibility of organized beings amply sufficient
> to explain their origin without having recourse to a design, *could* never be found
> buried among the secrets even of nature, were we able to penetrate to the principle
> upon which it specifies its familiar universal laws? This, in its turn, would be a
> presumptuous judgment on our part. For how do we expect to get any knowledge
> on the point? Probabilities drop entirely out of count in a case like this, where the
> question turns on judgments of pure reason. On the question, therefore, whether
> or not any being acting designedly stands behind what we properly term natural
> ends, as a world cause, and consequently, as Author of this world, we can pass no
> objective judgment whatever, whether it be affirmative or negative. This much
> alone is certain, that if we ought, for all that, to form our judgment on what our
> own proper nature permits us to see, that is, subject to the conditions and restric-
> tions of our reason, *we are utterly unable to ascribe the possibility of such natural ends
> to any source other than an intelligent Being.* This alone squares with the maxim of
> our reflective judgment, and, therefore, with a subjective ground that is nevethe-
> less ineradicably bound to the human race. (Ibid., 228, my emphasis)

What a magnificent rigor, which preserves the noumenal possibility of an
integral mechanism while denying that we could do without an intentional
and intelligent organizing principle in our understanding of things. But one
must see clearly where this ban comes from. The fragments I have italicized
in the above quotations (like many others to be found in Kant), clearly indicate the origin of this ban: the vitalist conception of the nature of living
beings considered as the ultimate theoretical framework for any scientific
inquiry. Kant cannot be criticized for this position in 1790, at the time of the
publication of his third *Critique*. On the contrary, today we cannot ignore
how bold it was, excessively so, to claim that not even the production of a
simple blade of grass could ever be understood through the simple mechanism of nature. Despite the support of giants such as Kant himself, Claude
Bernard, Louis Pasteur, and Henri Bergson, vitalism collapsed, not because
a "new Newton" experienced a flash of inspiration but under the repeated
blows of the biochemical and biophysical techniques that enable us today to
understand the mechanical production of a simple blade of grass, and of a lot
of other things.

It may, of course, rightly be remarked that, so long as these techniques did
not exist, the opposition between vitalism and mechanism did not have the

same meaning. A figure such as Claude Bernard, for instance, today may be equally branded a mechanist and a vitalist. The meaning of his invocation of the specific complexity of living beings has changed, now that a nonintentional physicochemical complexity can be recognized and studied. The vitalism of Kant and even that of Bergson resulted from a correct critical appraisal of the scientific reason of their time and not from a blind belief in a vital spirit. But that is exactly what clouds the issue with regard to the seemingly clear vision of bygone days: new techniques teach us to look at natural mechanisms through different logical and philosophical categories. What used to be logically obvious about what is necessary and what is contingent is not that obvious anymore. What was contingent in nature could be considered as some sort of disorder, as chaos, whereas the organization of living beings, by contrast, could not be reduced to what was necessary, what was deducible from mechanical causalities. Self-organization, already conceived as being what made living beings peculiar—"organization with an internal purpose"—was perceived as an organized contingent, a singularity that is not causally necessary, since its determinations are governed by its *internal* purpose. The self-organized was singular and therefore contingent because it might have been different and therefore was not necessary if it only happened to be the product of mechanical causes. The logical identification between contingent and random, the opposition between organization through causal necessity and chaos, made it impossible to see in natural mechanisms anything other than the repetitive product of a priori necessary laws—modeled after the Newtonian mechanics governing the movements of the stars. For a reason armed only with the model of mechanical causalities—"sufficient reason"—nothing was conceivable apart from this repetitive order of physical laws and chaos. The organized singular, which was not necessary because it was governed by its internal finality, could only be perceived through the projection of our own experience of freedom, while recognizing the noumenal, "supersensible," character of that experience. One could not, then, do without final causes, although it was clear that, for all that, the eventual union of final causes and mechanical causality in nature—Leibniz's preestablished universal harmony—exceeded the capacities of our understanding.

Yet today this contingent aspect of singularity, which characterizes the production and evolution of organized beings, is no longer incompatible with the chaos that sometimes characterizes the inorganic. The study of complex phenomena through methods of physics and chemistry reveals that what is random has a share in what is organized. The emergence of organization is no longer a mysterious property of Life; it can be observed in physicochemical systems in which our knowledge allows us, at least to a certain extent, to control it—of course, only with regard to its global properties and not to what makes it peculiar, for we would then no longer be dealing with emergence. An *a posteriori necessary* may be recognized in physical systems whose dynamic equations present bifurcations: several solutions are possible and one among them is necessary, but the one that will appear is the result of random fluctuations. In other situations that physicists call "chaotic" or "deterministic chaos" (here the mathematical meaning of the term is very different from its usual meaning), the imbrication of determinism and singularity is even greater and leads to a degree of predictability approximating zero (diminishing exponentially from the time when the prediction is made), even though all the mechanical determinations that govern those situations are well known. It is only in the hypothetical, physically unachievable case of a perfect knowledge of the initial conditions of the evolution of a system, and with *no* experimental error *in the least* being tolerated, that long-term predictability is conceivable: contrary to classical physical phenomena, in which the order of magnitude of measurement errors remains the same through time, here it increases exponentially.

Accordingly, we have enough elements to represent the phenomena of self-organization of the living as the result of chance encounters between molecules, which direct the effects of physicochemical laws in an irreversible manner. The latter determine, at any moment, a wide range of possible products resulting from the interactions between huge numbers of different molecules. The actual occurring of these encounters, which is contingent to the extent that it might have been different, contributes to the creation of a temporal evolution, a history that may be narrated as if an intention had directed it.

Today, one cannot but be struck by the reading of Kant, who without further ado rejected as absurd the idea, attributed to Epicurus or Democritus, that chance might be granted a role in the purposive mechanisms of nature, so firmly convinced was he that the reconciliation of chaos and organization was impossible.

The understanding—which indeed is still limited—that we can have today of the organizing role of randomness as a process that, at least for us, creates meanings takes into account the limits of our knowledge. But this does not turn it into a subjective kind of knowledge—thanks to mathematization (notably in its probabilistic form) and to the objective definition of the conditions of measurement and observation.

But all this is still too recent to have changed radically the unconscious philosophical aspirations we have inherited from the Enlightenment and from the critiques of reason. In this vision of a nature produced in the same way as a work of art is produced, we still find a collusion between the beautiful—turned into the sublime—and the good. Of course, truth is not entirely confused in this vision, since the distinction between objective "determining reason" and subjective "reflecting reason," between sensible and supersensible, is still preserved, even if we do not use these concepts that often anymore.

In brief, we have lived, and keep on as if we were still living, in this fragile equilibrium in which scientific rationality's universalism seemed to be able to coexist with rational ethics, in which, without being confused with one another, the true and the good were indeed reconciled in the enlightened man's worldview.

But this equilibrium has broken down, and the foundations of the magnificent edifice—the necessity of our teleological judgments with respect to living beings—have collapsed. And the proper goal of the building of this edifice, namely, "moral theology," appears more and more as what truly determines its worldview. It was a secular form of theology (just as the cult of Reason was the secular form of religion in the West) that, far from establishing an authentic universality, finds itself confined to the limits of a defensive ideology. Its nobleness can still be felt, but felt as the nobleness of an aristocracy whose world has been destroyed by a revolution. In this worldview, the judgment passed upon the moral freedom of man as an ultimate end of nature cannot *not* be articulated together with the judgment passed upon living beings in general—that is, upon Life—as ends of nature. It was under cover of the Kantian distinctions between different types of reason that a mechanist science could develop. But, at the same time, it was under cover of the vitalist distinction between living beings, endowed with purposes and intentions, and everything else, that moral freedom, albeit supersensible and constituting the subject matter of a transcendental philosophy, could apparently remain within the province of what was natural and *universally intelligible*. Nowadays, contradiction breaks out almost daily, every time *Western* science addresses problems that *Western* morals find nearly insoluble. Whereas the scientific techniques invented by Western humanism have succeeded in treating what is living as a machine, to solve the identity and social problems resulting from this success, this very humanism can only rely on what is left of natural law,[1] in which the dignity of man depends on the sacralization of Life. It is just as if Life had remained inaccessible to scientific instrumentation—whereas, after having been mechanized and molecular-ized, it has long disappeared from where it was thought to be found. It is striking that African, Middle Eastern, and Far Eastern societies seem to have fewer difficulties in solving these problems than our societies do. That is perhaps because their own vitalism was never articulated with scientific rationalism. Originating from traditional animism, it could and still can coexist with science and with techniques imported from the West. These remain foreign cultural products, about which the very issue of theoretical articulation is never raised, even though, operationally speaking, such tech-niques are perfectly assimilated. For us, the end of vitalism occasions the collapse of the construction upon which modern scientific humanism rested. We are right in the midst of the "great agony of Kantianism" Alexis Philo-nenko spoke of twenty years ago.[2] We are left with only two possibilities: either we extend the mechanism of objective causalities, in which there is no room for freedom, to man, or we draw a new distinction between, on the one hand, nature, including organized beings (be they living beings or not, i.e., organized machines), and, on the other hand, moral subjects, consid-ered, as such, to be "supernatural."

We have suggested elsewhere how we can represent "man as a game" so as to escape the alternative between "man as a machine" and "man as a god" by recognizing the place of play in being. This unveils a real-unreal that characterizes, it seems to us, specifically human phenomena such as the devel-opment of our *Homo sapiens* psyche, our practice of natural languages, and, more generally, our knowledge games, those of scientific knowledge included. Given this realization, we must relinquish some philosophical, the-oretical, and a priori universality: specifically, we must relinquish the "moral law as a formal rational condition of our use of freedom." But we thus also reap the benefits of having more room for practical universality, like that of human rights, which were recognized and accepted in the twentieth century, in the midst of and despite ideological, religious, and cultural clashes. This kind of universality is indeed fragile, always in the process of being built, for it has no other foundation than the good will of people who happen to be

---

1. The oft-expressed stances of the churches are spectacular illustrations of this point.
2. See Atlan 1991a: 239.

determined by moral laws, which are often antagonistic. But perhaps that is less illusory. In order, however, to regain the certainty and security (which quickly turn dogmatic) provided by "foundations," some have been led to a precritical regression. Blithely skipping over the three last centuries, they feel at ease within the world of alchemy, natural philosophy, and theology, out of which modern science emerged. But they forget that modern science could do so only by breaking with the world of its origins. That is how the disillusions of scientific humanism, which can no longer be vitalist, too often lead to a regression to fundamentalist religion. There, God, astrology, and the use of new scientific theories take the baton from political ideology and nourish redemptive messianisms in which all uncertainties are swept away once again.

Thus some, for example, thanks to a new vocabulary, try to convince themselves that the movements of the stars animated by Aristotle's active intellect are not that different, everything considered, from rational mechanics. Some people are ready to believe that we are dealing with different ways of referring to the same thing, and that final causes can be reinterpreted within the framework afforded by rational mechanics and physical principles of evolution. In other words (those of Leibniz in his *Discourse on Metaphysics*), people think as if formal purposiveness and intentional purposiveness could be identified with one another.

Of course, that is not so. To realize this, one need only analyze the new, sophisticated forms purposiveness takes today, thanks to the sciences of complexity, both in physics and biology.

Complex Systems: New Aspects of Formal Purposiveness

Mechanics has always accommodated principles of evolution in which the end state plays a crucial role in the description of temporal processes; in other words, it fits well with a mode of reasoning that also seems to be teleological. By contrast, why has biology, ever since the first attempts to apply physicochemical and mechanical methods to living beings, always defended itself against finalism? For very clear reasons:

1. Finalism in physics is always formal, calculable, and nonintentional. By contrast, biology had to struggle against the noncalculable and intentional finalism it inherited from the animism of the cosmic consciousness that suffused the philosophy of nature up to the seventeenth century.

2. Finalism in physics is not, in general, defined as such, for it is expressed in the form of optimizing principles, that is, of maximum and minimum, which state that such and such a magnitude expressed by a mathematical function has to reach a maximal or a minimal value at the *end* of a process: for example, that the free energy of substances likely to react chemically will be minimum at the end of the reaction; or that the entropy of an isolated spontaneous process tends toward a maximum. These physical principles allow us to predict, by way of computation, the evolution of some processes and in no way assume that a consciousness or an intention would govern them. Furthermore, these optimizing principles in physics are always restricted to domains of validity that are neatly circumscribed: for instance, the principle of entropic increase concerns only closed and isolated systems.

Today, one might imagine a physical principle of complexification of matter based on the cosmological theories of the Big Bang and the evolution of matter that would lead to the appearance of living beings and to the evolution of species up to man. But we are still far from it, because there is no universal and objective measure of complexity that is applicable to all fields. Even if a consensus were to emerge on this point, complexity does not necessarily imply meaning and signification. Sometimes, one is even tempted to argue the opposite: a random series deprived of meaning achieves maximum algorithmic complexity if it is reproduced as such. We are currently working on the definition of a measure of complexity that would remedy this shortcoming and that would take meaning into account—a measure according to which a random set would be given zero value.[3] But the meaning must always be defined relative to an observational situation.

That is why optimizing principles in physics—unless they are misinterpreted and, in a sense, wrongly spiritualized—have no animistic or intentional connotations. The ambition of biology has been and still is to imitate physics in that respect, that is, to account in a nonintentional way for living processes that obviously appear to be end seeking, such as the development of an embryo from an egg, or even, to a lesser degree, the evolution of species.

Thanks to molecular biology, this ambition has to a large extent been achieved. Of course, all the problems are not yet resolved, but research in that field is headed down a path that also seems, because it is the most efficient and fruitful, to be irreversible.

---

3. [See the notion of sophistication introduced in Chapter 5, above.—Eds.]

Vitalist finalism can no longer really have advocates today, in a time when one cannot fail to refer to the molecular and cellular substratum of living phenomena as a solid base, even if one must take into account phenomena of cooperativity and emergence. The notion of a genetic program was conceived to replace the purposiveness of a hidden intention that seemed to direct the development of living beings toward an ever-increasing complexity, through a mechanical, nonintentional end-seeking, comparable to that of a computer program.

Because the notion of genetic program has already been heavily criticized, there is no need to draw out here its character as a heuristic metaphor, which becomes deceptive when its explanatory value is taken too seriously. But these criticisms, made by biologists and physicists, were not intended to revive, so to speak, vitalist theories—quite the opposite. They were intended to strip the program metaphor of the leftovers of intentional purposiveness it inadvertently contains: for, in general, a program is written by somebody in order to achieve a given task.

It is this need to resort to a programmer and to an end intentionally set beforehand that the most recent theories of self-organization have tried to eliminate. In other words, these theories do not tend to make biology less physical—quite the opposite. They aim to explain the development of the structures and functions of the living by referring to the dynamics of interaction and coupling that take place between chemical reactions and transport phenomena, which, *in themselves and each on its own*, are not at all fundamentally different from the reactions and transports observable outside of living organisms. In embryology, the meaning of the notion of morphogenetic field, once used to describe the effects of mysterious "vital" forces responsible for the orientation of an egg's development toward its adult form, has changed: this notion now designates a field of chemical, electrical, and mechanical forces resulting from the interactions between various molecules. Just as in a network, whose global behavior is determined at least as much by the nature of connections as by the individual properties of elements, these interactions can create phenomena of cooperativity, in which forms and functions appear at the global level of dynamics but which cannot be reduced to a mere additive superimposition of the individual properties of molecules. These phenomena of emergence often have a spectacular character, for they may be counterintuitive and may have surprises in store even for those who model them with the help of programs integrating, in various ways, determinism and randomness. This is perhaps why work on self-organization can still be misleading, and it is worthy of further attention.

In very different fields—the physics of disordered systems, geology, meteorology, chemistry, population dynamics, and so on—it happens that several processes may interact with one another, either because they share the same sources of energy or because they exchange regulating signals. There, one observes phenomena of emergence that are called by different names: cooperativity in chemistry, catastrophe in the theory of the same name, dissipative structures, self-organization, and so on. Research into these phenomena has highlighted the importance of levels of organization and the possibility of the emergence of specific properties at a global level that cannot be observed at the level of individual constituents.

Hence the idea, which has now begun to circulate, that consciousness, desire, and intentionality might also be the results of emergent phenomena at the level of the human brain, which would of course not exist at the level of molecules, not even at the level of individual neurons, but would result from cooperativity between the thousands of neurons that make up the brain. It's important to point out a confusion here.

Of course, phenomena of functional self-organization may occur on the basis of networks of interacting elements (cellular automata, genes . . .). These phenomena enable one to analyze, almost step by step, the mechanisms of emergence—at a level that is global and integrated in comparison to that of base elements—of structures and functions that simulate some of the cognitive functions of our brain. One may refer, for instance, to the classification of and discrimination between forms based on "self-generated" (i.e., not explicitly programmed) criteria.[4] One thereby simulates the emergence, within a network of automata, of a semantic function capable of creating meanings by the very fact that it sets criteria of recognition. In a similar vein, one can imagine that intentional phenomena, in which "goals" are "created," might also appear as emergent phenomena in processes of functional self-organization, and that a certain distance of the system from itself would appear, simulating something like "self-consciousness."

Translated by Vincent Guillin

---

4. See Atlan 1987a: 563–76 and Atlan 1986.

EIGHT

Ectogenesis and Reproductive Cloning: Reasons For and Against

(2005)

When we consider the state of current debates on the transposition to the human species of reproductive techniques that would dispense with fecundation (techniques now being developed), we are struck by how these techniques bring to light the ever-determining role of female uterine implantation in the process of procreation.[1]

Uterine implantation remains a benchmark and a barrier between, on the one hand, a very large set of potentialities (including many things and many goals that are quite different from a personal human existence) and, on the other, engagement in a process that must lead to a personalized birth. A prohibition against having children born by cloning, for whatever reason, can thus, rather simply, be articulated as a prohibition concerning uterine implantation.

This rather simple situation is what gestations outside of women's bodies unsettle.

---

1. [See Atlan 2005, chap. 4.—Eds.]

Indeed, unlike the opposition to reproductive cloning, whose motivations are largely phantasmal, opposition to the artificial uterus, like opposition to the contraceptive pill and to abortion, will have trouble resisting the right women claim to have control over their bodies. The debate about cloning will begin again because the conditions, both technical and symbolic, under which cloning can be achieved have changed. One of the rare biologists to envision the theoretical possibility of cloning mammals, Nobel laureate Joshua Lederberg, predicted in 1966—at a time, that is, when most thought of this as a radical impossibility—that human reproduction by cloning would have to wait until ectogenesis had been established as a mastered and socially accepted technique in order to benefit, in a way, from its fallout (Lederberg 1966). This prediction is justified by the fact that each of these two techniques constitutes the crossing of a qualitative threshold in the denaturalization of reproduction. Debates about the acceptability of crossing the threshold of asexual reproduction, seen in relation to techniques of medically assisted procreation already employed, will very probably be of a different nature in societies where the other threshold, that of ectogenesis, has already been crossed. Once ectogenesis were to become a normal mode of gestation that would make it possible to do without a mother's uterus to give birth to a child, the question of what an embryo is and of its legal status would be posed in a new way, even more dramatically.

As long as ectogenesis remains in the domain of futuristic predictions bordering on science fiction, implantation in a natural uterus remains both a benchmark and a lock that enables the control of technology and the prevention of the possible application to the human species of asexual reproduction, such as cloning by transferring the somatic nucleus and perhaps eventually parthenogenesis. This lock, which is both technical and symbolic, will be cracked once ectogenesis has become a familiar practice—if not a family one—adopted by a non-negligible number of women. While today there is little motivation to give birth to children by cloning—phantasms of immortality projected onto the genome aside—it is possible to conceive of stronger motivations and social pressures for ectogenesis if a significant number of women come to want to have children but also avoid the servitude of pregnancy. In turn, it is possible to conceive how a foreseeable fallout for family structures and for the diversification of forms of filiation could then contribute to normalizing modes of biological reproduction, such as cloning and parthenogenesis, which today are considered socially unacceptable.

Once artificial uteruses have been used to give birth to normal babies of nonhuman mammals—cattle, sheep, pigs, pets—nothing will be able to prevent the application of the technique to the human species. One could imagine that ethical arguments, which public opinion takes over from national and international law, could eventually be mobilized with the goal of achieving absolute prohibition. The precedent of reproductive cloning could be invoked; at the moment it is likely to be prohibited at the level of research and development that would be necessary to transfer to the human species a technique that is still at an experimental stage for animals. But this precedent is misleading, because human ectogenesis would pose ethical, social, and political problems very different from those of reproductive cloning or, for that matter, of parthenogenesis and grafting [*bouturage*], given the—still hypothetical—case that these techniques could easily be put into practice in mammals.

The Fetishism of the Gene

It may be useful briefly to recall certain elements in the debate about human reproductive cloning, which largely work within the boundaries set by the hypothesis—as yet unverified—that the technique would be biologically safe for the animal, producing neither abnormal pregnancies, nor malformations of embryos, nor offspring with congenital diseases. The first reactions, whether of horror or of fascination, to the birth of Dolly the sheep, the first clone of a mammal, were in fact the result of a misunderstanding. Since a clone is produced from the same genome as the organism that is cloned, a clone was seen as an identical copy of that organism. This view of things was the result of thirty years of uncritical teaching and bad popularization of molecular biology, which had reduced the biological to the genetic. The genome, like a computer program, was thought to be charged with determining the "essence" of each individual's biological identity. Two individuals having the same genome thus became biologically identical, even if some, good dualists that they were, were ready to admit possible, albeit limited effects of the environment and of culture on mental and affective development. Forgotten were well-known observations of monozygotic twins, who carried the same genes and were nonetheless biologically different, thanks to epigenetic mechanisms in their development that led to different immune systems and different brains.

Attempts have been made to correct this massive misunderstanding by replacing the simplistic view "everything is genetic" with a more nuanced idea of the complex molecular and cellular interactions that constitute the individuality of an organism by means of evolving processes that continue throughout its existence. In fact, the structure of genes is just one, relatively static part in the set of an individual's biological determinations.

In this context, cloning a mammal does not mean producing a copy, because two genetically identical individuals are not identical. Once this misunderstanding has been cleared up, one would think that the main reason for prohibiting—or desiring—human reproduction by cloning would have been eliminated.

But the culture of "everything is genetic," even a certain fetishism of the gene, is so strong that the reproduction of human beings genetically identical with others, whether by cloning or by splitting embryos (i.e., artificially producing what nature produces in making twins) has remained prohibited. Such, at least, was the recommendation of the French National Advisory Committee on the Ethics of the Life and Health Sciences, followed by numerous national and international ethics panels. Although the framework of international legislation still remains to be put into place, it seems that there currently exists an international consensus on prohibiting any practice of human reproductive cloning. But this consensus rests on a new misunderstanding. Many biologists, medical professionals, philosophers, and even people of faith would be ready to permit this practice, within limits and in well-circumscribed cases, if it were biologically without dangers. All agree on the fact that, at the current stage of the technique as it is practiced on animals, and taking into account the very weak rate of success (successes that themselves are relative and uncertain), to experiment on humans, in this case, women, would be criminal from the viewpoint of the common rules that frame all human experiments for therapeutic or research purposes. It is therefore fortunate, with regard to a possible conclusion of the current legislative debate, that animal reproductive cloning has not been perfected since what would be at stake is the possible launch of experiments on women that would seek to transpose this technique to the human.

But underneath this immediate consensus, debate about a hypothetical case continues, namely, the case that human reproductive cloning might become available as a new technique of medically assisted procreation, an addition to artificial insemination, in vitro fertilization and its variants, such as the intracytoplasmatic injection of spermatozoids or even of nonmature spermatozoids taken from a testicle.

There are reasons other than the erroneous invocation of identical copies for prohibiting human reproductive cloning, even if it were biologically "safe," at least as safe as other techniques of medically assisted procreation. Hypothetically, these reasons would not be biological reasons so long as this practice remained limited to rare cases and did not become the normal mode of reproduction for the human species. (In that case, indeed, a certain danger of a damaging loss of genetic diversity in the species would exist, a risk that is due to the asexual character of this type of reproduction.) It is hard to see, too, how metaphysical or religious reasons could apply, unless those who put them forward belong to a naturalist religion or morality that prohibits all human intervention in the natural course of things. But there are equally strong social reasons for prohibiting such practice, reasons we would be wrong to neglect, even if they seem less "absolute." One could name problems of filiation (a son or daughter would at the same time be a brother or sister, a twin sister or twin brother), the dangers of anti-clone racism, and, generally, all forms of instrumentalizing a child to be born, conceiving it as a means of perpetuating a fetishized genome that is to be reproduced.

Not qualifying these cellular artifacts as embryos does not imply, as it has been taken to, that the children who would be born following uterine implantation, whatever the licit or illicit circumstances, would be less human than others. It is the desire to define at all costs, a priori, to label once and for all what must be the "essence" of a human being as the "essence" of an embryo, that leads to this absurd conclusion. On the contrary, as a great witness in the American National Bioethics Advisory Commission's debate on human reproductive cloning affirmed in 1997: despite the offense to human dignity that the possibility of this practice constituted in his eyes, no kind of diminished dignity must be attributed to a child born from it.

What we can learn, among other things, from the diversity of the living about its substantial unity and evolutionary continuity is (1) that what is not a living thing can become one, starting from (molecular) elements that are not alive, and (2) that what is not an embryo can become one, what is not a human being can become one, just as what is not a tree—a sprout—can become one. After all, we cheapen nature's (and, in its wake, agriculture's) efforts in growing a tree if we identify, purely and simply, the tree with its "essence," said to be contained in its sprout.

There is a certain danger that human "clones" would be ostracized and excluded, even if such behavior were to be based on totally erroneous conceptions. The inanity of any alleged biological foundation of racism does not prevent racism from manifesting itself.

In the vast majority of cases, these reasons for not taking the path of human reproductive cloning should be opposed to phantasmal motivations of immortality or genetic reincarnation. Put differently, the possible reasons for taking that path are weaker than the reasons for prohibition, even if the latter could, in certain particular cases, be avoided.

Recalling these elements of the current debate about human reproductive cloning lets us see that the stakes of the future artificial uterus lie elsewhere. They do not concern future children, even though such children will be affected by the conditions of their birth, nor do they concern the representations of these children in the desires of third parties.

We could also imagine social reasons for prohibiting the artificial uterus, but they would not long resist the reasons for engaging in ectogenesis, since the reasons for replacing pregnancy with development in an artificial environment would not be on the order of a more or less phantasmal biological "desire for a child" at all costs. We have seen that the first justifications would probably be medical ones, to save the embryos of spontaneous abortions, then to permit women without a uterus to procreate . . . Yet very quickly a demand would develop from women who want to procreate while sparing themselves the constraints of pregnancy. From this point of view, the stakes of the artificial uterus must be compared not with the stakes of medically assisted procreation but with those of the contraceptive pill and the liberalization of abortion. It is not a more or less contestable "right to a child" that will be invoked, but the right of women to take control of their bodies. Once it will become possible to procreate without pregnancy, in what name could women's claim to be able to choose this mode of gestation be opposed? Women's right to take control of themselves and to refuse a pregnancy by making use of the services of an artificial uterus will be stronger, in a different way, than the right, sometimes invoked, to "satisfy the desire for a child," even if such a desire were there as well.

There are several sorts of individual rights. Some are limited only by the duty not to harm anyone else and do not imply any duty on the part of society (except regulatory agencies) to allow them to be exercised. The right to go on a dream vacation, spending a fortune on it, does not imply a duty on the part of society to finance the exercise of this right. By contrast, rights to education, work, or health imply duties for society if they are not to remain purely theoretical. The more individualistic a society is, the more the domain of rights of the second sort is extended. The right to have children at any cost, despite natural impossibilities, can be considered part of the right to health and thus implies a duty for society to permit its exercise. That is what is at issue in mobilizing the medical community and employing it in the service of this right through medically assisted procreation, even more so when it is covered by health insurance plans. But are there no limits to this right and the duty of the collectivity it implies? In particular, does society not have to ask what effects the exercise of this right would have on society's own relations to the children born in this way? From this point of view, we can analyze the similarities and differences between claims to medically assisted procreation, by whatever technique, including perhaps cloning, and claims to women's right to take control of their bodies, including such similarities and differences in questions of procreation as would exist if the artificial uterus were available.

Translated by Nils F. Schott

NINE

Weak Reductionism

(1986)

An overview of the problems posed by multilevel natural organizations indicates that we must accept both a unity of processes that we do not (yet?) know and the diversity of the disciplines that subdivide them and allow us to learn about them, progressively and partially. This leads to the perception of a unity of science that can be only the unity of its *practice*, motivated by its *successes*. The successes are technical, relating to mastery and control; the practice is that of weak reductionism, for which physics remains, to a certain extent, the fundamental discipline, even though the other sciences cannot be totally reduced to it.

We have seen how even the question of the relationships between matter and mind can be approached from the angle of the brain-thought problem in a de facto reductionist process that characterizes scientific *practice*.[1] In this

1. [This process is what Atlan calls "weak reductionism" and proceeds to discuss in this chapter. For his discussion of the brain-thought problem (and its relation to scientific practice), see Atlan 1993, chap. 2, esp. pages 46–53.—Eds.]

procedure we have encountered the influence of the information sciences and paradigms of the computer and artificial intelligence, which allow us to explode the alternative between strong reductionism and spiritualist holism by demonstrating that properties of the whole can emerge *in machines.* As we have seen, however, this influence creates new theoretical difficulties that must always be borne in mind, for lurking in this process is the danger of going no further than the provisional syntheses that are always possible—if only for purposes of pedagogy or popularization—at the price of tentative extrapolations, generalizations, and global explanations. Weak reductionism consists, inter alia, of not giving in to this temptation; at the same time, the unity of the processes continues to be postulated and physics—because of its successes—continues to be the paradigmatic discipline, the one whose method should be emulated. This reductionist praxis often leads the practitioners of each discipline, anxious to preserve the prudence and rigor without which their discipline would not exist, to fall back on themselves—at least in their practice as scientists—and renounce the need for total explanation.

But when we study complex systems organized into different levels of integration, this excessively hermetic compartmentalization is no longer possible, because it destroys the very object of study. Then reductionist practice resumes its unifying garb and discourse, though guarding itself against falling back into a strong, metaphysical reductionism. *The causal search for physical phenomena* remains indispensable, but its meaning differs from that spontaneously attributed to it by the strong reductionism of physicalist metaphysics on at least two points.

The first point concerns the transition from one level to another, such that the properties of the higher level are considered to be determined by those of the more elementary level. This transition cannot be seen as a simple causal determination of the same type as cause and effect relationships in a temporal sequence of events, which determine one another on the same level of observation and organization. Neither can it be seen as a simple relationship of spatial inclusion of the parts in the whole, nor, as Oppenheim and Putnam would do,[2] as an association of these two relationships, the causal and the inclusive, where parts *spatially contained* in the whole *causally* determine its properties. As we have seen, even in artificial machines the determination of the whole by its parts implies a qualitative leap, with the emergence of new properties that cannot be observed in the parts and cannot be described by a

---

2. See the Appendix to this essay.

simple association of the properties of these parts. What is more, this determination takes place thanks to relationships among the parts that are not merely spatial contiguity but also functional connections. The latter create spaces (reaction spaces, more or less complex topological spaces) that are more relevant than the usual Euclidian space (Bienenstock 1985).

The second point in which the meaning of reductionist practice applied to multilevel organizations differs from physicalist metaphysics concerns the nature of the physical phenomena that analysis makes it possible to place in causal relationship: these are physical in that they are described by physics, not because they are immediately available through direct sensory apprehension of the physical reality—in the material sense—of which the universe is composed.[3] In other words, reductionist praxis derives from physics, when

---

3. Richard Feynman cannot avoid the question of reductionism on the basis of physics, "the most fundamental and all-inclusive of the sciences . . . the present-day equivalent of what used to be called *natural philosophy*, from which most of our modern sciences arose." Significantly, those other sciences (described as "sisters" rather than as "daughters") are chemistry, biology, astronomy, geology, and *psychology*, whereas the "relation of physics to engineering, industry, society, and war, or even the most remarkable relationship between mathematics and physics" is merely mentioned, due to "lack of space" (Feynman, 1963: vol. 1, chap. 3, p. 1; the last of these is discussed at length in Feynman 1967). For Feynman, the question of the how physics theoretically grounds these sciences returns on its own as an imperative or an explanatory ideal. But he adds himself to the list of authors already cited, philosophers and scientists, who defend a nontriumphalist position, comprising the pragmatic coexistence of a de facto reductionism as the moving force of scientific explanation in the process of generation, along with an acute awareness of the obstacles to elaborating a theory of this reductionism and having it accepted as the metaphysical foundation of a unitary theory of science. These obstacles are of two sorts. Some are connected with the aforementioned difficulty of translating the phenomena studied by the other disciplines and the laws they have established into the language of physics. Even the case of its nearest relative, inorganic chemistry (see Bunge 1982) shows how these difficulties are not totally overcome in practice, even if they are in principle. The other kind of obstacle is even more important: it is related to physics's property of not integrating history, in that physical laws (including those of evolution) are supposed to be themselves eternal and outside of time. For Feynman, only a physics that takes into account a *possible history of physical laws* by asking the question "Here are the laws of physics; how did we get there?" could speak of the same problems as sciences that incorporate history, such as astronomy (the history of the universe), geology (the history of the earth), and biology (evolution). This is a very strong indication that, unlike strong reductionists, as well as spiritual physicists, Feynman does not fall into the realist error of assimilating the description of reality given by physics to reality itself. This appears quite clearly in his conclusion, which, *taken out of context*, might be charged with being antiscientific or even mystical. In effect, for

necessary, the concepts it finds useful for describing the properties of atoms and molecules, or measurable energy exchanges between well-defined systems composed of atoms and molecules, or mathematical formulas that have solved certain physical problems. But it does not infer, from the operational success of its concepts, a theoretical discourse on the unity of reality. (Even less can this unity be described on the apparently uncontestable basis of macroscopic material reality, that of objects directly perceived by our senses, as a truly materialist metaphysics would do if it could ignore the problems posed by mathematical abstraction in physics.)

The nonrenunciation of reductionism, although limiting it to a praxis that conditions science without deriving any unitary theoretical consequences, is evidently troubling if one can conceive reality only through a unified knowledge whose field would be coextensive with that of the natural sciences. Such pragmatic reductionism circumscribes science's domain of legitimacy; it indicates the limits of the scientific process, which can progress only by *forcing itself* to be reductionist, only by "playing the reductionist game," whereas "believing in it" would be evidence of great naïveté[4]—the naïveté of believing in the objective truth, in some fashion or other, of the "fact" of the

---

Feynman, if there is a unity of nature (and not of science), it is that of a glass of wine, in which one can certainly describe various phenomena: physical (fluid mechanics, optics, atomic physics), astronomical and geological (of the glass), of course biological (fermentation), and even psychological (the pleasure of inebriation). But these sciences are only the effect of our minds, which, for the sake of convenience, divide their objects of investigation into segments—physics, biology, geology, astronomy, psychology, and so on—although nature knows none of that.

4. If science implies belief, let it be what Feynman suggests as its definition: "Science is the belief in the ignorance of experts. . . . Science doesn't teach anything: experience teaches it. If they say to you: 'Science has shown such and such,' you might ask, 'How does science show it? How did the scientists find out? How? What? Where? . . . And you have as much right as anyone else, upon hearing about the experiments (but be patient and listen to all the evidence), to judge whether a sensible conclusion has been arrived at. In a field which is so complicated that true science is not yet able to get anywhere, we have to rely on a kind of old-fashioned wisdom, a kind of definite straightforwardness. I am trying to inspire the teacher at the bottom to have some hope, and some self-confidence in common sense and natural intelligence. The experts who are leading you may be wrong. . . . I think we live in an unscientific age in which almost all the buffeting of communications and television, words, books, and so on, are unscientific. As a result, there is a considerable amount of intellectual tyranny in the name of science" (Feynman 1969: 320).

reducibility of the real to some unique (ultimate) reality, whether material or not, on the basis of scientific theories.[5] The same naïveté is found among spiritualist physicists who discover Spirit at work in the reduction of the wave function. In their defense we may cite, without condoning their approach, the oft-noted ambiguous role of mathematics in science.

Translated by Lenn Schramm

Appendix (Atlan 1993, pp. 80–82 n. 26)

In the first wave of double-helix biology, Oppenheim and Putnam wrote an article on the unity of science (Oppenheim and Putnam 1958) that is a veritable manifesto of triumphant strong reductionism. The language of physics, the basic discipline, will, according to them, come closer and closer to replacing those of the other disciplines. Significantly, even though they thought they had found in the infant fields of molecular biology and cybernetics justification for the purest unitary materialism, they described the different levels of organization of reality as "levels of reduction." One of their presuppositions was that the relations between levels could be envisioned only in the form of spatial inclusion. This may be why the levels of mental process and of the disciplines that deal with it—psychology, linguistics, and so on—did not even appear among the six levels of reduction they proposed. These six levels were limited, "from bottom to top, to those of elementary particles, atoms, molecules, cells, multi-cellular organisms, and social groups; i.e., those among which a relationship of spatial inclusion can indeed be observed. By contrast, thought and language, not to mention feelings and beliefs, cannot be situated vis-à-vis living organisms either as the whole or as parts, from the perspective of the *spatial* relationship of parts to the whole. . . . As a result, psychology and psycholinguistics cannot even have the status of scientific disciplines, even though they, quite as much as sociology, can be characterized by their objects of observation and their theories and are therefore no

---

5. Hilary Putnam, criticizing theses he had once defended [see the Appendix to this chapter], rightly observes that the fact "that these two sides of psychology [that which is "extremely close to biology," and that produced mainly by "societal beliefs and their effects in individual behavior"] are not distinguished very clearly is itself an effect of reductionism" (Putnam 1973: 146).

less scientific than the latter." For Oppenheim and Putnam, though, they seem to have the status of angelology or demonology, bent on studying the properties of things that do not exist.

Somewhat later this strong reductionism came under attack by Popper's disciples and other practitioners of analytic philosophy, as is evidenced by an article on the same subject but with quite different inspiration and content (Agassi 1969). Finally, some years later, Hilary Putnam himself (Putnam 1973), making honorable amends, described his previous position, which he had shared with most philosophers of science of his generation, as "wrong." Revealingly enough, the analysis of the status of psychology stood at the center of his argument. The reductionist thesis he had formerly defended implied that the human brain could be assimilated to a Turing machine and that "psychological states of a human being are Turing machine states or disjunctions of Turing machine states" (Putnam 1973: 136). This is the thesis he now rejects, because, inter alia, it ignores "societal beliefs and their effects on individual behavior" (ibid. 146). In general, his main criticism concerns the explanatory power of reductionist theories, even when they work, which decreases in proportion as they are situated on a level of organization that is more remote from (i.e., more elementary than) the more global level that is to be explained, for "from the fact that the behavior of a system can be *deduced* from its description as a system of elementary particles it does not follow that it can be explained from that description" (ibid. 131). This is because the explanation relies on the characteristics that are relevant to a situation to be explained; and these characteristics appear only on the level where the phenomena to be explained are perceived. [See note 2 to this chapter for Feynman's objections to such a unitary metaphysics.]

In still more recent work, Putnam carries his break with reductionism farther and exposes the weakness of functionalism viewed as a computational theory of thought that takes the computer metaphor too seriously. This theory resurrects a reductionism of mental states, no longer to physicochemical states but rather to computational states (of some hypothetical software), which cannot themselves be reduced to the physical states of the hardware. In this vein, Putnam criticizes Chomsky and his idea of an innate organ of language and even more so Fodor and his theory of localized mental representations. He relies, inter alia, on the central role he ascribes to interpretation and the modes of formation of the convictions on which our appreciation of reality are based—as much through our scientific language as through our habitual way of using language. He abandons the last vestiges of a mentalist theory of linguistic meaning, which persisted in his 1975 papers, and, like many contemporary philosophers who have been disappointed by logical positivism, sees no more than a social construction in the meaning of words. For a review of his later position, see Putnam 1986.

PART THREE

Spinoza

TEN

The Spinoza Path

(2005)

ROGER-POL DROIT: When we started, you were talking about Spinoza's thought as occupying a place apart, difficult to situate in the development of Western thought. This, maybe, is the point to return to that in order to see if it can offer us some help, a model, or simply an indication for our project of looking elsewhere. It seemed that the first Kantian antinomy—of the finitude or infinity of the world—dissolves when we take, as Spinoza did, infinitely infinite substance as our starting point.

HENRI ATLAN: Let's not go too fast. At the beginning of the *Ethics*, in the definitions, Spinoza does not, properly speaking, posit substance as being infinite. At that point, that is not demonstrated; we're not dealing with anything more than the definition of a word, very different from what Kant understands by "substance," which he then shows to be the cause of itself and to be unique. I'd rather tend to say that this question, if we take it as Kant formulates it, in the form of a finite or infinite world, is more an empirical question.

For the moment, it has not been resolved, even in astrophysics, but people try to resolve it by measuring the acceleration of galaxies, by measuring constants in order to know which one is the most adequate cosmogonic model. Is ours a universe in expansion, that is, infinite, or, on the contrary, a universe in contraction that will collapse, that will once more become a singularity? In any case, today astrophysics considers this question to be of an empirical order, even if astrophysics has not yet found a response.

RPD: Why does Kant present it as a dilemma of reason? Is it because he couldn't yet conceive that one day one would be able to extend, if I dare say so, the field of experience to the dimensions of the universe?

HA: That is in fact the case, and this is also, by the way, where Kant sees the error of dialectical reason, which believes that it can conceive of the existence of absolute totalities (the world, all the composed substances that constitute it, all the causes of actions and events) in the same way it conceives of the existence of phenomena, objects of sensible experience. He finally resolves the antinomies of this "pure" reason thanks to critical reason, which carefully distinguishes the phenomenon, accessible to experience, from the thing in itself, inaccessible. The critical solution makes him affirm that the world "exists neither as *a whole, infinite in itself,* nor as *a whole, finite in itself.*" Neither finite nor infinite, it is nothing but indefinite.

Nonetheless, I'd tend to say, for my part, that the motif of these antinomies lies in the question of knowing whether the world was created by God or not. To admit that the world is infinite means that there is no creation because "infinite" is to be understood in terms both of time and of space. The arguments Kant develops rest on a certain vision of the spatial infinite and the temporal infinite, a vision that today can evidently be contested, given the ways of representing both space and time.

Finally, let's not forget that "practical" reason, which seeks to found morality, makes Kant opt, despite everything, for the thesis of the third antinomy, that is, free will, which participates in a suprasensible domain of liberty specific to the human being, such that the human is, in a way, conceived of "in itself."

RPD: Nevertheless, for our purposes here it is more interesting to show that important changes are of a philosophical and not just an empirical order. In order to see how Spinoza and "infinitely infinite substance" put this first antinomy aside, render it unarticulable, or disqualify it, we can investigate to what degree this story plays itself out between philosophical problems or ways of asking questions, and not between solutions of the empirical type.

What change does the definition "By God, I understand an infinitely infinite substance" bring about when compared to the ping-pong game of the first antinomy?

HA: It's evident that the antinomy dissolves here in that Spinoza is as much on the side of the thesis as on that of the antithesis as they are defined by Kant. He does in fact introduce God as an impersonal infinite, but the world is coextensive with him, thus infinite as well. For Kant, on the contrary, if God exists there can be only a finite world, and if the world is infinite, there is no creator-God. The infinite God who is not a creator unites, in a way, the thesis and the antithesis.

RPD: God as substance is as infinite in space as in time. He is *sive natura*: God, or if you prefer, Nature. The two overlap and coincide.

HA: Monotheistic theological tradition absolutely did not conceive of God as nature. What dominated was always the idea of a transcendent creator-God, with the exception of the Stoics and certain currents of esoteric Jewish thought. There, an immanentist current survived, expressed schematically by an equation of *elohim* ("gods") with *hateva'* ("nature"). But this current of Kabbalah became more and more a minority opinion as time went on, especially after the Renaissance. The Kabbalah has continued, despite everything, to make its way through the centuries, even though it has often been looked upon with suspicion because of its attachment to immanence and has been accused of "pantheism" or "atheism."

There were some very straightforward relationships between the Kabbalah, Neoplatonism, and Gnosis, and a certain number of Christian heresies, including, by the way, relative heresies, such as those of Jacob Böhme or even Pico della Mirandola, which, after the fact, looks like a Christian Kabbalah. Leibniz, too, inherited some of this but was suspicious of it. He was attracted by the proto-scientific elements he found there, but he wished to repel its consequences.

RPD: A path of transmission must pass through the work of Ramón Llull and his legacy, in Leibniz, in the question of combinatorics.

HA: Exactly. These thinkers were attracted by the curious idea—which has remained common to these prescientific traditions and, afterward, to scientific procedures properly speaking—of a sort of linguistic (or abstract, in any case formal) texture of reality. The idea that the world is created by the Word suddenly becomes operative in these currents—alchemistic, kabbalistic, hermetic, Gnostic, or otherwise. Afterward, this becomes truly operative, in an effective way, in modern science.

RPD: We also find in Leibniz's explanation of the universal characteristic the famous formula *calculemus* ("let's calculate"). Leibniz imagines that one day we could forge some sort of conceptual algebra that would make a sign correspond to each concept. One would then write questions as one poses an equation, and they could be resolved by calculation. To the question "Does God exist?" the first answer would be: "Let's calculate!"

HA: Let's not forget that the conciseness of this idea as well as its operational success in certain branches of the natural sciences (above all, in physics) have pushed twentieth-century philosophers—the Vienna Circle, for example, and all of the analytic philosophy that sprang from it—to take up once more this project of a calculus of propositions. The calculus of propositions, which would allow you automatically—and today with computers—to resolve all those questions, is completely Leibnizian.

RPD: We mustn't forget, either, that the answer obtained at the end of the process evidently depends on the data entered at the beginning.

HA: Absolutely. Hence the interest, despite everything, of the critical method inaugurated by Kant.

RPD: It would be possible to get back to Spinoza via the anecdote alleging that Leibniz donated the money necessary to publish Spinoza posthumously. There is no certainty, but, after Spinoza's death, an anonymous donor disbursed a sum sufficient to have all of his works, among them the *Ethics*, posthumously printed. This volume appeared in Amsterdam, one or two years after Spinoza's death, still anonymously, or almost, since only the mention "BdeS" figures on the title page. At Oxford, when I was twenty, I was lucky enough to hold in my hands one of the very rare copies of this edition. There must be only three left in the world.

HA: Leibniz was so complicated that anything is possible. Perhaps he wanted to facilitate the diffusion of Spinoza's works. Perhaps he himself wanted to have access to these works and this was the only possible solution. Did you know that one of his friends pointed this out to him, and Spinoza finally decided not to let Leibniz read the manuscript of the *Ethics*?

You undoubtedly know the terrible judgment Leibniz pronounced on Spinoza's thought: "It is a monstrous doctrine that pushes to the extreme Cartesianism and the Kabbalah of the Hebrews." From my point of view, I think he's right, but in a congratulatory way. If we admit that there are monsters that bear our hopes—like Goldsmith's hopeful monsters in evolutionary theory—then we can admit that Spinoza was one of these hopeful monsters.[1]

RPD: In the eighteenth century, people still said about Spinoza that he'd been "a very bad Jew who wasn't, for all that, a good Christian," because the Jews who left Judaism at the time became Christians. He appears to be the exception.

HA: All the more so since he had taken the inverse path, being the son of a *converso* who became an Orthodox, even official, Jew of the Amsterdam community. In effect, he abandoned this status without, for all that, becoming Christian.

RPD: Does this position out of sync with his time suffice to explain the hatred directed at him? Few philosophers have been so insulted, slandered, and vilified in all ways imaginable. In nineteenth-century Germany and France, Spinoza was seen only as a dangerous pantheist or an atheist, as if one had fabricated a "Spinoza scarecrow."

HA: Being out of sync was not sufficient to trigger this hatred, but it certainly was a contributing factor. Spinoza was out of sync in a provocative way, concerning points situated at the heart of theological power, while claiming to speak in the name of theological truth. It is understandable that this irritated many people.

I have also been struck by the at once moving, exasperating, and desperate character of his correspondence. He tries to convince his Protestant interlocutors, who are, as he says, reasonable Christians. The audience he addresses

---

1. [Atlan extends this discussion of Spinoza as a "hopeful monster" in Atlan 1999b: 322–28, 335.—Eds.]

are Protestants. They are certainly not Jews or Catholics; they are very much the Protestants of the official churches (those of England and the Netherlands). You can see how sincerely he would like to persuade them that he brings them the means to be better Christians. But he does so by denying free will, by saying that God is nature.

RPD: That's also why he also needs to convince them that he's not an atheist . . .

HA: All the time. The entire tradition of Spinoza commentary can be separated into those who say "He really was an atheist but he hid his game for political reasons" and those who affirm: "He wasn't an atheist at all. The proof is that he always said he wasn't, he always defended himself against that." On the contrary, he was a true believer, if one can say that . . . except that he wasn't a "believer" but had "the true knowledge of God," as he himself said.

Over and against these two traditions of commentary, Spinoza . . . does not respond to the question "Atheist or not atheist?" He says, "No, I'm not an atheist," but otherwise, he talks like an atheist.

RPD: If we return, starting from Spinoza, to our antinomies, then there is one where he is not just outside or out of sync, but very much on one of two sides. In the opposition between free will and determinism, he is purely and simply—and entirely—on the side of determinism.

HA: That's not so certain. Evidently, from the viewpoint of adequate knowledge, Spinoza is well on the side of absolute determinism. But from the viewpoint of the inevitable conditions of human knowledge, which cannot eliminate inadequate ideas, the situation is less clear-cut. The only adequate knowledge belongs to the third and the second kinds—that is to say, knowledge by way of reason for common properties and knowledge by way of intuitive science for singular beings. But we mustn't forget that, however many adequate ideas you might be able to arrive at, you will never, for all that, get rid of inadequate ideas.

You will thus, by means of these inadequate ideas, always have the experience of the illusion of time and therefore also of the illusion of free will, since you won't know all the causes of your volitions. From the viewpoint of this double belonging of the human condition, to adequate knowledge, on the one hand, and, on the other, to inadequate knowledge—which is inevitable—the problem is displaced. It is no longer an antinomy.

RPD: It's no longer "either . . . or . . ."

HA: In the end, it's two experiences. We have the experience of the timeless, especially through mathematics. We thus know what Spinoza calls eternity. We have this experience, and that leads us to the experience of absolute determinism, but we also have the experience of time, and that leads us to the experience of the illusion of time and free will.

This double face, in Spinoza, is often poorly understood. You can see how all sorts of misunderstandings come in. Even among scientists, some, otherwise confronted with the determinisms that we research and that we increasingly discover to be at the origin of behavior we thought to be free, nonetheless think: "No! That's not possible! There must be free will, because if not, there is no longer any morality, there is no longer any responsibility."

Translated by Nils F. Schott

ELEVEN

Immanent Causality: A Spinozist Viewpoint of Evolution and the Theory of Action

(1998)

Spinoza's monism is seldom taken into consideration when the mind-body problem is stated and the solutions for it that have traditionally been proposed are discussed. Sometimes Spinoza's philosophy is referred to as a "parallelism" that is tantamount to a revised Cartesian dualism. Sometimes it is considered to be a materialist monism, thereby making Spinoza a precursor of the eighteenth-century materialist philosophers, such as Diderot and the Baron d'Holbach. In fact, Spinoza's theory of psychophysical identity is neither of these. I will argue that it is particularly well adapted to a discussion of the mind-body

NOTE: This work was supported by a Ishaiah Horowitz Scholarship in residence in Philosophy and Ethics of Biology, at the Hadassah University Hospital Human Biology Research Centre, Jerusalem. [Atlan puts together his own, sometimes rather free, translations of Spinoza—drawing on the Latin and on available translations, for this essay the 1955 Dover edition of the *Ethics*, translated by R. H. E. Elwes, or the 1982 and 2002 Hackett editions of the *Ethics* and *Complete Works*, respectively, translated by S. Shirley. Throughout this essay and the following ones, we have maintain Atlan's own renditions, following them with the page number in the 2002 *Complete Works*.—Eds.]

problem in the framework of the present-day natural sciences. In fact, Spinoza's philosophy can only be understood if one takes into consideration his notion of immanent causality. The cause of itself, *causa sui*, which pertains to substance, is distributed in the modes through their essences or *conatus*, although the modes are produced by one another, come to existence, and are destroyed in their infinite chain of efficient causes. Given such a notion of immanent causality, evolution can be seen as the unfolding of a dynamic system or a process of complexification and self-organization of matter, produced as a necessary outcome of the laws of physics and chemistry. In this process, new species come into existence one after the other as effects of mutations with stabilizing conditions working as their efficient causes, whereas their particular organizations are particular instances of the whole process. This view of evolution is compatible with the idea of a dynamic evolutionary landscape with peaks of local stability. The whole dynamics created by the physical constraints of composite organized bodies is an *intemporal* theoretical description of possible organisms. The actual peaks of stability are populated one after the other in a historical, partially contingent fashion, which constitutes *temporal* evolutionary processes. The latter may be oriented by adaptive natural selection, but that is not always necessarily the case. Temporal evolution is also a self-organizing process, driven by randomness, whereby more "sophistication" (Koppel and Atlan, 1991b, Atlan 1995b) or "logical depth" (Bennett 1988), in the sense of functionally meaningful complexity, can be committed to memory and accumulated.

Spinoza's "parallelism" is based largely on proposition 7 of part 2 of the *Ethics*, where Spinoza deals with the relationship between matter and mind: "The order and connection of ideas is the same as the order and connection of things." However, Spinoza himself never uses the word *parallelism*, and he explicitly states his stance as an ontological monism underlying conceptual dualism: "thinking substance and extended substance are one and the same substance, comprehended now under this attribute, now under that. So, too, a mode of Extension and the idea of that mode are one and the same thing, expressed in two ways" (pt. 2, prop. 7, scholium; Spinoza 2002a: 247).

By contrast, Spinoza's "materialism" is based on statements like: "No one will be able to understand this [the union of the mind and the body] adequately and distinctly unless, at first, he is sufficiently acquainted with the nature of our body" (pt. 2, prop. 13, scholium; 2002a: 251); and "Each one judges concerning things according to the disposition of his own brain or rather takes for things that which is really the modifications of his imagination" (pt. 1, appendix; 2002a: 238–43). However, the attribute of thought seems to prevail, in a way, over all other attributes, including extension, in that it produces not only "ideas" but "ideas of ideas." Ideas are not only modes of Substance as it is expressed by the attribute of thought, in the same way that bodies are modes of substance as it is expressed by the attribute of extension. Ideas are also ideas of all other modes of substance as it is expressed by all other attributes. As such, they contribute to the activity of the "infinite understanding," which is itself the infinite mode of the attribute of thought. In other words, thought entails an epistemological superiority over matter, in that matter is understood and "intellected" by means of ideas with a reflective nature, that is, "ideas of ideas." If an idea is the mode in thought associated with a body in extension, the idea of this idea expresses the reflective nature of thought through the capacity of "intellection," that is, knowing and understanding. In addition, the *Ethics* ends with several propositions concerning "those points which appertain to the duration of the mind without relation to the body" (pt. 5, prop. 20, scholium; 2002a: 373), such as "eternity" (pt. 5, prop. 23; 2002a: 374). This part of the book seems so strange and foreign to a materialist stance that many philosophers view it as a kind of enigma in the work of Spinoza, who, all of a sudden, appears to be a mystic. In fact, close to the end Spinoza warns the reader against an idealist misinterpretation by recalling that "He who has a body capable of many things, has a mind of which the greater part is eternal" (pt. 5, prop. 39; 2002a: 380). However, seeing this statement as an expression of a materialist ontology would also be a misinterpretation.

In fact, Spinoza's monism is an original one, as it is neither materialist nor idealist. It is difficult to grasp within the framework of existing ontologies, both monist and dualist. That is probably the main reason why Spinozism is seldom considered to be a serious alternative to the conflicting solutions to the mind-body problem.

The purpose of this essay is to suggest a way of understanding Spinoza's original stance on psychophysical identity, starting from what is called his "physics," that is, the small set of axioms, lemmas, and postulates between propositions 13 and 14 in the second part of the *Ethics*. First, we shall see that these few statements are more relevant to what we would today call a biophysical theory of the organism than to physics per se. Then, based on a remark of Hilary Putnam (1981), I shall refer to the idea of the *synthetic identity of properties* to show that a similar way of thinking, which is different from the usual philosophical analytical conceptualization, is at the basis of both psychophysical identity as it is conceived by Spinoza and the identity of microphysical and macrophysical quantities as it is conceived in contemporary physics. This will help us solve a puzzle encountered by Donald Davidson (1999) when he explicitly referred the "anomalous monism" at the basis of his theory of action to Spinoza's psychological monism.

Spinoza's Physics

In Spinoza's own terms, his physics is "poor," since it is limited to "a few [*pauca*] premises on the nature of bodies" (pt. 2, prop. 13, scholium; 2002a: 251–52). His purpose was not "to lecture on the body" but only to mention what he needed in order to deduce more easily what he wanted to demonstrate regarding the "Nature and Origin of the Mind," to which the second part of the *Ethics* is devoted.

In the eyes of a modem Newtonian and post-Newtonian physicist, these few statements are not only irrelevant but uninteresting, even from a historical point of view, and therefore contrast the physics of Pascal, Descartes, and, especially, Huygens and Leibniz. However, this opinion is based on a misunderstanding, since Spinoza's concern in his "physics" is not mechanics, that is, a physical theory of motion, which could be considered a precursor of today's physics. Rather, these few statements in the *Ethics* outline a theory of "the nature of simple and compound bodies." These premises of a theory are more relevant to what we would consider today as chemistry or biophysics, that is, a physical theory of composite individuals, with no basic difference between living and nonliving bodies. As Hans Jonas (1965) has pointed out, for example, one can easily recognize metabolism, growth, internal activity, and locomotion in lemmas 4, 5, 6, and 7, respectively. These are various ways in which "an individual can be affected and, despite this, preserve its nature." The "nature" of an individual is defined by its "form," that is, by the fact that its constitutive parts "remain united with one another" despite the changes that they can undergo, provided that "a certain *ratio* of movement and rest" is retained. This invariant *ratio* of movement and rest is the "form" or "nature" of the individual, and it works like a law of organization, which defines a given individual. However, this form is not imposed on the body by a soul, as it is, for example, in Leibniz's conception. Nor does it work like a machine, as it does in Descartes' *animal-machine*, based on the model of an hydraulic or mechanical device, where an individualized source of energy, like an internal fire, is necessary to activate the parts. The law of organization results only from the way in which the parts are in contact with one another, and what is necessary to maintain the entity invariant is produced by the metabolism and activity of the parts themselves. According to the scholastic tradition, in which movement meant not only displacement in space but also change of any sort—like, for example, a chemical transformation—the "certain *ratio* of movement and rest" defining a given individual must be understood as a certain *ratio* of change and invariance. This is reminiscent of what today we call dynamic systems, such as those made up of several chemicals that are able to react with one another, where a global law of organization is defined by local kinetics of chemical reactions. It is easier for us to define precisely such laws of organization, because we can now make use of differential calculus, which did not exist in Spinoza's time. From this point of view, there is no doubt that we are indebted to Leibniz and Newton, who discovered it, although their respective ontologies and metaphysics are hardly acceptable to us nowadays.

At the end of this brief summary of Spinoza's physics, let us remind ourselves that it applies equally to all individual bodies, living or not. As a result, one should not be misled by Spinoza's statement about the generality of the definition of the human mind as pertaining to all things and "no more to man than to the other individual things, which are all, though in various grades, animate" (pt. 2, prop 13 and scholium; 2002a: 251). This statement should not be confounded with universal animism or panvitalism. It is the very opposite, as is clear from context. The last quotation follows the remark that "in truth, no one will be able to understand [the union of the mind and body] adequately and distinctly unless he is first sufficiently acquainted with the nature of our body" (ibid.). The whole scholium is a commentary on the definition of the human mind as the idea of the body that actually exists (pt. 2, prop. 13; 2002a: 251). Being applied to any individual, living or nonliving, this definition therefore implies that a stone has a mind. But the mind of a stone must not be confounded with a soul that would keep the stone alive. This mind, like the human mind, is merely the idea, as a mode of thought, that is always united in nature with an existing material object, as a mode of extension of which it is the idea. Therefore, the "mind" of a stone must not be confounded with the soul of a human being or another "animate" being. It is just the opposite. The mind of a human being is the idea of his body, as is so for the idea of a stone or any other thing. The differences between minds reflect differences between the bodies of which they are the ideas. Our distinction between animate and inanimate beings results only from our insufficient and inadequate knowledge of the nature of composite individual bodies.

Of course, the more complex bodies are, made of various parts that in turn are composite bodies, the more complex are their ideas (pt. 2, props. 14 and 15; 2002a: 255). The ideas of human bodies, assumed to be the most complex, exhibit a greater capacity for understanding and reasoning, that is, for producing ideas with the reflexive nature of "ideas of ideas." In other words, consciousness and reason are properties of the idea of the human body, which accompany the complex nature of that body, namely, the large number of ways in which the human body can be affected and can do things without losing its "form" or nature. Contrary to the idea of a stone, whose object is relatively less complex than a human body, "the human mind is apt to perceive many things, and more so accordingly as its body can be disposed in more ways" (ibid.).

Thus, under certain conditions related to the complexity of the body—that is, the many ways in which it can be disposed—the mind may not only be an idea but may also produce and have ideas, like the infinite understanding of which it is a part.[1]

In this way, Spinoza shows that the "soul" or, more correctly, the "mind" (*mens*), should not be confused with the classical soul as the origin of movement and "fire" for the living body. Moreover, and contrary to what our inadequate immediate knowledge usually leads us to believe, even in living and moving bodies *"the mind cannot determine the body to be in motion, or at rest, or in any other state (if there be any),"* and similarly, *"the body cannot determine the mind to think"* (pt. 3, prop. 2; 2002a: 279; my emphasis).

This proposition is at the core of Spinozist psychophysical theory, and it is certainly difficult to understand. Davidson has written an important article (1999) in which he tries to reconcile this statement with his own theory of action and perception. This endeavor is significant, because Davidson's theory is based on his "anomalous monism," which, according to him, is closely related to Spinoza's monism and "parallelism."

---

1. The "idea of God" is one of the human mind's ideas. It is the "infinite Understanding of God" (in both the sense of understanding God and that of being understood by God) that is the same idea in the human mind—which, being finite, is a part of this idea—as it is in itself as the whole of this idea. This conception, which lies at the basis of Spinoza's theory of knowledge, is, however, outside the scope of this essay.

Before turning to this thesis, I will summarize Putnam's remark on the synthetic identity of properties in physics. I will refer to it later in my discussion of Davidson's theory of action in light of Spinoza's theory of psychophysical identity.

The Synthetic Identity of Properties

According to Putnam (1981), the notion of physical magnitudes, which we employ in physics, implies the existence of a "synthetic identity of properties," which is to be distinguished from an analytical identity, or equivalence. The physical magnitude "temperature," for example, is identical to the "mean molecular kinetic energy." However, this identity is not analytic, because the two sentences "A gas has a temperature T" and "Its molecules have a mean kinetic energy equal to $\frac{3}{2}kT$" are not synonymous, even though we learn from the kinetic theory of gases and statistical thermodynamics that they represent two different ways of expressing the same property. If two identical physical properties were linguistically definable as predicates, then we should be able to describe them by means of synonymous enunciations, that is, sentences with identical meaning or, in other words, sentences that are analytically equivalent. The example taken from the kinetic theory of gases shows that this is not the case, since two physical properties (having a particular temperature and having a certain molecular mean kinetic energy) are identical, despite the fact that their descriptions are not synonymous. Thus, we must admit that physical properties can exist that are "synthetically identical" without being analytically identical, that is, without being conceptually identical, in the usual sense of a concept expressed by a meaningful unequivocal enunciation.

Therefore, we can safely say that such a synthetic identity may exist between mental states and brain states, although we may not be able to describe it in a unique, nonequivocal enunciation. A brain state may be identical to a mental state and also to the sensation of a given qualitative state, even though we need nonsynonymous expressions to describe each of these. In other words, according to Putnam, "what the physicist means by 'physical magnitude' is something different from what philosophers call a 'predicate' or a 'concept.' . . . Properties, as opposed to predicates, can be 'synthetically identical.'" Therefore, "if there is such a thing as *synthetic identity of properties*, then why shouldn't it be the case that the property of being in a certain brain-state is *the same property* as the property of having sensation of a certain qualitative character (very much in line with Spinoza's thinking)—even though it is not a conceptual truth that it is, even though, in fact, it seems to many to be *a priori* false?" (Putnam 1981: 84–85).

Spinoza himself often takes another example from geometry. The definition of a triangle entails that the sum of its angles equals the sum of two right angles. This property is a consequence or an effect of what Spinoza calls the generic definition of a triangle, which is its reason or its cause. However, this property is identical with what is expressed by "the sum of the angles of a triangle equals 180 degrees," although the two expressions are not synonymous.

For the purpose of my discussion of Davidson's interpretation of Spinoza, it is important to notice that these examples raise the problem of a possible causal relationship between the two properties, which are synthetically identical but analytically different. This problem is solved simply by noting that such a causal relationship between identical properties would be meaningless (a cause and its effect cannot be identical), even though it may seem meaningful linguistically. Although described in two different ways, the two properties cannot be cause and effect, since they are identical: temperature is neither the cause nor the effect of the molecular mean kinetic energy, because it is this energy. Similarly, "the sum of the angles in a triangle is equal to two right angles" is neither the cause nor the effect of "the sum of the angles in a triangle is equal to 180 degrees" because they are two different enunciations of the same property. (This geometrical example may seem less obvious than the previous, physical one. It seems that the sum of the angles equals 180 degrees *because* the sum of two right angles equals 180 degrees. In fact, the two properties can be demonstrated independently by means of two right angles or one single angle of 180 degrees. In any case, the example of temperature is more instructive because it can be easily applied to all thermodynamic quantities—such as pressure, volume, free energy, and entropy—that can be defined at two different levels, macroscopically and microscopically). In the same way, a mental state is not the cause or the effect of a given brain state, since it *is* this brain state, even though we cannot describe the mental state and the brain state by synonymous expressions.

This view is exactly the same as that of Spinoza, who states explicitly not only that body and mind are two different aspects of the same but also that no causal relationship can exist between them, which is obvious in view of the aforementioned discussion.

Action and Perception: Davidson's "Anomalous Monism"

However, this lack of causal relationship is hard to reconcile with our imme-diate experience of perception and action: we feel spontaneously that our voluntary movements are caused by some mental state that corresponds to our experience of making a decision. Conversely, it seems that our percep-tions are effects on our mind of modifications of our bodily sense organs. Davidson has devoted much of his work to elaborating a naturalist theory of action and perception that does not assume a kind of dualist interactionism imbued with the same mystery as Descartes' "solution" of the pineal gland. In this, Davidson invokes Kant and the necessity of establishing the reality of human actions caused by voluntary decisions in order to save the reality of ethics based on free choice and responsibility (Davidson 1970). By contrast, in a kind of summary of his theory, Davidson presents it as a modern version of Spinoza's "parallelism." The association of Kant with Spinoza here is strange, since it is well known that Spinoza's *Ethics* is a search for freedom that denies the reality of free will. Therefore, in a Spinozist, non-Kantian universe, there is no need to establish the reality of bodily movements that are caused by decisions of our mind in order to build a "model of human nature" that would lead us to freedom and happiness (pref. to pt. 4; 2002a: 320–22).

In any case, Davidson's argument goes as follows.

Let us consider two individuals, or two consecutive states of one individ-ual, I and II, such that I causes II. Each of them is a union of the mental and the physical. Let A and B respectively be the mental and the physical in I, and C and D the mental and the physical in II, as is represented in the accom-panying diagram.

In Spinozist terms, A and C are, respectively, modes I and II of substance under the attribute of thought, and B and D are modes I and II of substance under the attribute of extension. Proposition 7 in the second part of the *Ethics* states: "The order and connection of ideas is the same as the order and con-nection of things," which means that the causal relation between A and C, or rather, the production of C by A, is the same as the causal relation between B and D, through which D is produced by B. (This becomes obvious in the reformulation of proposition 7 when it is used to demonstrate proposition 9: "the order and connection of ideas is the same as the order and connection of *causes*" (my emphasis). In other words, the causal connection between A

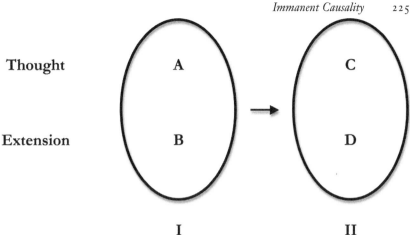

Thought

Extension

I   II

Fig. 11.1. Davidson on the Union of the Mental and the Physical

and C, through which an idea A is the cause and the reason of an idea C, is the same as the causal connection between B and D, united with A and C, respectively.

However, the two attributes of thought and extension are only two different expressions of the same substance, which is only one. Contrary to Descartes, Spinoza holds that "thinking substance and extended substance are one and the same thing, which is comprehended now through this attribute, and now through that attribute. Thus, also a mode of Extension and the idea of that mode are one and the same thing, but expressed in two manners" (pt. 2, prop. 7, scholium; 2002a: 247). This means that A is identical to B, and C is identical to D. Incidentally, it is for this reason that A as a mental event cannot be the cause of B as a physical thing: it is not because A and B are two different substances with no interaction between them, but because A and B are one and the same. The one cannot be the cause of the other, since cause and effect must be different (except for God, i.e., nature as a whole, which is *causa sui*, the cause of itself). This is stated explicitly by Spinoza in part 3, proposition 2: "The body cannot determine the mind to think, and the mind cannot determine the body to be in motion, or at rest, or in any other state (if there be any other)" (2002a: 279). The scholium to this proposition contains an explicit reference to proposition 7 of part 2.

However, Davidson questions the meaning of this proposition when it is applied to our scheme. Whereas it is clear to him that A cannot be the cause of B, because they are identical, A, nevertheless must be the cause of D, because C and D are identical and A is the cause of C. The same holds for physical event B, which must be the cause of mental event C, because it is the cause of physical event D and D and C are identical. Thus, it is clear that body B determines mind C to think by means of this cross-causality or transitive causality: "B causes D, and since C is identical to D, B causes C." The same also holds for mental event A, which causes physical event D. Therefore, how can Spinoza say that the body cannot determine the mind to think and the mind cannot determine the body to motion?

As Davidson puts it: "The difficulty, described in this way, is so apparent that one must assume that we have misinterpreted Spinoza on some essential point, or have failed to make a distinction that was crucial to his position" (1999: 100).

After having quoted conflicting interpretations of Spinoza that obviously miss the point, he suggests his own solution, which fits his theory of action and perception. A usual misinterpretation is to refer to Spinoza as a materialist monist. Davidson only mentions Edwin Curley, but several commentators adopt the same line of thought. Their reasoning goes as follows: being a materialist monist, Spinoza denies the reality of the mental, and therefore for him the interaction between mind and body does not exist and does not pose any problem. However, it can easily be shown that Spinoza's denial of interaction does not stem from any denial of the reality of the mental and, hence, that he cannot be seen as a materialist in the modern sense of the word. Another misinterpretation is the distinction between the two attributes and their so-called parallelism. This amounts to denying that Spinoza is a monist, referring to the fact that the mind and body cannot be causally related to one another because the attributes of thought and extension are separate. Davidson attributes this interpretation to Alan Donagan, but it can be traced back to Leibniz, who invented the term *parallelism* to describe the relationship between mind and body described in Spinoza's *Ethics* (pt. 2, prop. 7; 2002a: 247). As Davidson rightly comments on Donagan's interpretation, "If Spinoza is not a genuine monist, mind and body are not really identical, and the apparent contradiction [to which he] pointed disappears. But then so would the apparent contrast between Spinoza's and Descartes's metaphysics also disappear, and Spinoza would need some substitute for the pineal gland" (Davidson 1999:101–2).

Leibniz's substitute for the pineal gland is his principle of universal harmony, according to which the mysterious action of God compensates for the lack of interaction between mind and body, maintaining harmony and adequacy between the two parallel realms of the mental and the physical by ways unknown to men. At this point, Spinoza departs from both Descartes and Leibniz, because they resort to the mystery of an unintelligible God, thereby entering into what he calls the "sanctuary of ignorance" (pt. 1, Appendix; 2002a, 241).

What Spinoza said about Descartes' theory of the pineal gland and animal spirits is well known: "I cannot sufficiently wonder that a philosophic man, who clearly stated that he would deduce nothing save from self-evident bases of argument and that he would assert nothing save what he perceived clearly and distinctly—one, moreover, who so many times reproved the scholastics for wishing to explain obscure things by means of occult qualities—should take up a hypothesis far more occult than all the occult qualities" (pref. to part 5; 2002a: 363–65). Of course, Spinoza could not say anything about Leibniz, who succeeded him, but we can easily imagine that he would have reacted in a similar way to the principle of universal harmony.

Having discarded these misinterpretations of Spinoza, Davidson addresses the important question of how we are to understand the concept of causation, more specifically: "What did Spinoza mean when he wrote 'The body cannot determine the mind to thinking and the mind cannot determine the body to motion'? Did he mean by 'determine' what we mean by 'cause'?" (1999: 101).

Davidson's answer is based on the idea that a causal explanation of something is not always the same as the actual cause of that thing. In other words, he states that a logical reason (*ratio*) for something may be different from its physical cause (*causa*). This distinction is, of course, not an ontological one, since this would lead us again to an ontological dualism. It is merely the result of our frequent inability to have an adequate explanation of something or some event, that is, one that fully "deduces the occurrence (or probability of occurrence) of the event to be explained from the laws of nature and a statement of the prior conditions" (ibid. 102–3). Although we may not have a full explanation and therefore know only partial causes (which Spinoza himself calls "inadequate"), the "inadequacy of our knowledge of the cause and the effect does not throw in doubt the causal connection" (ibid. 103). The mind-body problem is an example of inadequate knowledge, since we do not have a vocabulary that would allow us to describe mental events in terms of physical ones and vice versa, and we do not know psychophysical laws that would allow us to reduce mental events to physical ones. Such a situation is not inconsistent with ontological monism. This position is Davidson's own "anomalous monism," that is, an ontological monism of mind and body "compatible with the failure of nomological reduction, that is, with the absence of strict psychophysical laws. . . . Monism, coupled with the failure of nomological connections, implies that a complete or adequate explanation of a mental event cannot be given in physical terms, and a complete and adequate explanation of a physical event cannot be given in mental terms" (ibid. 104–5). For Davidson, this anomalous monism is akin to Spinoza's ontological monism coupled with a dualistic (or multiple) explanatory apparatus.

The question I want to address now is whether Davidson's reading of Spinoza is a correct interpretation of Spinoza's concept of causation. His reading allows him to save the reality of causal relations between the mental and the physical, despite the assertion that the body cannot determine the mind to think and vice versa. This assertion is understood as belonging to the realm of causal (inadequate) descriptions or explanations and not to the realm of causes. "Causal relations in nature are indifferent to how we describe them. . . . Causal *relations* as [Davidson] conceives them are between events however described; causal *explanations*, on the other hand, depend on the vocabulary or concepts used to describe events and to formulate laws" (ibid. 106).

Action and Perception in the Light of Spinoza's Monism

In what follows, I want to question Davidson's reading of Spinoza and to suggest a related but different way to solve the apparent contradiction he diagnoses. As he mentions, this contradiction stems from a misinterpretation of Spinoza. However, this misinterpretation does not concern the question of determination by causes versus explanation by reasons. I will argue, first, that the contradiction disappears within the system of the *Ethics* if we keep in mind the immanent and nontransitive nature of the causation of all things produced as modes of substance by substance itself (see pt. 1, prop. 18; 2002a: 229). Then, in the following section, I will make use of the idea of the synthetic identity of properties to draw an analogy with today's physics, which can help a modern reader to better conceive the kind of union between mind and body implied by Spinoza's doctrine. In effect, it seems that, for Spinoza, even if the *existence* of this union were to be asserted and demonstrated by deduction from the unity of the unique substance, its *nature* would still be an object of investigation that one would be "able to understand adequately and distinctly" only "by being sufficiently acquainted with the nature of our body" (see the discussion above of pt. 2, prop. 13, scholium; 2002a: 251), that is, through further physical research and knowledge.

Insofar as the question about the meaning of "determination" in proposition 2 of part 3 is concerned, it is difficult to accept Davidson's answer that it belongs to the realm of inadequate knowledge of causes, since the demonstration of this proposition makes use of the basic idea in the *Ethics* that ideas are produced and can be known (adequately) as modes and effects of nature only under the attribute of thought and that physical events are produced and can be known (adequately) as modes and effects of nature only under the attribute of extension. Moreover, the distinction between *causa* and *ratio*, or between causal relation and causal explanation, is contradicted by Spinoza's insistence that they are the same, albeit viewed through different aspects, as noted in proposition 7 of part 2 and his own commentary. Davidson himself acknowledges that there is some difficulty here, which leaves him with a "remaining doubt" about the correctness of his interpretation. Nowhere does Spinoza seem to imply that the absence of a causal relation between the physical and the mental is due to our lack of knowledge. Indeed, the opposite is true. Our usual experience that a decree of our will can cause a bodily movement is due to our inadequate knowledge of the real causes of both our bodily movements and our ideas, thoughts, emotions, and unconscious and conscious feelings, which, *being identical to those movements*, always accompany them. Thus, if it is true that "a complete or adequate explanation of a mental event cannot be given in physical terms, and a complete and adequate explanation of a physical event cannot be given in mental terms," this does not mean that we must accept the necessity of real causal relations between the mental and the physical, even though we can describe them only in partial, confused, and inadequate ways. It is clear that, for Spinoza, even if we had a perfect, infinite knowledge of nature ("God's infinite understanding"), we would still maintain that there cannot be causal relations between the mental and the physical. Then how can we solve the apparent contradiction that was analyzed schematically above: If A causes C, and C is identical to D, how can we say that A does not cause D? (The same applies for B, D, and C.) What is at stake here is not the nature of the causal explanation, which, when adequate, cannot be distinguished from the causal relation. What is at stake, rather, is the nature of the identity between C and D, and the transitivity of the causal relation between A and D that we deduce from this identity. A transitive cause is such that its effects are produced beyond itself. The effects do not occur in the same entity as the cause itself. That is why, for Spinoza, "God is the immanent and not the transitive cause of all things" (pt. 1, prop. 18; 2002a: 229).

The effects of nature (i.e., *natura naturata*) are produced within the same entity (namely, "God, i.e., Nature") as that of the causes (i.e., *natura naturans*). According to modern logic and mathematics, a relation r is transitive if from A r C and C = D, we can deduce A r D. This itself results from our axiomatic agreement that equality is a transitive relation: if X = Y and Y = Z, then X = Z. Our reasoning in the above problem implies that the causal relation between A and C is transitive. This, however, means that we consider D to be not *really* identical to C, but merely related to C by a relation of equality or equivalence. D and C are, therefore, two different entities. Again, we come back to a dualist position, which is certainly not consistent with the Spinozist context of our analysis.

The only way one can circumvent this is to postulate a real identity between C and D, which are united in II, and the same for A and B, united in I, as required by Spinoza's monism, and to consider the causal relation between A and C to be nontransitive. Thus, we should not say: A causes D because A causes C and C = D, because of the transitivity of the relation between A and C. Rather, apparently, we should say: A causes C or D indifferently (similarly, B causes C or D indifferently) because C and D are one and the same. But can we say so? In fact, this would suggest that the *descriptions* of C and D are synonymous and that they can be replaced by one another in the description of their causal relation with A or with B. In other words, to go back to Putnam's distinction, this would be possible only if the identity between mental C and the physical D (or likewise, mental A and physical B) was analytical. Or, to use Davidson's terms, it would imply a possible reduction of the mental to the physical by means of a common vocabulary and strict psychophysical laws. Since that is not the case, and since we must consider the identity between C and D to be synthetic, we cannot replace C by D in the relation A causes C (the same holds for replacing D by C in the relation B causes D).

It seems that we are arriving at the same conclusion as Davidson, since our reasoning apparently implies that a causal relation still exists between A and D, although we do not have the means to describe it as a causal explanation. This is not quite the case, however. The causal relation existing in nature that we are unable to describe is not between A and C, or between B and D, but between I (i.e., A and B together) and II (i.e., C and D together).

It is simultaneously a causal relation and an explanation, that is, *causa* and *ratio*. The relation between A and C and that between B and D are *both* different ways of conceiving and describing the same relation between I and II. Thus, both the relation between mental events A and C and the relation between physical events B and D can describe the only causal relation that exists, namely, between I and II. The fact that this causal relation can only be described through its expressions in the two attributes has nothing to do with inadequate knowledge or hidden reality. The situation results from the fact that we can only have access to the essence of substance and to its causal production of its modes through the attributes. This does not preclude the possibility of adequate knowledge in either one of the attributes. In addition, even if we knew strict psychophysical laws allowing mental and physical states to be equated, this would not necessarily imply analytical identities between them. The same holds for microscopic and macroscopic descriptions of physical quantities, which are synthetically but not analytically identical. Causal descriptions in the mental and the physical domains can be adequate without there being any other cross-causal relations. Replacing mental state C with physical state D would result in mere confusion and meaninglessness, since the identity between C and D is synthetic and not analytic.

The difference from Davidson's position is that we introduce the dualism of descriptions, not within the causal relation—which remains as one and which includes both the causal relation and the explanation, assumed to be adequate—but within the identity of the events, where mental C and physical D, although identical, need different descriptions, which cannot replace one another when related to A and B, respectively.

The Analogy with Physical Magnitudes

A comparison with synthetically identical physical magnitudes can help one to visualize the situation better. Let A be the temperature of a gas and B the mean kinetic energy of molecular motion of that gas. As mentioned above, A and B are synthetically but not analytically identical. Similarly, the pressure of a gas is synthetically identical with the exchange of momentum of the molecular collisions of that gas with the walls containing its volume. Let C be the pressure and D the force of the molecular collisions corresponding to that pressure.

As is well known, an increase in the temperature A of the gas causes an increase in the pressure C. Similarly, an increase in the molecular motion B causes an increase in molecular collisions. However, there is no sense in talking about cross-causal relations between A and D and between B and C. The only thing we can say is that an increase in temperature causes an increase in pressure, and that this is accompanied by an increase in molecular collisions. This is so because pressure and molecular collisions are identical, even if their definitions are not exchangeable in the description of what is going on, since the latter refers to molecular properties and the former to thermodynamic properties of a macroscopic sample of matter. In other words, the causal relation between A and C acts *as if* it were transitive and passed on to D because of the identity between C and D, but it is *not* transitive because the identity between C and D cannot be described as an epistemic (analytical) identity.

Our solution to the puzzle presented by Davidson is closer to the letter of Spinoza's *Ethics*. It is also closer to its spirit, in that it does not dissociate explanation (*ratio*) from cause (*causa*) and does not assume that the apparently contradictory propositions (pt. 3, prop. 2 and pt. 2, prop. 7; 2002a: 279, 247) concern only inadequate causes, which is obviously not the case, as can be judged from their contexts and demonstrations.

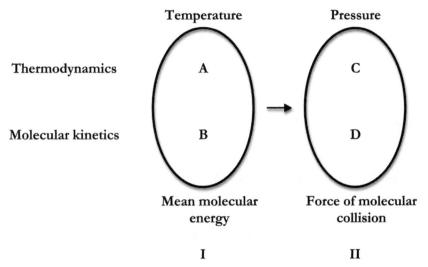

Fig. 11.2. An Example of Synthetic but Not Analytical Identity

Thus, Davidson was forced to depart from Spinoza in order to save his causal theory of action and perception, although he is obviously very sympathetic to Spinoza's unique association of ontological monism and epistemic dualism.

Of course, nothing forces us to be orthodox Spinozists. We must try to weigh the advantages and disadvantages of this radical denial of a causal relation between the mental and the physical, which is associated with a synthetic identity that needs different vocabularies with different analytical meanings.

An obvious disadvantage of this position is that it contradicts our commonsense experience, which tells us that a decision of our will may be the cause of our bodily movement, and that a physical affection of our body may be the cause of our perception and thinking.

However, in order to save our commonsense experience—and also, maybe, the Kantian moral a priori of the existence of free will—we must pay the price and assume the reality of relations depicted by crossing arrows in our schema. Is this price too high? As mentioned above, these arrows represent causal relations that are supposed to be real but that are hidden and can be described only by means of inadequate causal explanations. How could inadequate explanations then refer to something real? If there is no adequate causal explanation between the mental and the physical, then what is the point in talking about a hidden causal relation that we can only conceive of confusedly, simply because it is suggested by our commonsense experience? Do we need to save a causal theory of action and perception at any price?

I will not go into the classical moral argument that states that if we do not assume free will, there will be no responsibility and no ethics. This argument has been dealt with by all proponents of absolute determinism, such as the Stoics and Spinoza himself, who have clearly dissociated the possibility of moral conduct and responsibility from the necessity of free will. After all, Spinozist ethics is a path to freedom and joy based, among other things, on the knowledge that *free will is illusory*.

Instead, I will make use of present trends in cognitive sciences that support a noncausal theory of action and perception compatible with ontological monism. In this context, the Spinozist stance allows us to make better sense of what is implied by these trends and to overcome the difficulties stemming from our commonsense experience. Contradictions between theoretical physics—quantum theory and relativity—and common ideas about space and time have already taught us to be cautious about common sense as a source of insights into reality. Pragmatic theories of cognition have shown how acquisition of knowledge and action in the world are mutually dependent insofar as their attribution of meaning is concerned (see Shanon 1993 for a review and discussion).

Functional Self-Organization

With the help of models of functional self-organization (Atlan, Ben Ezra, et al. 1986, Atlan 1987a), it is possible to understand, at least in principle, how meaning and intentionality can be created by a more or less frequent repetition of events leading to self-observed behaviors of dynamic systems. This model is made by associating memory devices with self-organizing networks of neuron-like automata. The memorization of self-organizing behaviors generates meaning and intentionality in a process that we call intentional self-organization (Atlan 1992b, 1995b; Louzoun and Atlan 2007). In the light of this model, we can understand that pattern recognition and decision making are basically the same process, even if the former seems to be a recollection of the past and the latter a projection into the future. The differences between the two are only a matter of emphasis in a unified process of creation of meaning by self-observed self-organization of behavior. If, for instance, we are more sensitive to the aspect of observation, we experience what we call perception or pattern recognition. By accentuating the behavioral aspect, by contrast, we experience what we call will and decision making. However, as Spinoza said in his concise way: "Will and Understanding are one and the same" (pt. 2, prop. 49, corollary; 2002a: 273). In other words, our usual distinction between perception and knowledge, on the one hand, and conscious decision making or voluntary action, on the other, is due to our "imaginary knowledge," that is, to our confusion between an "idea, or a conception of the mind, and the images of things which we imagine . . . which are found in us by the concourse of bodies" (pt. 2, prop. 49, scholium; 2002a: 273–77). More generally, this confusion is due to our limited knowledge of the true nature of the union of our mind and body. This is what makes "men think themselves free on account of this alone, that they are conscious of their actions and ignorant of the causes of them; and, moreover, that the decisions of the mind are nothing save their desires, which are accordingly various according to various dispositions [of their and other interacting bodies]" (pt. 3, prop. 2, scholium; 2002a: 280–82).

This quotation is taken from Spinoza's comment on proposition 2 in part 3, which denies the determination of the body by the mind and vice versa. Here, he makes his point even more clearly: "Now all these things clearly show that the decision of the mind and the desire and decision of the body are simultaneous in nature, or rather, one and the same thing, which when considered under the attribute of Thought and explained through the same we call a decision, and when considered under the attribute of Extension and deduced from the laws of motion and rest we call determination" (ibid.). The last sentence shows that Davidson's suggestion that, for Spinoza, "determination" means explanation and not causal relation is hard to accept, since what he calls determination here is both a causal *production* of a bodily movement, that is, something "considered under the attribute of Extension," and its causal *explanation* by laws of physics, that is, something considered under the attribute of thought.

The scholium continues with the development of "another point, . . . namely, that we can do nothing by a decision of the mind unless we recollect having done so before" (ibid.). This additional point exemplifies the role of memory in decision making. Its further development will lead to the identity between deciding and knowing, or "will and understanding" (pt. 2, prop. 49; 2002a: 272).

From all this, we can see how consistent and radical Spinoza's stance on psychophysical monism and causalism is. As a matter of fact, his philosophy inspired a "model of man" in the form of what he called a "spiritual automaton," which is more adapted to the kind of mechanistic models we can build today than to most ideas of his time (Spinoza 2002b: 24).

Conclusion: Support from Neurophysiological Data?

Although such a stand may be difficult to reconcile with some of our usual experience, it provides the best theoretical framework for understanding spectacular and apparently paradoxical neurophysiological data on the chronological order of decision making and action. In a set of experiments involving subjects undergoing brain surgery who had to remain conscious, B. Libet (1985) consistently found that a conscious decision to act corresponds to an electrical brain event that occurs two to three hundred milliseconds *after* the initiation of action in the motor cortex. This finding triggered a lot of controversy (Libet 1992), because it does not fit a theory of action in which the conscious decision to act is an efficient cause of the action. However, this observation can easily be understood in the context of our model. It means that the action is triggered by some neuronal unconscious stimuli and thus supports the Spinozist statement that the cause of a bodily movement can only be a physical event in the body. A conscious observation with an understanding of our action accompanies that action but is not its cause. We can interpret it as a decision of our will that determines the action because we do not know the unconscious events in our body that are its real causes. Moreover, there is a slight delay between the triggering of the action and our being conscious of it, because consciousness and understanding take time. They need to be retrieved from memory. As Marc Jeannerod (1992) so nicely puts it in his comments on the neurophysiology of consciousness, "the where in the brain determines the when in the mind."

Thus, we can make sense of Spinoza's radical denial of the causal relation between mind and body entailed by their identity, since we can conceive of how "understanding and will are one and the same." This can be achieved if we have a better understanding of what is meant by causality and identity in the framework of Spinoza's *Ethics*.

Causality must be understood not as an abstract relation between events but as a process of production, which is nontransitive from mental to physical (and vice versa) because they are synthetically but not analytically one and the same. Following Putnam's suggestion, an examination of the kind of synthetic identity of properties exemplified today by physical quantities apprehended at two different levels allows us to grasp the kind of mind-body identity at the core of the Spinozist system.

TWELVE

Spinozist Neurophysiology   *Henri Atlan and Yoram Louzoun*

(2007)

In this chapter, we will approach the mind-body problem using a monist ontology. Spinozist philosophy is certainly the most radically monist attitude toward this problem. This is apparent, for example, in propositions such as "The body cannot determine the mind to think, and the mind cannot determine the body to be in motion, or at rest, or in any other state different from these (if there be any other)" (pt. 3, prop. 2; Spinoza 2002a: 279), where Spinoza denies the possibility of causal relationships between the mind and the body, not because they would pertain to two different substances, as in Descartes, but precisely because they are "one and the same thing, expressed in two ways" (pt. 2, prop. 7, scholium; 2002a: 247).

Analyzing some aspects of this psychophysical monism will help us better understand the philosophically counterintuitive implications of our model for the emergence of goals in a self-organizing network, as well as the neurophysiological data on voluntary movements.[1]

---

1. [For the model, see the full version of this chapter (Louzoun and Atlan 2007).—Eds.]

Let us first note that this Spinozist denial of a causal relationship between mental and bodily states implies that the cause of a voluntary bodily movement must always be some previous bodily (brain) event or set of events, and not a conscious decision, seen as a mental event as described in subjective reports of conscious experiences. The difference between voluntary and involuntary movements is the nature and degree of conscious experience that accompanies them. In either case, a conscious mental event may accompany the brain event *but not be its cause, being in fact identical with it*, although not describable in the same language. Results from neurophysiology support this view: unconscious initiation of voluntary action precedes the conscious decision to trigger the movement. Thus, our model may provide Spinozist monism, however counterintuitive, with some theoretical and philosophical grounding.

This kind of counterintuitive identity between different properties or events that are identical but cannot be described in synonymous enunciations has been called a "synthetic identity of properties" (Putnam 1981: 85, 206–7), to be distinguished from analytical identity, where synonymous descriptions can replace one another. Hilary Putnam found an example of synthetic identity in the notion of physical magnitudes, which we employ in physics, such as "temperature" and "mean molecular kinetic energy": these are synthetically but not analytically identical. In the same context, Putnam explicitly related Spinozist psychophysical identity to such a synthetic identity, as a way to overcome many well-known difficulties in understanding this approach to the mind–body problem.[2] Similar results for affects and emotions, indicating a lack of causality between body and mind, have been proposed by Antonio Damasio, with the same reference to Spinozist monism as its philosophical interpretation (Damasio 2003).

This stance, as well as Wittgenstein's view of intentional descriptions (Wittgenstein 2001: 635–63), has been neglected by most philosophers and cognitive scientists, mostly because it contradicts our commonsense experiences and the commonly accepted ethical implications of free will that accompany them.[3] Thus, due to the influence of mentalist theories in psychology, intentions are viewed as some kind of conscious mental states, able to cause bodily movements whenever an intentional action is executed. These theories raise several difficult questions, such as:

---

2. [See also Chapter 11, above.—Eds.]

3. For analysis and criticism of this point, see, e.g., Anscombe 1957, Chalmers 1995, Davidson 1970, Fodor 1981a, and Shanon 1993.

1. How can a mental state be the cause of a physical movement?
2. More generally, what is conscious intentional experience made of?

The first question has been addressed, more or less successfully, by several philosophers. Among their approaches, Donald Davidson's theory of action may be the most comprehensive (Davidson 1970; see also Davidson 1999), especially in view of his definitely monist attitude, which he explicitly relates to the *Ethics* of Spinoza. However, his retention of commonsense conscious subjective and ethical experiences prevents him from overcoming serious difficulties in trying to reconcile Spinoza's explicit denial of a causal relationship between subjective states of mind as such and objective bodily movements with his "anomalous monism."[4]

The second question covers several problems related to different aspects of what we call consciousness. According to David Chalmers (Chalmers 1995: 200), some of these problems are "easy," although not trivial: they deal with specific cognitive aspects of consciousness, related to objective mechanisms that account for cognitive properties, such as memory, learning, adaptation, and so on. However, what he calls the "hard problem" is the question of how physical processes in the brain "give rise to subjective experience." This question is the reverse of that concerning intentional actions, where subjective intentions are supposed to cause physical movements.

In our work, we take leave of mentalist causal theories of action in order to seek a more objective approach to the question of causality.[5] The model presented here exhibits one of the main features outlined by G. E. M. Anscombe (1957) in her attempt to circumvent the logical difficulties of these theories, namely, an approach to intentionality through the study of intentional *actions*. This implies that intentions and actions are not dissociated to start with, and that the normal state of affairs is the execution of an intention. Dissociation, which may occur when an intention is not accompanied by an action, is the result of an obstacle to or inhibition of the execution of that action.

In this view, the "hard problem" of causality between the mental and the physical is eliminated: there is no causal relationship between intention as a mental state and action as a bodily movement, because "roughly speaking, a man intends to do what he does" (Anscombe 1957: 45). Because this view

4. [See also Chapter 11, above, and Davidson 1999.—Eds.]
5. [See Chapter 2, above.—Eds.]

seems counter-intuitive and raises new questions in the wake of Wittgenstein's investigation into the status of intentional statements in language games, Anscombe feels compelled to add: "But of course that is *very* roughly speaking. It is right to formulate it, however, as an antidote against the absurd thesis which is sometimes maintained: that a man's intended action is only described by describing his *objective*" (ibid.). In many instances, the objective of the agent is a description after the fact, aiming to answer the question: "Why did you do it?"

Let us conclude with several features of the nonmentalist model of intentions we wish to present, which appear almost literally in Spinoza's writings, to the point that one can speak of a "Spinozist neurophysiology."

1. Decision to act and previous knowledge allowing prediction are two different aspects of the same process associated with voluntary actions, although the former seems directed toward the future and the latter toward the past. That is so because intentions are described by means of intentional actions and not intentional mental states as causes of actions. "Will and Understanding are one and the same" (pt. 2, prop. 49, corollary; 2002a: 273) seems to state this counterintuitive concept.

2. In our model, general sets of goals are memorized by learning from experience. The knowledge acquired results from interaction between the internal structure of the network and the history of its most frequent encounters with classes of stimuli from its environment. In the context of the classical controversy about the reality of "Universals," we read:

> these general notions [called Universals] are not formed by all men in the same way but vary in each individual according to the things whereby the body has most often been affected and that the mind most easily imagines or remembers. For instance, those who have often admired the stature of man will by the name "man" understand an animal of erect stature; those who have been accustomed to regard some other attribute will form a different general image of man, for instance, that man is a laughing animal, a featherless biped, or a rational animal. Similarly, in other cases, everyone will form general images of things according to the conditioning of his body. (Pt. 2, prop. 40, scholium; 2002a: 266–68)

Thus, this "conditioning of his body" comes about by the way the cognitive system (mind–body) is assembled and also by the way it has been most frequently affected.

3. According to the neurophysiological data on voluntary movements, as well as our model, voluntary action is triggered by some unconscious stimulus, accompanied but not caused by a conscious state of the mind. A conscious

observation with an understanding of an action accompanies that action but is not its cause. We can interpret a decision of our will as determining the action because we do not know the unconscious events in our body that are the real causes.

> Now all these things clearly show that the decision of the mind and the desire or decision of the body are simultaneous in nature, or rather one and the same thing, which, when considered under the attribute of Thought and explained through the same we call a decision, and when considered under the attribute of Extension and deduced from the laws of motion and rest we call determination. (pt. 3, prop. 2, scholium; 2002a: 279)

4. There is, however, as B. Libet's observations show, a slight delay between the triggering of action and our being conscious of it, because consciousness and understanding take time: as in our model, they need to be retrieved from memory. In other words, "we can do nothing by a decision of the mind unless we recollect having done so before" (ibid.).

5. In adopting the stance that we have, we obviously *lose* something, namely the commonsense assumption about free will and the causation of actions by decisions of a nonbodily mind. However, we *gain* understanding of intentional actions without resorting to hidden causal properties of mental states. We do not necessarily deny the reality of free will, although we modify its content. According to Libet, free will can be located in a kind of *veto* function, that is, the possibility of inhibiting movement after it has been initiated. In addition, we say nothing about the possible effects of long-term deliberation and decisions to act "in principle," with a more or less extended period of time until the decision initiates action. Spinoza's stance on free will is more radical:

> men believe themselves free on account of this alone, that they are conscious of their actions and ignorant of the causes of them; moreover, the decisions of the mind are nothing more than their desires, which vary according to the various conditionings of their and other interacting bodies. (Pt. 3, prop. 2, scholium; 2002a: 281)

6. Finally, the picture of intentional action we present helps us better understand what "desire" in a practical syllogism is about: an unconscious drive with awareness of the goal to which one is driven.

Spinoza extends this definition of desire to the realm of moral judgments:

> Desire is appetite accompanied by the consciousness thereof.

It is thus clear from what has been said, that we do not strive for, wish for, long for, or desire anything because we deem it to be good. On the contrary, we deem a thing to be good because we strive for it, wish for it, long for it, or desire it. (Pt. 3, prop. 9, scholium; 2002a: 284) ·

THIRTEEN

Knowledge, Glory, and "On Human Dignity"

(2007)

The idea of dignity seems inseparable from that of humanity, whether in its universal dimension of "human dignity" or in the individual "dignity of the person." We do not (yet?) speak of animal dignity or of the dignity of nature, even though some are tempted to invoke—erroneously—rights for animals or rights of nature. In truth, whenever we try to define them, these notions of dignity, humanity, and individuality all reflect back onto one another. What is more, such attempts at definition often come up against almost insurmountable linguistic obstacles when they seek to lay down "values" or a universal ethic.

In order to overcome such difficulties, which are inherent in all intercultural dialogue where differences and the linguistic idiosyncrasies that characterize them are taken into account, it is better to proceed "from the bottom up," starting with case-by-case analyses of situations where these notions come into play, rather than following a "top-down" approach, starting from abstract definitions, which are necessarily influenced by the languages and cultural traditions in which they are expressed.

The ethics governing the sciences and technology, in particular, the biological sciences and biotechnology, provide examples of concrete situations where such case analyses can be made. Some observers, drawn into excessive generalization, perceive the danger that a posthuman environment will emerge, ushered in by the constantly accelerated pace of technological development throughout the twentieth century and its effect on the human condition through the more or less catastrophic social, cultural, and moral repercussions that seem inevitably to accompany them.

Granted, the present-day explosion of knowledge and technical invention and its unbridled exploitation, fueled by the imperative of the market, keeps throwing up new problems in relation to future dangers that it allows us to glimpse. But these dangers are those of a rise in inhumanity, rather than of the "disappearance of man" or of his being replaced by some form of posthumanity.

Let us immediately recognize that in all times inhumanity has characterized the human species. In fact, only human beings can be inhuman or experience the inhuman. Mineral, vegetal, and animal existence can only ever be assigned to the nonhuman. It is precisely because science and technology are among the most characteristic spheres of human activity that the everlasting question arises of whether their products are human or inhuman.[1]

Ever since man mastered fire and invented the wheel, science and technological invention have exercised both fascination and dread, for they have served simply to increase man's power over nature and over his own condition, including his tendency toward inhumanity. The traditional expressions of humanism and the "values" that they have encapsulated have not managed to prevent sudden outbreaks of inhuman behavior on a wide scale. Such values have even been invoked in order to justify these. One can think of the massacres of native peoples and of forced conversions and other excesses of colonization, together with the millions of victims sacrificed to ideologies, both secular and religious, at the altar of human salvation by any means, whether these populations wished it or not. We are still not immune to inhumanity of this type, which is always associated with totalitarian ideologies, even if, like the road to hell, they are paved with the best of intentions. Our only practical option for diminishing the probability that these excesses will continue to spread is to hold fast to democracy as a form of government and

---

1. [See Chapter 1, above.—Eds.]

to the Universal Declaration of Human Rights as a barrier against the practice of inhumanity, independently, to a certain degree, of any theoretical adherence to any particular idea of man or nature.

In fact, the danger of inhumanity is consubstantial with humanity and the human species itself. The English language has a wider range than French for labeling the human characteristic whose opposite is the inhuman; it distinguishes the *humane* from the simply *human*, whose opposite is not the inhuman but simply the nonhuman, the nonbelonging to the species. Thus humaneness, of which inhumanity is indeed the opposite, along with the notion of dignity, with which it is closely associated, is not reducible simply to its biological components. It results, to be sure, from biological evolution, but also from cultural evolution, whose mechanisms are not the same and which does not necessarily obey the same laws. It is perhaps by reflecting on this notion of human dignity that one can delineate what is understood by this humane dimension of man, which is subject to threat. But the notion of dignity remains obscure, even though omnipresent in discourse as a criterion of ethical demarcation. It leads us to rate certain practices as not acceptable, such as reproductive human cloning, or experimentation on human subjects without consent, or other treatments considered inhuman, such as torture or slavery: these we disqualify on moral grounds as being contrary to essential human dignity.

Therein would appear to lurk a vicious circle arising out of an essential tautology: inhumanity being defined as an offense against human dignity, and an offense against human dignity being defined as inhumanity.

But this is in fact not so, and to help understand it, a short detour via analyses of the equally obscure but related notions of honor and "glory" may assist.[2]

---

2. [The French use of the term *gloire* in this essay poses certain difficulties of translation. *Gloire* in French conveys primarily the meanings "high renown, honor, noble reputation, fame," that is, a great admiration and respect bestowed by others, and only secondarily that of "prestigious splendour and majesty," especially when referring to the "glory of God." (See *Le Robert, dictionnaire du francais*, where the second meaning is listed as "dated or literary"). Atlan's essay makes use of both these meanings, but apart from the section discussing the notion of human dignity in Pico della Mirandola, the sense of *gloire* as "high esteem or reputation" tends to predominate. While the range of meanings of *glory* in English generally parallels those of *gloire* in French, the weighting of the English meanings appears to be the reverse of French. Though the sense "exalted praise, honor, or admiration accorded by common consent to a person or thing; honorable fame, renown" is the first one given by the *Oxford English Dictionary*, modern usage does not use "glory" in

❖

Let us first recall that the notion of "glory," both human and divine, infuses many texts of the Middle Ages and the Renaissance, a pivotal era that ushered in the scientific revolution in Europe. In the writings of Pico della Mirandola, for example, "glory" is explicitly associated with dignity, to which he devoted the whole of a small treatise, as an introduction to his "Nine Hundred Theses," a vast compendium "on the sublime mysteries of Christian theology, on the loftiest questions of philosophy, on unknown teachings."[3] Dignity and glory belong to creatures who dwell "beyond the chambers of the world [in] the chamber nearest the most lofty divinity. There, as the sacred mysteries reveal, the seraphim, cherubim and thrones occupy the first places." But the human condition, higher than that of all other creatures, leads us to "compete with the angels in dignity and glory."[4] Even more, though this effort, the dignity of man, to whom freedom is given as a doorway to the possible and a capacity of self-realization, elevates him above the angels and allows the world to reach perfection. Our dignity is thus no different from the glory of God: "aroused with ineffable charity as with fire, placed outside of ourselves like burning Seraphim, filled with divinity, we shall not now be ourselves but He himself who made us."[5]

We do not have to limit ourselves to this theological and mystical vocabulary, which can seem rather old-fashioned today. But we can bring this terminology down to earth, so to speak, by stripping it of its mystic aura. Depending on context, these notions of glory, honor, and dignity are interchangeable. As we have seen, today such notions are difficult to define, even though they play an essential role in the definition of particular moral and juridical norms. To take an example, Article 2 of the Universal Declaration of Human Rights declares that "all human beings are born equal in rights and dignity." The notion of equality sometimes serves to provide substance, at least on the political level, to that of dignity, therein constituting one of the foundations of democracy. The notion of dignity is also present in considerations of biomedical ethics as a value that must be respected in all circumstances. And the very concept of a crime against humanity implicitly

---

the sense of "reputation" except in limited circumstances ("he won glory on the battlefield").—Trans.]

3. Pico della Mirandola 1965: 19.
4. Ibid., 7.
5. Ibid., 14.

contains, as its inverse, the right to this "indefinable human dignity." According to Mireille Delmas-Marty, this usage provides a definition of "humanity" that would go beyond mere belonging to the species, namely, the humanity that is destroyed in crimes against humanity.[6] Conversely, this moral and social definition of humanity, whose opposite is the inhuman rather than the nonhuman, in turn allows the notion of dignity to be defined. We thus arrive at a circular, but not tautological, definition where the "not inhuman" is defined by dignity and dignity by the "not inhuman." We also find in Spinoza a moral definition of *humanitas* in that particular sense, where what "is habitually called *humanitas*" consists in that "we also endeavor to do whatever we imagine men to regard with pleasure, and . . . we shun doing whatever we imagine men to regard with aversion" (pt. 3, prop. 29; Spinoza 2002a: 293). We find in this a form of the classic and supposedly universal "golden rule" of not doing to others what one would hate having done to oneself, but *modulated here by the role of the imagination*. Indeed, nothing proves to us that others have the same desires and aversions as we do. We are left with having to imagine this, which markedly reduces the altruism of the rule. Only in a society where all live within the realm of reason can the reciprocity of the rule truly function, since all would have the same aversions, if not the same desires.

But in this comment about humanity in the form of affect or feeling we also hear an echo of one of the definitions of "glory" that Spinoza gives elsewhere, in association with the praise of others: "Honor [*gloria*] is pleasure accompanied by the idea of some action of ours which we think that others praise" (Spinoza, "Definition of the Emotions," 2002a: 316). This definition reveals the ambivalent character of this human sense of *gloria* (*gloire*), something not particularly "glorious," given that it comes down to a reputation derived from public praise. Spinoza emphasizes this ambivalence in the context of what he calls *humanitas*, as it relates to "this *conatus* to do, and also to avoid doing, something simply in order to please men" (pt. 3, prop. 29, scholium; 2002a: 293). This, he says, is then called "Ambition." We are here very far from the lofty heights of the intellectual love of God and of human freedom. But these forms of humanity and dignity, which are reflected in each other, in effect describe a minimum degree of these qualities, attainable by all through the imagination. This Spinozist "first type of knowledge"—

6. Atlan, Augé, et al. 1999: 81–82, 99–109.

through the imagination—however confused, distorted, and occasionally illusory it might be, is granted to all human beings. Contrary to what might be said, imaginative consciousness, based to a greater or lesser extent on illusion, is probably more equally distributed within the human species than is reason. Therefore, retaining the function of the imagination in this definition of "glory," we may define human dignity as *"the minimum quantum of recognition [gloire] without which an individual would be excluded from human society; that is, following these definitions, a minimum level of self-esteem and satisfaction, as well as of recognition and acknowledgment by others, in the absence of which the condition of a human being would be inhuman."*

But "glory" is also the name that, for Spinoza, the Bible "not without reason" gives to beatitude, or Spinozist freedom, the highest perfection that the philosopher is thought to attain through the trained exercise of his understanding.[7] Taking into account Spinoza's familiarity with the Hebrew Bible, this sends us to the Hebrew word *kavod*, meaning "glory, honor, or dignity," which shares the same root as *kaved*, meaning "heavy." We are thus led to consider *kavod* as a weight, ontological and moral in nature, such that possessing this dignity gives a human being an irreducible "weight," in other words, an intrinsic value in him- or herself. This is the minimum "weight" accorded to a human existence, without which it would become inhuman.

This interpretation can be linked to the formal sense of the "power" of a number accorded in earlier times to the notion of dignity in mathematics: the nth "dignity of a number" specified, in Leibniz, for example, n "continual multiplications" of this number by itself.[8] It is worth noting that this link between dignity and humanity is not tautological in that it is not a case of simple identity. This can be seen in actions and behavior. An inhuman action consists in removing or denying the essential dignity of a human being. But an action that falls short of the standards of human dignity [*une action indigne*] is not necessarily inhuman. Such behavior or action is unworthy or dishonoring for the perpetrator himself to the extent that it diminishes others' appreciation or recognition of him. This is associated with quasi-universal experiences of the absence of dignity, such as shame and, indirectly, humiliation.

It is here that the two senses of the human and the humane can come together, as one would expect from a truly monist conception of body and

---

7. See Atlan 2003: 132ff.
8. Leibniz 1991: 41.

mind as two different aspects of the same thing. One cannot set aside the human body in any definition of human dignity. The question of ontogenesis and first beginnings, the question of where the limit is drawn nevertheless remains hanging. At what moment does a body become a human body? That this question clearly underlies the debates about the nature and status of the human embryo. It is posed against the backdrop of the unity of nature and the gradualism observable in the continuity of development, as in the evolution of species. In such a context, a response to the question concerning the point at which a human body begins to exist cannot be founded upon an illusory essentialist definition, in the sense of an abstract essence of man that would be infused in him once and for all, whether through the genetic heritage—at the moment of conception as marking the constitution of the genome—or through a more or less arbitrary appreciation of certain degrees of consciousness appearing over the course of development or evolution. There remains the possibility of a definition based on emergence, whereby the humanness of the human being becomes established progressively, as his or her body forms. The threshold at which this body starts to be human is therefore that at which *its human form can be recognized*, including, most obviously, the face, the "glorious" form above all. This would correspond to an ancient Aristotelian definition, incorporated into Jewish and Muslim tradition, as well as Christian tradition in its Thomist form, which followed the thesis of "late animation" before this was superseded by the doctrine of "early animation" that is current Catholic orthodoxy.

Whatever judgment might be brought to bear on these questions and, in consequence, whatever pragmatic decisions might be taken concerning thresholds and barriers that cannot be crossed in concrete situations, the issue is not a threat to human nature per se, but the emergence of new forms of inhumanity, in the awareness that the danger of inhumanity is consubstantial with the human dimension of the human species.

The humanity of *Homo sapiens* is in danger, as it always has been, and each advance in the progress of knowledge brings new dangers. Knowledge is intrinsically ambivalent, at once both good and harmful, for it disturbs the order of things, opening up new possibilities out of old certitudes. Like all creative activity, it is both destructive and constructive. And when it is accompanied by an increase in mastery over nature, its ambivalence is multiplied by that of nature itself. For nature is not solely good, as certain naive ecologists seem to believe. It is at once beneficent and harmful, a source of both prosperity and suffering, as is every transformation imposed upon it. Yet we cannot halt the march toward knowledge. The wish to know is consubstantial with the entire human condition. And the lucidity brought by what rigorous and disinterested research teaches us is constitutive of what creates human dignity.

In conclusion, we should not dread false fears or pursue mistaken targets. The danger lies not in the disappearance of the humanity of man but in the appearance of new forms of inhumanity, following on from the inhumanity of ages past. Drawing lessons from past experience, we will do well to recall that inhumanity has always prospered from illusions about discoveries that were thought to be definitive or final, heralding the dawn of a new age or a long-anticipated salvation. Far better to retain the uncertainty of not knowing, which we keep discovering even as new knowledge appears.

For narrow is the way between rigid adherence to immutable beliefs and the intoxication of new discovery, which always lurks behind the arrogance and illusion of omnipotence.

Translated by Colin Anderson

PART FOUR

Judaism, Determinism, and Rationalities

FOURTEEN

Sparks of Randomness

(1999)

Adam knew his wife again, and she bore a son and named him Seth
["gotten"], because "God has got me another seed in place of Abel,
for Cain killed him."

—Genesis 4:25

When Adam had lived 130 years, he begot in his likeness after his image,
and he named him Seth.

—Genesis 5:3

The *sparks of randomness* are the drops of semen that Adam, according to
legend, emitted during the 130 years when he was separated from Eve. The
phrase is a literal translation of the Hebrew *nitzotzot keri*. The word *keri* is
derived both from *mikreh*, a random or chance event, neither regular nor
planned—an accidental occurrence—and from the involuntary emission of
sperm that the Bible denominates *mikreh layla*, a "nocturnal event."

A lapse of 130 years separated the birth of Seth from Cain's murder of
Abel, these two sons having been born after Adam and Eve had eaten from
the Tree of Knowledge and been banished from the Garden of Eden. According to a talmudic legend, developed at length in the midrashic literature and
Kabbalah, during these 130 years Adam lived apart from Eve; the sperm he
spilled during that period created and nourished demons, the origin of the
"lost generations" of the Flood and the Tower of Babel.

This story of knowledge and sperm is difficult to understand today, after
two thousand years in which the knowing subject has been separated from the
body, one of the objects of its knowledge. Whether we refer to the biblical

knowledge of Genesis, the spermatic knowledge of the Midrash, or the seminal reason of the Stoics, it is the forgotten source of our present intuition of a physical union between soul and body. This forgetfulness has been indispensable to the autonomy of a wide-awake knowledge, detached from the fusion of dream and illusion. But can forgetfulness do its part without our being aware of it?

It seems that we must restore the voice of the knowledge gained through sex and the fecundity of this concept so that we can understand what biology, the cognitive sciences, and psychoanalysis are trying to tell us, perhaps clumsily, *in addition to* what they *may* be telling us explicitly.

Knowledge, sexuality, generation, concepts and conceptions, birth and abortion, angels and demons, aging, disease, and death—science and technology constantly bring us back to these eternal problems, inherent in the human condition, while refashioning the terms in a way that is sometimes dramatic and unprecedented. Does a new science have a new morality? Who can decide this, how, and with what tools? With what concepts and conceptions of the world, of existence, of what is good and what is bad? What words can we use to talk about them? What style? The empirical and logical mode of science and technology, whose terms are indispensable for posing the problems? The narrative mode of literature? The talking heads on television? We should recall that today the standard response to the question of how virtue can be taught is Protagoras's rather than Socrates': good and evil are not taught exclusively by means of scientific knowledge, but by means of images drawn from epic poetry, in which moral problems are raised. In most cases, the heroes and antiheroes of these situations are the source of our acceptance or rejection of what we imaginatively identify as good or evil. Critical scientific and philosophical analysis makes it possible for us to delve ever deeper into technical and conceptual subtleties. But science and philosophy themselves are not sources of universally accepted standards. Nor can religious dogmas, for all the help their authority may provide to those who embrace them, produce norms acceptable to everyone and suited to the complexity of specific situations.

In fact, myth has always seized upon these questions and expressed them in its own synthetic and oneiric mode, which supports visualizing and associating and which, perhaps better than science, enables us to uncover the hidden threads of a concealed fabric woven in different registers of experience and knowledge that analysis strives to distinguish: not only Prometheus and Oedipus, but also the biblical myth of the Trees of Life and of Knowledge, the Flood, and the Tower of Babel. To speak and write about living well requires a certain style in which diverse languages—scientific and technical, legal and philosophical, poetic—can coexist without being confounded; where the perception of reality is always pregnant with the imaginary, but the latter never takes the former's place; where the rationalities of science and of myth can subsist side by side without being confused and can criticize each other.

Thus new ethical reflections would seem to require inventing a new form of discourse. Its birth was registered when we came to the realization that modern science, contrary to Condorcet's dream, not only fails to resolve all social and political problems but in fact creates new ones, because it spawns new possibilities without providing the means to settle them. What is more, scientific discourse is not always free from dubious extrapolations. Sometimes, unknown to those who conduct it, myth still manipulates it, and ancient stakes, thought to have been left behind long ago, return to the surface. The Big Bang restores creation, if not the Creator. As for the celebrated "human genome," with its poorly defined contours, referring to it as a patrimony and sanctuary is no less imaginary than seeing the heart as the seat of the passions and the bile as the medium of anger. In the new form of discourse that we must construct, we must burn whatever fuel is to hand. We must not hesitate to stoop to case-by-case legalism to argue about what is permitted and what forbidden. But to analysis of technical details and examination of basic principles we must juxtapose complicated plots, real or mythological, and interpret them on several levels, where what is not said and not thought, yet always present, can at least be made visible, so as to become, even if only for the moment, partly said and partly thought.

This form may not be as radically new as it seems. No doubt we can draw inspiration from the dialogues of the schools of antiquity, in which myth, science, and philosophy were not yet separated, neither from one another nor from the experience of right thinking in pursuit of living well—with oneself, with others, and with nature. Alongside the works of the ancient schools of philosophy, the inquiries conducted in the rabbinic academies in Palestine and Babylonia have come down to us in the unique style of the debates and narratives of the Talmud. The legal disputations aimed at establishing just laws on the basis of multilevel interpretations of biblical myths and statutes constantly rub shoulders with new legends, the *aggadot* of the Talmud. Midrash and Kabbalah took up these accounts and developed them into new myths, reenergizing and amplifying their interpretive power. We, in our own turn, can be inspired by this form, without necessarily adhering, of course, to the literal sense, for at least two reasons. The social, scientific, technological, and philosophic context of two thousand years ago is incompatible with that of today, even if, in the interim, what we can call "human nature" does not seem to have changed very much, at least in the biological sense. Above all, however, it is the nature of mythical narrative to be taken up again, generation after generation, in a recursion that amplifies it and in which the letter of the commentary, and of commentary on the commentary, serves as a new text to be interpreted, a pretext for new interpretations.

What is the status of the randomness of birth, chance, and the ignorance of causes that we call "destiny" in a world that we are increasingly able to control, where uncertainty itself seems to be programmed by probabilistic estimates of risk? Is it not the vocation—or the destiny—of our species to use its inherent capacities, its large brain and its cognitive and linguistic abilities, to order and control the rest of nature? But does what applies to the rest of nature also apply to the human species? Is it the destiny of humanity to suppress destiny by means of planning? Are human knowledge and technology a violation of some natural law, a rape of Nature, on which they are imposed like some monstrous exception? Could they be a curse to which humanity is subject, generation after generation, massacre after massacre, always increasing in number and intensity, proportional to humanity's control over everything that is not itself? Or are they merely one product among many of that same Nature?

Are we the children of Prometheus only? Are we not also the children of Adam, who was enjoined to fill up the earth, to occupy it and dominate it, along with the fish in the sea, the birds in the sky, the terrestrial animals, and every living thing that creeps on the Earth? The human domination of nature is not, clearly, the product of the cultures that preserved this myth. On the contrary, the narrative merely expresses, in its own manner, the dominion of the human species as it has always been experienced, in all latitudes and by all cultures.

In the same way, the narrative of Genesis is also the account of a curse. The narrative winds like a serpent around the two poles of the human exception: knowledge and morality, *Homo faber* and *Homo sapiens sapiens* at one end, and the suffering they inflict on others and on themselves at the other. At one end is the evil that they do, that they suffer, that they imagine, and that they plan; at the other is the good, the happiness, the bliss that they also imagine and try to plan—in short, the angels and demons with which they fill their universe, inside and outside themselves. The myths of Genesis and of the sparks of randomness make it possible to explore all this.

Curiously, for bourgeois morality and for law, the English term for an infant born out of lawful wedlock is *natural child*, as if the institution of marriage endowed children with the superior status of "artificial children." In fact, today we are not far removed from the production of true artificial children, not through social institutions but through biotechnology. Will this be bad or good? For humanity today? For the humanity of the future, which may consist at least in part of such children?

When we refer to the biblical account of Adam and Eve's transgression and the curse that followed it, we will do our best to forget received ideas about "original sin." In the Western world, the story of Adam, Eve, and the serpent is generally associated with the Augustinian interpretation imposed by Christian orthodoxy since the fifth century: human beings are doomed to unhappiness and suffering because of the sin of the flesh committed by our first parents, Adam and Eve. Sex is fundamentally evil; holiness requires abstinence and celibacy. We inherit their sin and guilt at the moment of conception, the result of that very same sin; this explains all the unhappiness and suffering with which human beings are afflicted from the moment of birth, including infants, who have not yet had the occasion to do something bad of their own free will. For some Augustinian theologians, this universal predestination to evil is the negation of free will; the only way to escape it is divine grace and obedience to the authorities of the city and the Church. In this form, which centuries of catechism have made familiar to and inculcated in millions of children, the story has played and continues to play a decisive role in the moral and religious conscience that is almost consubstantial with Western civilization. It has shaped notions that still hold meaning, even for those outside the Church, of male and especially female sexuality, of the family and the body, of birth and death, of guilt, holiness, and innocence. But the early Christian church, subversive and persecuted, before the Christianization of the Roman Empire placed it in the saddle, did not always hold this interpretation. Augustine himself was able to impose it only after long theological and political controversies. Elaine Pagels's *Adam, Eve, and the Serpent* (1988), which offers a contemporary and critical Christian perspective on the history of early Christianity, clearly depicts the protagonists and matter of these controversies. It is a matter of divergent interpretations of the first chapter of Genesis, where this foundational story is told. In the year 418, at the end of the Pelagian controversy, the Augustinian reading came to be considered orthodox. All others were condemned as heretical, and their proponents were excommunicated. Their interpretations, less misogynistic and less appalling, highlighted not the sexual aspects of the story but rather its message of individual liberty and responsibility in obeying or transgressing divine law, in a world that is fundamentally good, in the image of its Creator. These interpretations, from that time forth deemed heretical, even though they had been defended by bishops of the Church, had been preceded by other interpretations of Gnostic inspiration, which were, if possible, even more heretical and were condemned even more quickly. Today, especially since the discoveries of the library at Nag Hammadi, we know that Gnostic interpretations of the Bible and Gospels were an integral part of early Christianity, at least during its first two centuries. One of their characteristic traits is the symbolic nature, cosmic rather than anthropological, that they attribute to the biblical protagonists. In some of these interpretations, for example, Eve is an icon of wisdom, the mother of the universe, rather than a woman of flesh and blood. In a similar vein, the virginity of the mother of Christ was not understood literally (Pagels 1989). In general, these extraordinary stories, including the first chapters of Genesis, were understood less as edifying tales with a moralizing bent, intended to nurture a particular social and religious doctrine of the Church about the relative values of celibacy and marriage, for example, than as myths of origin, like the Greek or Egyptian traditions they had supplanted. This difference must be underlined. For some Christian Gnostics,

> the story was never meant to be taken literally but should be understood as spiritual allegory—not so much *history with a moral* as *myth with meaning.* These gnostics took each line of the Scriptures as an enigma, a riddle pointing to deeper meaning. Read this way, the text became a shimmering surface of symbols, inviting the spiritually adventurous to explore its hidden depths, to draw upon their own inner experience—what artists call the creative imagination—to interpret the story. . . . Consequently, gnostic Christians neither sought nor found any consensus concerning what the story meant but regarded Genesis 1–3 rather like a fugal melody upon which they continually improvised new variations [gnostic interpretations of the creation story], all of which, Bishop Irenaeus said, were "full of blasphemy." (Pagels 1989: 64)

But this was nothing new. Philo had employed and greatly expanded this method of allegorical interpretation. Similarly, for some philosophers, mainly Stoics, the *Iliad* and *Odyssey* were not to be understood according to their surface meaning as accounts of the rivalries and loves of the gods. Their surface meanings concealed deeper truths of natural philosophy, which a symbolic reading could uncover. The rabbis of the Talmud and the Midrash were raised in this type of interpretation, associated and superimposed on the plain meaning of the biblical text, to the extent that it allows itself to be grasped. Kabbalistic interpretations that uncover and develop the "hidden" meaning of the text merely amplified and systematized this tendency, already found in the Talmud. It is not surprising that the Kabbalah sometimes evinces a familiarity with gnostic themes, as in its doctrines about what preceded creation, to which we shall return. Gershom Scholem saw this as reflecting the direct influence of Gnosis on some sources of the Kabbalah.[1] Moshe Idel, by contrast, suggests that it was Gnosis that drew on ancient Hebraic influences, or at least that the influences were mutual (Idel 1988: 115–57). The interpretations of the story of Adam and Eve, such as the legend of the sparks of randomness that inspired the present work [Atlan 1999b], are much closer in method, if not in content, to gnostic construals than to those derived from Augustine.[2] They differ, nevertheless, in at least two points. First, the symbolic interpretation does not cancel out the literal meaning but is superimposed on it. Adam is at the same time the first man, an archetypal figure of the human nature in each person, "male and female," and *Adam Qadmon*, the primordial human being, a cosmic and divine figure that fills the universe, simultaneously creator and created. (From this point of view, the kabbalistic readings are often less rigorously allegorical than those of Philo, for example.) A second and decisive difference is that, contrary to Christian gnostic interpretations, which were swiftly condemned by the Church and banished from its doctrine, the kabbalistic interpretations, with all their diversity, have remained an integral part of orthodox rabbinic Judaism. They are less well known than interpretations that are easier to understand and to employ in religious instruction whose goal is edification and enlightenment, but over the centuries, until the beginning of the twentieth century, their authors were frequently prominent teachers of rabbinic orthodoxy, sometimes community rabbis. For all these reasons, we will use the biblical story of Adam and Eve and the rabbinic commentaries upon it as a myth that is pregnant with multiple meanings, discarding the received notions of original sin and hereditary curse, the inevitably evil character of physical nature that derives from them, redemption through celibacy, and mortification of the flesh as the road to

---

1. See Scholem 1987 and 1974: 10–14 on "rabbinical gnosticism."
2. See, e.g., Tardieu 1974.

salvation. We will read it, not as a story with a moral, but as an album of images of various and contrasting aspects of the human condition, associated in particular with the protracted period of childhood and maturation that follows birth and with the long interval between sexual maturity and intellectual and emotional maturity: the Tree of Knowledge is assimilated before the Tree of Life (although we can easily imagine how the inverse chronology might have had better consequences). If, all the same, there is a moral to be drawn from the story, it must involve the search for some sort of redress or reparation—that is, a way to ameliorate this condition by identifying the harmful effects, the sources of pain and suffering, in order to eliminate, attenuate, or transform them.

Translated by Lenn Schramm

FIFTEEN

Nature's Ultimate Trick: The Parable of the Divine Intrigues ('*Alilot*)

(1999)

Given what we have seen about certain currents of kabbalistic thought that are to be strictly distinguished from spiritualist theology, it is not astonishing that we can find in it—as among the Stoics, but surviving to the present day within the context of Hebrew monotheism—more than one kinship with the notion of the nature *of* human beings and of nature *in* human beings, simultaneously actor and acted on by the human form of the "image of god(s)" (*tzelem elohim*).[1] Here we encounter, as in every monism in which the distinctions between body and mind are not essential but depend on the perspective we adopt to observe their activities, the games that determinism plays with itself. In general, instead of an essential duality of matter (passive, created) and mind (active, creating), these distinctions are part of an "infinite chain of being" (*hishtalshelut ha-havayah*): what is body and opacity, when contrasted to a soul or a light that fills it, is itself mind and light when contrasted to the vehicle that it fills and animates in its turn. One consequence is that the soul

---

1. See Atlan 1999b, chap. 3.

is not thought of as an immaterial entity. In the Hebrew tradition, the notion of soul has its own history; the soul was not always "spiritual" in the sense of incorporeal. In Genesis, the human body molded from the earth is animated in different ways by different souls, created from other elements: the *nefesh hayyah* ("living soul"), common to animals and human beings, is identified with the blood (*ha-dam hu ha-nefesh*). The *ruah* ("spirit") is breath and wind. The *neshamah* or intellectual soul (related to *neshimah*, "respiration"), perhaps immortal, is literally an in-spiration of life (*nishmat hayyim*), blown in through the nostrils.

Naturally enough, in this tradition we find the forgotten vision of an absolute determinism of nature in human beings, which is not expressed here, of course, in Greek or in Latin, but through the images and characters of the Hebrew Bible, amplified by the parables of the Talmud and Kabbalah. Among these images we have already met the *tzelem elohim*, the form of the human body or divine form in human beings, which makes them simultaneously actors and acted on; we will also encounter the image of the game or toy, the *sha'ashua'*, the plaything of the Creator, the game that Creator and creation play with themselves and with Wisdom, and especially the terrible image of a crafty deity (or nature) that employs *'alilot*—pretexts—to make human beings responsible for the sins that he or it made them commit (or that they were determined to commit). This radical thesis is developed by Shlomo Eliaschow in his *Book of Knowledge*. That author is heir to the long line of rabbis (including the Gaon of Vilna) for whom the teachings of the Talmud and Midrash, as well as Lurianic Kabbalah, concern a world governed by necessity, in which everything that happens must happen and is foreseen from all eternity. A passage in Tractate *Avot*, classically invoked to affirm free will, is inverted into an assertion of absolute determinism: *ha-kol tzafui ve-ha-reshut netunah*—"everything is foreseen but permission [*or* the possibility (to choose)] is granted" (M *Avot* 3:15.) The classical reading of this dictum by idealist theologians, following Maimonides, begins with the second half of this statement: "We are granted the possibility of choosing"; that is, we have free will to decide the sequence of events, for which we are thereby responsible. As for the first part, "everything is foreseen," it is taken as referring to the mystery of God's omniscience.[2] Eliaschow, by contrast,

---

2. Maimonides develops this thesis in the introduction to his commentary on *Avot*, known as the *Eight Chapters*, as well as in his *Sefer ha-Madda'* (*Book of Knowledge*; see Atlan 1999b, chap. 2, n. 28). The mystery of the omniscience of God, who is the creator ex nihilo of nature and its laws, derives from the fact that, for Maimonides, God's essence is

begins with an affirmation of uncompromising determinism: "Everything is foreseen, including what is done by our choices." For there is no denying that we have the possibility of choosing each time we make a choice. But this is merely a game and in no way modifies the "everything is foreseen" character of rigorous determinism; free will, which is perfectly real as an internal and subjective experience, is an illusion if we believe that it determines the sequence of events. Eliaschow expounds an absolute determinism in which "everything is foreseen and decided for all eternity, including what is done by the choices of human beings" (Eliaschow 1977, §8, pp. 295–96). In this vision of things, human responsibility and, once again, guilt are the result of one last trick played by creation, a plot, a terrible 'alilah that toys with human beings, beginning with the first couple, Adam and Eve, and thereafter with all of their descendents, past, present, and future.

To develop this idea, Eliaschow borrows a commentary from Midrash *Tanhuma* (on Genesis 39:1) of Psalms 66:5: "Come and see the works of God, terrible in his deeds [*nora' 'alilot*] among men." This phrase, "terrible [or terrifying] in his deeds" (*nora' 'alilah*), quoted in one version of the Yom Kippur liturgy,[3] anchors the comment, itself terrifying.

According to the context in which it is used, the same word, 'alilah, "deed, maneuver," may mean "feat," "brilliant deed," or "pretext" (in the sense of perverse manipulation or bad faith). When associated with God, it is usually

---

identical with his knowledge, which is creative, as he is, unlike human knowledge, which is distinct from the knower. In consequence, "the knowledge attributed to [God's] essence has nothing in common with our knowledge, just as that essence is in no way like our essence" (*Guide* 3.20, Maimonides 1956: 293). "If someone asks us about the nature of his knowledge, we answer that we cannot grasp it, any more than we can fully grasp his essence" (*Eight Chapters*, chap. 8; Maimonides 1912, 102, translation modified).

This position has been criticized, notably by the Maharal of Prague in his introduction to *Gevurot ha-Shem* and in *Derekh Ḥayyim* (his commentary on *Avot* [5:6]), where he insists that "God's knowledge is not his essence" (*Tractate Avot with . . . Derekh Ḥayyim* [London, 1961], 233). Spinoza's criticism of the thesis of creative intellect is also a key point in his attack on Maimonides' ontology.

We should note that some of Maimonides' formulations, notably in *Sefer ha-Madda'*, may sow confusion and reinforce the thesis of those, like Leo Strauss and Shlomo Pines, who read the *Guide* as an "esoteric text" whose surface meaning conceals the author's true positions, as he alerts readers in the introduction to the *Guide*. (See, e.g., Robinson, Kaplan, and Bauer 1990). But for almost all of Maimonides' disciples and, over the centuries, mainstream Jewish theology, it is the surface level that carries the day.

3. [See Atlan 1999b, chap. 2, n. 64, and chap. 1, n. 72.—Eds.]

given the sense of "feat" or "great deed," a remarkable action that attests to its author's intelligence and power. ʿ*Alilah* is related to ʿ*alul*, "expected consequence," and to ʿ*illah*, "cause." (In Hebrew, the scholastic notion of God as First Cause is ʿ*illat ha-ʿillot*, "cause of causes.") But the same word, ʿ*alilah*, when applied to human actions, may mean "deceit" or "trick," the pretext a cunning and evil person employs to spin a web of lies to trap his adversaries and justify the harm he does them. There are many examples of this. In Deuteronomy (22:14 and 17), ʿ*alilot devarim*, "false charges" (lit., "pretexts of words") designates the wicked act of a husband who, wanting to divorce his wife because "he dislikes her," makes false accusations against her.[4]

In modern Hebrew, the word ʿ*alilah* means "plot"—both the plot of a novel or play and the devious scheme of a conspirator. In most cases, of course, the praiseworthy sense of a brilliant action is meant when the term is applied to God; that of perverse behavior when applied to human actions. In some places, though, the biblical text is ambiguous.[5]

---

4. The eleventh-century commentator Abraham Ibn Ezra (on Deuteronomy 22:14) glosses ʿ*alilot* as "causes" and *devarim* as "either true or false [sc. *words*]," i.e., good or bad arguments.

5. It would be impossible to analyze the many occurrences of the word ʿ*alilah* here, not even if we limited ourselves to those that may be ambiguous. David prays to be spared from doing evil, for example: "Incline not my heart to an evil thing, to practice deeds of wickedness [*le-hitʿolel ʿalilot be-reshaʿ*]" (Psalms 141:4). By asking God not to steer his heart towards evil, David is implicitly holding God responsible for any evil he himself might do. In fact, the expression *le-hitʿolel ʿalilot* is much stronger than simply "to do evil." It reinforces the notion of ʿ*alilot* as deceitful words or actions by making them the cognate accusative of a verb with the same stem, which already includes the connotation of a bad action—"to trick, deceive" or "to trifle with, mock." It is in this sense that the biblical text employs it with God as the subject and the Egyptians as the object ("I made a mockery [*hitʿallalti*] of the Egyptians"; Exodus 10:2).

Another example, remarkable because of the interpretive technique to which it has given rise, occurs in Hannah's prayer of thanksgiving: "YHWH is a god of knowledge and by him actions [ʿ*alilot*] are weighed" (1 Samuel 2:3). The ambiguity of the meaning of these ʿ*alilot* associated with YHWH is underlined by an exegetical tradition that employs the technique of *qere-ketib* (the text as read versus the text as written). In this verse, the written text is *loʾ* with an *aleph*, which expresses negation; thus "ʿ*alilot* [plots and pretexts] are not appropriate to YHWH, the god of knowledge." But we are invited to read instead *lo* with a *waw*, "to him," which transforms the meaning into "ʿ*alilot* [brilliant feats] are appropriate to YHWH, the god of knowledge." Here the exegetical technique is used to attach both senses to ʿ*alilot* and apply them both to God, negating his relationship with the pejorative meaning and affirming his relation with the positive one. But the very recourse

There is at least one notable exception to this. When God speaks of his treatment of Egypt and Pharaoh, he uses a verbal form of this same stem, *le-hit'olel*, whose regular connotation is inflicting harm or doing evil. In Exodus 10:1–2, God tells Moses that he has hardened the hearts of Pharaoh and his servants so that they will refuse to allow the Hebrews to depart and so that the full scenario of the ten plagues, with all the signs and prodigies accompanying them, can play out to the end. "You will recount for your children," he tells Moses, *hit'allalti be-mitzraym*, which Rashi invites us to understand as "how I made a mockery of the Egyptians." This action spawned classical commentaries that seek, *grosso modo*, to salvage the morality of free will by observing that Pharaoh hardened his heart on his own after each of the first five plagues and was deprived of his free will only starting with the sixth plague, as if, with the escalation of his wickedness, he had decided for himself that he would no longer exercise choice. This type of exegesis, aimed at saving the morality of the simple sense (the *peshat*) of the narrative, vanishes in the midrashic reading of Psalm 66:5, *nora' 'alilah 'al benei adam*, "terrible in his deeds among men." As this verse is read by Midrash *Tanhuma* and developed by Eliaschow, Pharaoh's case is not all that exceptional. Human beings are the playthings of the events they think they determine. This can be extended to all human situations—not only to Pharaoh's ostensibly free exercise of his will during the earlier plagues, but also to all biblical characters and, more generally, to all human beings and the consequences of their choices. Nevertheless, in what seems to be a manipulative if not indeed perverse fashion, they are held responsible for them and accounted guilty for their consequences.

In this reading, *nora' 'alilah*, despite referring to God's actions in nature, is indeed understood as "terrifying in plots"—terrifying in pretexts and bad faith. The verse in Psalms is read, "Come and see the works of God, terrible in his *plots against* men." To avoid any doubt about its intentions, the midrash

---

to this technique, which leaves the written text without emendation while proposing a different way of reading it, shows clearly that the ambiguity persists and in fact invites reader-interpreters to delve more deeply into the text. This is precisely what the *Tanhuma* on the expression *nora' 'alilah* in Psalms 141, taken up by Eliaschow, does explicitly, unabashedly attributing the pejorative sense to God-Elohim.

It is remarkable that here this play on words is applied to a text that itself evokes, in self-referential fashion, the ambivalence of language. As at Babel, the perverse use of language is associated with perversion of "spermatic knowledge," which is then found to have been turned into the "seed of falsehood." [See Atlan 1999b, chap. 2, n. 55.—Eds.]

continues with a parable that gives us a clear picture of this terrifying behavior of God, who attempts to cast his own responsibility off onto the victim, using bad-faith argument, pretexts, and false charges, as in Deuteronomy. The tale is (by chance) that of a faithless husband who has already decided to divorce his wife (following the ancient judicial procedure, of course) and is looking for a pretext. He has already prepared the bill of divorce and only needs a reason to give it to her. He returns home and asks his wife to pour him a cup [of wine]; when she serves him, he rejects it as tepid and invokes this lapse as an excuse for divorcing her. She can only retort sardonically that the document had been ready when he entered the house, evidence that in his bad faith he was looking for the first available pretext.

This, according to the midrash, is the meaning of the verse in Psalms. The far from edifying parable of the false-hearted husband is invoked as a model of God's normal treatment of human beings, alluded to in this verse: God is "terrible in his plots—his pretexts and bad faith—against men." This thesis, heretical and scandalous on the face of it, is supported by other passages from the midrash and Talmud that deal with situations comparable to Pharaoh's, in which God plays with and abuses human beings. The victims are none other than Adam himself, Moses, Joseph's brothers, and finally, by extension, all human beings who are punished, ostensibly for their actions, even though their succession and consequences were foreseen from all eternity.

The first and clearest example is Adam, whose transgression, according to Genesis, brought death into the world. According to this midrash, Adam contests this charge, noting that death had been foreseen from all eternity by the Torah (which existed before the creation of the world), as proven by its institution of funerary rites of purity and impurity. If so, he cannot be held responsible for it. What is more, his very fall was inevitable, built into creation; it defines and shapes human nature just as much as his mortal character. To hold Adam morally responsible for his fall and for the death of his descendents is thus an unfair charge, an *'alilat devarim*, an accusatory pretext, which Adam himself, speaking through the midrash composed by his descendents, protests vigorously. The same applies to the other examples. Moses' sin, which prevented him from entering the Land of Israel, was a mere pretext: it was foreseen from all eternity that he would not enter the land. So too the sin of Joseph's brothers in selling him into slavery: it was a link in the chain of causes by which the earlier predictions of exile and slavery in Egypt, followed by the exodus and release from bondage—events foreseen from all eternity—were realized.

This notion rests strongly on a talmudic passage that is one of the most explicit in referring the guilt for the evil committed by human beings to the author of Creation (B *Berakhot* 31b–32a); it was he, as Rashi notes, who created the *yetzer hara*, the evil impulse, or, more precisely, the creative nature and dynamic of evil. Three verses are quoted that exonerate Israel of responsibility for their sins and thus permit them to escape a judgment that would otherwise certainly condemn them. The first of these verses is Micah 4:6, where God confesses to having abused (*hareʿoti*, from the stem *raʿ*, "evil") the exiled Israelites. Next comes a celebrated verse from Jeremiah 18:6, in which God admits that he manipulates the house of Israel like "clay in the hands of the potter." (Spinoza would take up this image to illustrate his thesis of absolute determinism and the illusory nature of free will.) The third verse is Ezekiel 36:26: "I will remove the heart of stone from your body and give you a heart of flesh." Even more explicit is the next verse, because it bears not only on inclinations—the heart of stone or flesh—but on the details of behavior: "I will put my spirit into you and cause you to follow my laws and observe my rules faithfully."[6]

This is certainly an appalling conception, scandalous to those who would hold fast to the classic moral theology of a personal God who punishes the evil and rewards the good that human beings freely choose to do. This, it seems, is the perspective of most of the book of Job and of a straightforward reading of the surface meaning of many verses of the Bible. But there is another way to read these texts, suggested by the first and last chapters of that same book of Job. Their lesson is that the questions raised by Job and his friends, however moving they may be, stem from a childish morality and magical thinking that expect God/nature to respond to human hopes and

---

6. In this talmudic passage, R. Eleazar has the prophet Elijah hurl a serious indictment against the Lord: "answer me, that this people may know that you are YHWH-Elohim; for you have turned [*hasibbota*] their hearts backward" (1 Kings 18:37). The word *hasibbota* expresses simultaneously the idea of *sibbah*, "cause," and *mesovev*, "reversed" (but also "effect," i.e., "caused"). The three verses quoted in the body of our text are then invoked to demonstrate that God admits the validity of the charge that he himself is the cause of the Israelites' sins. But R. Eleazar goes on to assert that Moses had uttered a similar accusation against "what is above," assigning ultimate responsibility for the sin of the golden calf to the God of Israel, who had lavished a surfeit of silver and gold on the people. Another sage, R. Hiyya b. Abba, likens the case of the people to that of a man who spoiled his son: "He bathed him and anointed him and gave him plenty to eat and drink and hung a purse round his neck and set him down at the door of a bawdy house. How could the boy help sinning?" (B *Berakhot* 32a).

behavior as if it too were a person, a moral subject in dialogue with another moral subject. In fact, the determining causes of the experiences of a human being, in this case Job, are to be looked for in a natural history that complies with its own necessities and that, for better or worse, mocks Job's choice of virtue rather than of vice.

Finally, we must insist that this radical and absolutely determinist reading of the passage in *Avot*—"everything is foreseen but the possibility to choose is granted"—is not theological. It must not be confused, for example, with a doctrine of predestination and grace, like those, Protestant or Jansenist, that are associated with elements of Augustinianism. Those are theological doctrines of salvation, whereas Jewish determinism is a theory of the sequence of cause and effect in the nature of things and events and in the knowledge we can have of them; in other words, it is a naturalist and epistemological theory of immanent causality. We should remember that the talmudic dictum that "everything is in the hands of heaven" is qualified by "except for the fear of heaven," a fear that is itself "the beginning of wisdom." As we will see, this implies a doctrine of salvation—namely, of what is just and what is wicked—that is realized neither by grace nor by works but by knowledge of one's works and self-awareness; that is, by acquiescence (in the Stoic sense) and the perspective on oneself that this knowledge makes possible.

Translated by Lenn Schramm

SIXTEEN

Mysticism and Rationality

(1986)

Both the adjective *mystical* and the noun *mystic* have two opposing connotations.[1] One of these, pejorative, designates the approximation, imprecision, and lack of rigor that take shelter behind the existence of some hidden reality or discourse. Only faith, in this reality or discourse, allows access to the mystical, because it is incommunicable and its origin is mysterious. In this sense, the hidden and mysterious paper over the unintelligible, the arbitrary, and, carried to the extreme, just about anything. They provoke a strong suspicion of unmastered error, of falsehood, even of trickery: here mysticism goes hand in hand with mystification.

---

1. The same applies to *mythic*. See Detienne 1984, 1986. The relations between mystical and mythological are complex, with both overlaps and differences. See, e.g., Eliade 1958 and Scholem 1965. For reasons that will be apparent later (see Atlan 1986, chap. 7), we would see in myth a collective expression that corresponds to the content of mystic experience for the individual. In other words, supposing that these two types of discourse refer to the same types of experience, the mythological is to the social and collective what mystical experience is to the individual.

But there is another, positive, connotation, that which has always been attributed to those called, often without any clear reason, "authentic mystics," who are supposed to have access to uncommon experiences and for whom the hidden and mysterious, although unintelligible, indicate something real—or at least an interesting, original, and "true" psychic experience.

This positive connotation of mystic experience has been reinforced in recent years by two related phenomena. During the 1960s, the psychedelic revolution in the United States provided tens of thousands of persons, generally young and raised in the positivist and pragmatic canons of Western civilization, with direct and rapid access, by means of hallucinogenic substances, to experiences that (as was soon noticed) reproduced at least in part the content of those described by mystics of all religious traditions.

Although these observations had their precursors, noted by writers and poets like Aldous Huxley and Henri Michaux, or by marginal experimentalists such as R. G. Wasson and Roger Heim, they had not strongly penetrated the daily life of our civilization. For at least two years in the 1960s, though, the systematic use of nonaddictive and non-habit-forming hallucinogens (LSD, mescaline, psilocybin, and so on) diffused through the most diverse social milieus, leading individuals who were in no way prepared for the discovery of "other" realities. The mass and reproducible character of these experiences gave them an "objective" reality, whereas formerly, in the best of cases, if their reality was not denied pure and simple, they were relegated to the practically impenetrable subjectivity of illuminati, poets, and artists.

This experience had an enormous cultural effect, including the critique and relativization of the philosophical and scientific tradition that Western thinkers had previously considered to be the sole reference, the unique standard against which the traditions of other civilizations were to be evaluated, if they were not simply ignored. In particular, the Far Eastern traditions of India, China, and Japan—soon followed by the West's own mystical traditions, Christian, Jewish, and Muslim, rediscovered and reevaluated—penetrated mass culture in the United States and then in Europe. The traditional foundations of the critical method (the law of noncontradiction, subject-object dualism, the postulate of objectivity, the reductionist materialism of the experimental method) were called into question and juxtaposed to the illumination of the mystics' cosmic consciousness, which almost anyone could henceforth discover thanks to hallucinogens and techniques of meditation.

Curiously enough, however, this new attitude toward existence, this new source of values, unifications, and exclusions (certainly new to the West) soon came to need to justify itself and itself became the object of rationalization: it is not easy to escape one's own sociocultural conditioning.[2]

While India, China, and Japan avidly absorbed the cultural products of Europe and America that accompanied industrialization and technological development, Western societies absorbed foreign traditions, but without ceasing to be American or European themselves. Here the capacity to metabolize everything, which characterizes the culture of these societies, is expressed by ambiguous rationalizations, rationalizations of the irrational that penetrate not only the masses and the media but also the more or less aristocratic circles of "enlightened" thinkers. Among the latter there seems to have developed a syncretism combining the rational elucidations of the "conscientious" scientist and the illumination of the mystic: F. Capra's *The Tao of Physics*, R. Ruyer's *Princeton Gnosis*, and, more recently, the proceedings of the Cordoba colloquium are among its most familiar manifestations. In defense of these confusions—perhaps explaining if not justifying them—may be cited the fact that the traditional teachings of the East, unlike Christianity, did not inherit a tradition of opposing and resisting empirical science and critical methods. What is more, access to these teachings is not embarrassed by a prior faith in a personal God.[3] Instead, these discourses about reality, the infinite, and nonduality, designated by exotic names (Brahma, the Tao, etc.) serve more as invitations to inquiry than as crystallizations of dogma. But this functions only at the infantile level of common catechisms. We need merely open the books and study them in their own language to recognize two facts.

1. All mystical traditions that attempt to rationalize their discourse come up against the same problems (even if they resolve them differently): infinite-finite, impersonal-personal, divine-human. Works such as those by H. Corbin (Jambet 1983), Gershom Scholem (Scholem 1954 and 1965), and contemporary Christian mystics who encountered Hindu spirituality attest to

---

2. The Catholic monk Henri Le Saux, who became an Indian swami long before these developments, was a pioneering witness to the difficulties and rents provoked by this sort of encounter for those who live it profoundly; even if, of course, it is an inexhaustible source of riches. See Davy 1981.

3. The influence of poet-philosophers like Alan W. Watts (1951, 1966) on the Beat Generation and the (counter) culture that came in its wake is undeniable. Too much poets not to be mystics (even independent of psychedelic experiences), but too marginal and

the common foundations and individuality of these traditions (Anonymous 1982). The differences in the solutions they propound are initially no more than different accents on one or another pole of a shared dialectic. But these different stresses lead, in the realm of practical behavior and legislation, to the enormous contrasts and distances that separate the sociocultural reality of the civilizations that are nourished by them (unless we consider their doctrines to be rationalizations of these sociocultural products as they relate to the discovery of this common ground of mystic experience).

2. The Reality involved here (Brahma, the *Einsof* of the Kabbalah, the theologians' *Deus absconditus*) has nothing in common, save for the word itself, with the reality uncovered by scientific research, even though, in both cases, the object is to expose a hidden reality. For mystics, Reality is the invocation of an infinite transcendence, one, absolute, compared to which the reality uncovered by scientific research is multiplicity and relativity, division, fall, and illusion (*maya*) for those who believe implicitly in the truth of its appearances. The unifying endeavors of physics, aimed at finding a unified theory and single formula, must not deceive us: far from designating an Absolute whose immediate and ineffable attainment is the province of the illumination of the saint (or prophet, or redeemed), they are the result of an empirical-mathematical confrontation in which nothing is supposed to escape quantitative discourse, determined by the conditions of measurement and measurability.

This does not rule out studying these traditions. They do not stop at the ecstatic experience of the ineffable but call on the intellect and on discursive reason, and have probably done so ever since their origins. They did not wait for the development of Western science and philosophy to make use of the human rational faculties, which were nourished by them and which carried them forward and renewed them over the centuries. This is why dialogue is possible between the Eastern traditions and the scientific and philosophical tradition—if it is carried out on the basis of the differences between them rather than their similarities, sustained by the light of reason rather than by the dazzle of illumination.

It is because facile comparisons of the syncretism of modern science and Oriental traditions are prejudicial both to Western rationalism and to Eastern

---

controversial for the institutionalized forms of mysticism and religion in the West (i .e., so-called Judeo-Christianity, which is much more Christian than "Judeo"), they quite naturally found in Far Eastern traditions what their native theistic culture could not give them.

traditions that we ought to unpick them. We should make these dialogues possible because they can be fruitful and profitable, but we must also respect the rules of the game that each of the parties plays on its own field, so that the metafield on which they meet is not merely a fusion where both disappear. This enterprise is clearly risky and unavoidably tentative in that it consists in laying down rules (metarules) for the game (metagame) in which games with different rules are set to playing each other. But the enterprise is also necessary, because, on the one hand, these games with their different rules (those of scientific research and those of mystical illumination) can no longer ignore each other, while, on the other hand, their shameless interpenetration can only overturn the chessboard and shuffle the cards, making it impossible to continue any game or even to begin one.

This is why it is important to locate, even if only approximately, the type of coherence admitted by mystical traditions and that by which their rationality, when manifested, is different from scientific rationality. The question does not reduce to that of the rational and the irrational, which is encountered, in varying degrees, everywhere, but rather to that of the relationship with reality, of concrete-abstract relations, of the rapport between reason and reality, between representations (theoretical, formal or analogical, artistic, revealed) and those things they represent and from which they flow: the belief in the reality of these representations and the question of the boundaries of delirium, illumination, and theoretical explanation.

Order and Chaos in Symbolic Rationality

The relationship between mystical traditions and reason is not always one of militant antagonism or involuntary perversion. Alongside these voluntarist currents, all of the great mystical traditions have intellectualist currents that value reason. Each in its own fashion has found other means to use the language of words and reason without being imprisoned by it. To perceive the effective rationality of these currents and how it differs from scientific rationality, a detour through our understanding of myth and poetry as languages may prove useful. We have learned to decipher, in what we in the West call myths and receive in poetic or symbolic form, a certain use of reason that is not necessarily conscious but is not necessarily unconscious, either.

Poetry has accustomed us to a usage of words in which what is actually said goes well beyond what is ostensibly said without its being, for all that, nonsense, unreason, or antireason. We know that the game has rules different from those of prose discourse. The studies on this point by the poet Claude Vigée (1982) are particularly illuminating. Whereas the language of prose aims at unicity of *meaning* and precision of content, culminating in philosophy, formal languages, and mathematics, poetic language seeks a purity of *form* that culminates in the unicity of the work, such that any modification of form would be a flaw or even destruction.

Nevertheless, these two types of expression remain closely related, since both involve the use of language. The relation between them is one of negative reciprocity: what is unicity for one is multiplicity for the other. The ideal of one, which at a certain level—whether of meaning or of form—is the anti-ideal of the other, reappears when the level changes. The unique form of the work of art speaks only thanks to a multiplicity (tending at its limit to infinity) of meanings and levels of meaning. What is more, this infinity of possible meanings can overtake pure form without meaning.

Conversely, the univocal meaning of formal language no longer requires any restriction of form—for all that it is called formal. It is called formal because it reduces meaning to pure syntax devoid of semantics: the meaning of its signs is reduced to that of the relations among them, in which the role of their significata is merely conventional and interchangeable. At this price, it obtains the univocality of meaning and absence of ambiguity that necessarily entail acceptance by every rational mind. This is what permits an algorithmic (i.e., mechanical and automatic) translation that is immediate and free of loss: unlike poetic language, the exchange of signs required by such a translation is a trivial matter. The form of the meaning, which constitutes the entire sense of a poem, going beyond its words taken individually and due to their polysemy, has no importance in formal languages, because the words have only a single sense, which is identical in all propositions and is independent of the form of these propositions. This is why, paradoxically, what is expressed in a proposition of formal language is independent of the particular form of that proposition: that form, by construction, is only a particular case of the unique general form that defines the syntax of the language—unlike the particular form of a poem, which is unique and cannot be reduced to that of a universal syntax.

These reciprocal inversions of unicity and multiplicity of form and meaning in poetry and prose may be at the origin of what Robert Pirsig calls the "search for Quality," in a novel of rare profundity (Pirsig 1974), in which

Zen, the motorcycle, and rhetoric are the occasion for an astonishing excursion to the frontiers of logical reasoning. He proposes an inversion of the roles of quality and reality, which makes him appear to himself and others as mad or mystical, both at once and alternately, continuing the line of ancient intuitions renewed today on various sides (Laing 1967, Roustang 1976).[4] One may also think about the Intense, intensity as such, invoked by Yves Bonnefoy (1989) when he wonders about an element that has not been taken into account "in the calculations today which attempt to situate exactly the significance of poetry" or when, for him, the (young) reader "understands everything in the polysemies through comprehensive intuition, through the sympathy that one unconscious can have for another, . . . in the great burst of flame which delivers the mind—as formerly the negative theologies rid themselves of symbols." After the roles of the unconscious in poetic creation and of the autonomy of signifiers in a text had been discovered, here too the "precision of study" has come to restrain and oppose the impatience of intuition by uncovering its illusion. For Bonnefoy, though, this is a question of a combat between "loyal adversaries."[5]

---

4. "Since the destruction of the Temple, prophesy has been taken from the prophets and given to fools and children," according to R. Yohanan in Babylonian Talmud, *Baba Batra* 12b.

5. It is in the same spirit of loyal combat that we oppose scientific knowledge and mystical rationality here. The Image lies, as criticism and disillusionment demonstrate. But it is also what the "young reader" finds in his intuition of the Intense. The loyalty of the combat for the poet, his concern to face it, to confront this aporia, leads him or her to a dialectic of dream and existence, illusion and reality, the third term of compassion at the height of passion and desire, which does not fail to recall the playing field, the intermediary between reality and illusion, for D. W. Winnicott and Eugen Fink (see Atlan 1986, chap. 7). This makes the poet accept two laws of literary creation *together*. One, the closure of the written word, demystifies the naïveté of the subject who would speak: "Hemmed in by the words he does not understand, by experiences whose very existence he does not suspect, the writer, and this is the element of chance which so distressed Mallarmé, can only repeat in writing that strictly limited particularity which characterizes any given existence" (Bonnefoy 1989: 163f). But the second "law" of literary creation bars him from staying there, because "the world which cuts itself off from the world seems to the person who creates it not only more satisfying than the first but also more real" (ibid. 164). The lie of the Image is only this impression of reality: a lying impression, in fact, if it designates some particular pseudo-reality erected into Reality, a particular Image absolutized and become Image. In the background, though, its truth, the "second level of the idea of poetry," is truth struggling against the abolitions, the closures—the "presence which opens"—of "this first network of naiveties, of illusions in which the will toward presence had become

Thus poets tell us (even in prose) about the type of positive relations that can be maintained between mysticism and discursive reason. Everything happens as if it were a question of dividing the world into two kingdoms that are to be found in somewhat different fashion, in the symbols of myth. As Dan Sperber clearly demonstrates (1975), symbolic language can be approximated by the use of quotation marks. These mean "your attention please, we are speaking from another point of view, at another level of perception and description of reality, where meanings are no longer the same." Above all, these other meanings suggested by quotation marks are not necessarily found in another lexicon already lying to hand. The most interesting cases are those in which they are merely hinted at by the use of quotation marks and the interweaving of levels they imply and are thereby created at the moment they are uttered. Deciphering symbols reveals their rationality in the form of these structures of relations, which have been illuminated by the studies of Lévi-Strauss and his disciples, in which the ideal of formal language reappears.

It is always a matter of finding an order or of making it appear (creating it) by a change of level in a discourse that is otherwise, and from another perspective, chaos, confusion, and contradiction. Yet something in the form of this chaos indicates that another order could be projected onto it. This is what I have defined elsewhere as nature's property of complexity (Atlan 1979) and richness: an apparent disorder in which we have reason to believe we can find a hidden order. The partition into two kingdoms, of order and disorder, overlaps that between rational discourse about reality and the experience of a reality with no reason.

Every scientific system rests on the postulate or faith in the ability of reason to disclose an order beneath the complexity and apparent chaos of our experience of the world. From this point of view, symbolic and interpretive thought goes as far as possible down the path of faith in the possibilities of reason: nothing is accepted as being devoid of meaning. Every phenomenon, even the most fortuitous and confused, every myth, even the strangest and most enigmatic, finds an explanation that renders it rational by means of a

---

ensnared" (ibid. 171). On this level, poetry "has denounced the Image, but in order to love, with all its heart, images. Enemy of idolatry, poetry is just as much so of iconoclasm" (ibid. 172). "These images which, if made absolutes, would have been its life, are nothing more, once one overcomes them, than the forms, the simply natural forms, of desire, desire which is so fundamental, so insatiable that it constitutes in all of us our very humanity" (ibid. 171).

change of level—the symbolic explanation—in which reason appears not only in what is said overtly but also symbolically, "between quotation marks."

Everything proceeds as if the fortuitous, the absurd, and the random were excluded from reality. A series of events or an a priori disparate aggregate of randomly associated facts is united by causal relations in one fashion or another. It seems as if our reason cannot bear the absence of order and reason in things. As Sperber puts it: "Symbolic thought is capable, precisely, of transforming noise into information" (Sperber 1975: 79). It is remarkable that poetic creation (and artistic creation in general) also ends up transforming noise into information, albeit at a different level, as we saw earlier. Vigée has provided a gripping description of the different stages of the poetic experience:

> It is in the violent and lucid flame of the initial emotion that the disparate elements among which the poet "discovers unheard relations" are welded into accurate images. . . . No apparent order as yet allows itself to be distinguished in this indescribable welter of sounds, images, undatable memories, which issue at the same time from the most secret regions of the soul and from its least intimate domain: that of learned signs. This experience rather resembles a throw of the dice: by means of everything that the past has accumulated in us, we would vanquish chance by means of chance, in order thereby to attain a superior reality of the spirit over which it no longer has any dominion. . . . A nascent coherence is suddenly established in the sounds and images of primitive chaos. "Useful" verbal and imaginative elements (i.e., those that can fulfill their representative function for affective consciousness) organize themselves according to the central rhythmic motion and forces that emanate from it like filings around a magnet. Thereafter unused materials become obstacles to creation. They are shunted off to the psychic frontiers and soon lapse into oblivion. . . . A perceptible structure is substituted for the void and chaos that existed previously. (Vigée 1960: 149)

Like Roger Caillois, for whom "it is a question of organizing poetry," for Valéry it is a matter not "of passing too simply from disorder to order, but of controlling self-variance . . .: awareness does not move toward unity, but on the contrary toward an organized multiplicity. . . . Distinguishing a 'psychesthetic chaos' from a 'form' that thwarts it while using it." And while his poetics searches for a benchmark in thermodynamic irreversibility, it derives from the "mystical element" of Wittgensteinian language, at once limitation and reunion (Oster 1981: 106 and 160).

I would say, rather, "*although* it searches for a benchmark in thermodynamic irreversibility." In fact, transforming disorder into order, thwarting chance while using it, making "order from noise" [see Chapter 3, above], and bringing about "order from fluctuations" (Prigogine and Stengers 1984) are recognized today as principles of natural organization based on the thermodynamics of open systems—of which living beings are particular cases. But is that really the same thing? Is the experience of the creation of information from noise, of order from disorder, in symbolic thought and artistic creation the same as that produced by scientific observation of nature and of the living world?

It certainly is not if one takes these expressions as literal descriptions of what is, of "reality." In one case, we are dealing with experiences of the operation of language and thought; in the other, with experiences of observations of nature interpreted in physical and mathematical theories (information theory, thermodynamics, and system dynamics). Yet these expressions probably do designate experiences that are quite similar, if we realize that both are descriptions of descriptions: that is, descriptions that take into account the position of the observer and describer, an individual endowed with reason who confronts the complexity of the brute data of his or her experience of nature.

Nevertheless, a more profound difference remains, again linked to the status of contradiction and paradox. These are to be eliminated from scientific discourse, whereas they are not only permitted but even endowed with particular "monstrative" virtues, as we have already seen, in mystical discourse. Even though symbolic thought and poets do not resort to active antireason and a quasi-systematic search for contradiction, the experience of contradiction and paradox, *despite* the effort to introduce order (or rules) and rationalize, acquires a positive content, at least provisionally, because it leads back to an irreducible originality of these attempts to introduce order: that of symbolic thought and the change of logical level indicated by the "quotation marks," or that of the poet and the poet's subjectivity.

For the scientific method, by contrast, contradiction and paradox are intolerable scandals, which threaten to undermine the entire structure. We are familiar with the role of the logical paradoxes that haunted the quest for the foundations of mathematics during the nineteenth and early twentieth centuries. They were not exploited to return to the ineffable or even to the irreducible, but rather to run them to earth, so as to make them surrender, if possible, and to eliminate them or, in total despair of success, to evade them, as Russell tried to do. Similarly, the discovery of organizational randomness by the nascent sciences of complexity would be profoundly denatured if it were to be understood as a manifestation and demonstration of an irreducible paradox that leads back to some "higher" or "deeper" elsewhere. Quite the contrary, it is a question of removing, by means of an appropriate formalism, the contradictions that appear when the notions of order, complexity, and organization are transposed unchanged from their usage in everyday language to scientific discourse concerning observations of nature. Natural language tolerates and even uses uncertain—because multiple and interlocking—significations, which in some cases rule out the use of univocal and adequate definitions. But what is no defect for everyday language is assuredly one for scientific research (even if, along with Wittgenstein, we hold that imitation of the natural sciences, including, inter alia, their "craving for generality," is the source of great confusion in *philosophy*).[6]

Scientific theories of complexity and organization cannot rest content with observing contradictions and underlining paradox. On the contrary, they must resolve contradictions and eliminate paradoxes. If, as in the example of order from noise, the apparent paradox is eliminated by taking the role of the observer into account, this must not itself induce error by suggesting that it involves a return to subjectivity. Whenever the role and status of the observer are taken into account in the natural sciences (an approach that began, at least explicitly, with quantum mechanics), we are dealing with the subjectivity not of an individual but of a theoretical being (the ideal physical observer) that is merely a shorthand reference for the totality of measurement and observational operations possible under the given conditions of the practice of a scientific discipline, and we are taking into account the corpus of knowledge characterizing that discipline at a given moment. The shift from the role of this ideal physical observer to one of individual subjectivity and consciousness is one of the main sources of misunderstanding and confusion in the spiritualist deviations of quantum mechanics and, of course, also in those of the new theories of order and complexity.

As we have seen, it is the individual, in his subjectivity and in the experience of his inner illumination (the "Self"), who is the point of departure for mystical and poetic experience—even if this interiority is subsequently extended to the All, in which no inside or outside is recognized, or the Self is expanded to the totality of being, or, going even further, one passes

---

6. "To think it is [a defect] would be like saying that the light of my reading lamp is no real light at all because it has no sharp boundary" (Wittgenstein 1965: 27); for a "craving for generality," see ibid. 17–18.

through the "I" and "Thou" (Buber, Rosenzweig) and the "face of the other" (Levinas) to a beyond of Being, an infinity that opens and shatters everything.[7]

Reason as a Complement of Illumination

Of course, mystical traditions do not stop at the experience of illumination, but take it merely as a point of departure. The possibility—indeed, the necessity—of not dwelling in the experience of illumination, of progressing and of speaking,[8] leads these traditions, at least in certain cases, to use reason as a tool of progression and discourse. Reason no longer serves as a perverse foil, as in koans, but is instead a valuable aid to be used correctly, with scrupulous observance of its rules.

Thus most great traditions admit a certain cohabitation or complementarity between mystical experience and rational discourse. A difference in accent may be observable in the traditions of India and those of the Mediterranean world: the role of rational discourse in the former, although undeniable (e.g., in some commentaries on the *Upanishads* and the *Bhagavad Gita*, or in a thinker like Sri Aurobindo[9]), is smaller than that of the experience and practice of the ineffable and of illumination. The contrary is true in traditions such as Judaism, Christianity, and Islam. The exoteric character of the mystical tradition of exercises and illumination in the Far East, where unity is lived outwardly in the mode of polytheism, while reflection and unifying consciousness are reserved for the esoterism of brahmans or monks, may correspond to this different accent—unlike Judaism, for example, where a

---

7. The question posed here is evidently that of the relations between philosophy and mysticism, whereas elsewhere those between science and philosophy are at stake. Perhaps Wittgenstein, again, can guide us by locating philosophy, as specializing in the meaning of words and aiming to resolve the contradictions that it itself creates by requiring natural language to have a rigor and generality it is not designed to have (Wittgenstein 1965: 17–29), in a quite specific hollow between science and mysticism. Like every hollow, it separates at least as much as it unites. This seems to be what one (lost) philosopher tried (but without great success) to state at Cordoba, lost in the All of spiritualist physicists, mystic predicators, and Jungian psychoanalysts (C. Jambet, cited in Cazenave ed. 1984).

8. The entire exoteric Jewish tradition (the Talmud) is grounded on the search for and constantly refined definition of the correct path (*halakhah*) to be followed in daily life, far from illumination, but not without some traces thereof.

9. Aurobindo Ghose 1955, 1977; See also Siddheswarananda 1977.

monotheistic and legalistic exoterism corresponds to a mystical and multiform esoterism, bordering on an overt polytheism reserved for masters and initiates.

Whatever the case may be, in some of the great texts of these traditions and the commentaries on them, one can find a certain rationality of mysticism, close to though different from symbolic thought, which it might be useful to analyze here. Even though we would then find ourselves confronting a certain rationality, in that the principles of identity and noncontradiction apply, the reality to which that rationality refers and that to which it applies constitute a domain peculiar to itself. Hence it is relatively easy to show that scientific rationality is radically different from it.

In other words, the existence of mystical tendencies in science, avowed or not, must be supplemented by the existence of a rationality of mysticism, it too conscious or not, which can amply explain the confusions that we regularly witness—explain but not justify them. We shall see that there remain irreducible differences that must be kept in mind in the interest of the fruitful pursuit of both processes.

Plato recognized a necessary complementarity between reason and divination; the latter was needed to supplement the former as a means of gaining access to truth, whereas reason's task was to test the content of this truth. A passage in the *Timaeus* is particularly suggestive here:

> Herein is a proof that God has given the art of divination not to the wisdom, but to the foolishness of man. No man, when in his wits, attains prophetic truth and inspiration, but when he receives the inspired word, either his intelligence is enthralled in sleep or he is demented by some distemper or possession. And he who would understand what he remembers to have been said, whether in a dream or when he was awake, by the prophetic and inspired nature, or would determine by reason the meaning of the apparitions which he has seen, and what indications they afford to this man of that, of past, present, or future good and evil, must first recover his wits. But, while he continues demented, he cannot judge of the visions which he sees or the words which he utters; the ancient saying is very true—that "only a man who has his wits can act or judge about himself and his own affairs." And for this reason it is customary to appoint interpreters to be judges of the true inspiration. Some persons call them prophets, being blind to the fact that they are only the expositors of dark sayings and visions, and are not to be called prophets at all, but only interpreters of prophecy. (Plato 2008, 71e–72b)

This complementarity resurfaces in diverse forms in the monotheistic religions—Judaism, Christianity, and Islam—where the exercise of reason is one source of knowledge, alongside revelation and experience (personal or that transmitted by trustworthy witness).

Within these traditions, however, it is interesting to study the separation between two forms of using reason: one properly theological, influenced by medieval philosophy and scholasticism; the other, probably the heir of gnostic traditions, considered to be more mystical and thus always somewhat suspected of irrationality. In fact, we can recognize, perhaps with greater ease today, that the more rationalist of these two streams is not the one you would think. In the theological current, reason complements, in the sense of merely adding to, dogmas and acts of faith posited or received a priori. The so-called mystical currents of these traditions, however, use reason as a tool to sort, verify, and express discursively, starting from the raw data of revelation. These two aspects are already present, but united, in the Platonic text just quoted. They seem to have been gradually separated until they wound up in opposition, split between the so-called rational philosophers of medieval theology and the mystical thinkers (kabbalists, sufis) of the same traditions.

Y. Jaigu opened the Cordoba colloquium by recalling that that city was the venue where this separation crystallized in Islam, between the rationalist Aristotelian philosopher Averroes and the mystic Ibn 'Arabī, as discussed by H. Corbin in "Creative Imagination in the Sufism of Ibn 'Arabī."[10] Given the mutual influence of Jewish and Muslim thought in that era, it is hardly astonishing that, on the Jewish side, the same separation between ostensibly rationalist and mystical philosophers coalesced around Maimonides, the great Jewish thinker born in Cordoba. This separation between "philosophers" asserting the rational tradition and "kabbalists" asserting the esoteric and mystical tradition became prominent after his time.

Before that separation, however, one could observe in the works of one and the same author, as in those of Plato, the two forms in which mystics use rational discourse: one explicitly mystical and suspected of irrationalism by the other camp, the other theological and scholastic, claiming exclusive title to rationality. Two important, albeit very different, examples of this are the poet-philosophers Judah Halevi and Solomon Ibn Gabirol. For the former, mystical content supported by poetic form is found in a philosophical work

---

10. Cited in Cazenave 1984: 3f. [In chap. 1 of *Enlightenment to Enlightenment*, Atlan discusses a 1979 conference in Cordoba, Spain, that centered on discussions of concepts of reality and consciousness in quantum physics and Eastern traditions (and their ostensible similarities).—Eds.]

(the *Kuzari*); for the latter, by contrast, the two are totally separated, by language (Hebrew for the mystical poems, Arabic for philosophical works), by style, and by readership.[11] The same cleavage appears subsequently in Maimonides' oeuvre, between the rationalist philosopher writing in Arabic and rejected by a large portion of the traditional Jewish public, on the one hand, and (rather than a Hebrew poet integrated into the mystical current of Judaism) the legal authority who occupies a central place in the post-talmudic juridical tradition. Contrary to appearances, there is no great distance between the foundations of this juridical tradition (which are certainly not Greek philosophy) and the mystical currents of Judaism, as later manifested by such rabbis as Joseph Caro, Schneour Zalman of Lyady, the Gaon of Vilna, or Joseph Hayyim of Baghdad, all of whom were simultaneously masters of Kabbalah and of the legal tradition. This is why, up to and including Maimonides, there was no clear opposition in Jewish tradition between "mystical" and "rational" currents; the line of demarcation between them more or less bisected the works of the authors. Only afterward do we find Jewish theologians, claiming descent from Maimonides the philosopher, author of the *Guide for the Perplexed*, clashing with the kabbalists (who rejected that Maimonides, even while fully accepting his juridical corpus). It is the content of this clash and what was at stake in it that interest us,[12] by exemplifying two opposing conceptions of the rational and the nonrational. The same opposing conceptions are found today, when the question involves not the use, properly speaking, of reason but a particular idea of the relations between reason and reality.

Translated by Lenn Schramm

---

11. See Gabirol 1970; the introduction by Jacques Schlanger clearly depicts the links between the mystic poet Ibn Gabirol, whose Hebrew *Keler Malkhut* has entered the canon of traditional texts, and the "Arab" philosopher Avicebron (none other than the same Ibn Gabirol), whose *Fons vitae* was quickly translated into Latin and practically ignored by Jewish readers.

12. Traces of this can be found in the general writings of kabbalists, such as *Vikuha Rabba*, an exchange of letters between a hasid and a mitnagged, or Moshe Hayyim Luzauo's *Hoqer Umequbal*, which pits a philosopher and a kabbalist discussing the accusations of incoherence and irrationalism made by the former against the latter, who manages to refute them and turn them back against their maker.

SEVENTEEN

Souls and Body in Genesis

(1982)

In the Western religious traditions, the soul and the body are viewed as two separate entities, sometimes conflicting, coming from two different origins: the earth and the natural for the body, God and the supernatural for the soul. These traditions have undoubtedly influenced the problem known in neuropsychological sciences as the mind-body problem, more precisely, the proposed solutions or pseudo-solutions to this problem, known as dualistic (leading to parallelism and spiritualism), on the one hand, and monist materialistic, on the other.

More recently, under the influence of modern cognitive science, computer science, and work in "artificial intelligence," a satisfactory way out of the alternative between monism and dualism has been proposed. It is known as functionalism (Fodor 1975, 1981a) and is also based on materialism, although it takes into account a kind of autonomy of mental life, its main idea, very schematically, being the distinction between structure and function.

The main contribution of computer science to this field has been the recognition of different levels in the description of organized systems, each level corresponding to different possibilities of function. The best example is that of a high-level computer language, in which very sophisticated functions of a computer program are written and implemented, while this implementation implies the functions of millions of electronic components, each of which is described in a completely different language. Between the two, there are several different levels of language, from so-called machine language to higher-level language. Of course, translation from the one into the other is always possible, but progress in computer programming relies on the fact that these translation languages can function as autonomous levels. Thus the functions to be implemented on a high level can be analyzed and described *as if* they were *independent* of the material support that actually is necessary (but not sufficient) to their implementation; in other words, the state of the electronic component during the implementation of a program is generally ignored and even considered irrelevant to the specific problems of programming in a high-level language. This is justified not only from a practical point of view but also because several different structures can perform the same function. The different levels of organization in the computer, although necessary and determined by one another, function within completely independent logical frameworks, at least apparently.

The existence of different levels of organization in natural integrated systems, such as living beings (i.e., molecule, cell, organ, organism), has been recognized long ago, together with the difficulties in going from one level to the other in a plain way in order to understand the structure and function of a higher level from that of a lower one, made of more elementary components. Here the basic difficulty comes from the fact that the techniques of observation and experimentation, together with the theories developed to account for these observations, are different at every level. In fact, one can say that the very existence of the different levels of organization in a living organism is created by differences in the techniques of experimentation that define different fields: schematically, the molecular level is observed by means of biochemical techniques, the cellular level by means of microscopy and cell physiology, the organ and the organism by means of biochemistry and general physiology, and so on. It is impossible to observe at the same time with the same accuracy more than one of these levels of organization, and in fact a cell, as we represent it to ourselves from manuals, has never been observed; it is a mental superposition of various representations of its chemistry, its spatial structure with its changes in time, and its biological functions, coming respectively from different techniques and fields such as biochemistry, electron and optic microscopy, cell physiology, and general physiology.

In computer simulations and in works in artificial intelligence that aim to simulate intelligent behavior and natural language, the same question of going from one level to the other has been encountered. There it is dealt with by a different approach: the existence of different levels in programming and function has been artificially produced as a kind of logically necessary process to undertake more and more complex tasks with more and more complex organization. Therefore there is no conceptual difficulty in moving from one level to another since they are, together with their articulations, man-made. However, the kind of practical difficulty one encounters in the analysis of natural organizations exists if one wants to describe in a causal way how the functioning of an elementary level determines that of a higher one. In this case, the solution of these difficulties is set as a goal for future work. As Daniel C. Dennett puts it:

> One learns of vast programs made up of literally billions of basic computer events and somehow so organized as to produce a simulacrum of human intelligence, and it is altogether natural to suppose that since the brain is known to be composed of billions of tiny functioning parts, and since there is a *gap of ignorance* between our understanding of intelligent human behavior and our understanding of those tiny parts, the ultimate, millennial goal of Artificial Intelligence must be to provide a hierarchical breakdown of parts in the computer that will mirror or be isomorphic to some-hard-to-discover hierarchical breakdown of brain-event parts. (Dennett 1978: 113–14)

In this essay, I aim to show that, when one is sensitive to this problem of multilevel organization, encountered in both biology and computer science, one is led to a better understanding of the verses concerning the soul-body relationship in Genesis, together with their traditional commentaries. In the light of this new approach, which seems to have been an old one, too, the question appears to be less theological and more naturalistic. From this point of view, the approach seems closer to the ancient concept of souls (vegetative, animal, and intellectual) than to the more recent idea of the soul in Western religion, with its connotation of being as mysterious as the God from whom it is said to originate.

First of all, the fact that the soul in Genesis is designated by several different names, some of them common to animals and humans, becomes meaningful, together with the traditional commentaries dealing with the exegetic problems of the verses that tell how these various souls originated.

Specifically, the main description of the creation of human souls is to be found in Genesis 2:7: "He blew into his nostrils a soul [*Neshamah*] of life and man became a living soul [*Nefesh Ḥayah*]." In this verse, at a first glance two different names are used for soul: *Nefesh* and *Neshamah*, in such a way that *Neshamah* appears to be the means by which man, starting from dust, becomes *Nefesh*. Therefore *Nefesh* seems to be at a higher level, since it is the goal that the insufflation of *Neshamah* is achieving. Or at least *Neshamah* is a necessary condition for making a human living *Nefesh*. However, *Neshamah* is specific to man, whereas *Nefesh* and even *Nefesh Ḥayah* have been described earlier in the biblical text in animals, being produced by the waters of the seas and by the earth in order to bring into existence fishes, birds, and terrestrial animals (Genesis 1:20–24). Therefore, *Neshamah* seems to be at a higher level, but *Nefesh Ḥayah* indicates a continuity between animal and human life. This problem is dealt with in a short commentary by Rashi on Genesis 2:7: "Animals also were called *Nefesh Ḥayah*, but that of man is more alive [*ḥayah*] than any other because knowledge and speech were added to it." There are also short and enigmatic commentaries on the verses in chapter 1, where the *Nefesh Ḥayah* of animals is mentioned. Rashi explains it on one occasion (Genesis 1:20) as a *Nefesh* with vitality (*ḥayuth*), which *will* dwell in it, and on two other occasions (Genesis 1:21 and 24) as a *Nefesh* with vitality (*now*). This indicates clearly that *Nefesh* cannot be understood (at least by Rashi) as a source of vitality, since at first it exists without it,[1] when it is only the product of the waters (Genesis 1:20).

Writing on the same verse (Genesis 2:7), Nahmanides summarizes the two kinds of positions taken by previous thinkers, according to whom the Bible talks of three different souls in man or one single soul endowed with three

---

1. Only at a later stage, by a special act of creation ("He created," Genesis 1:21) that vitality is infused into the *Nefesh* of fishes and birds. And then the earth brought forth the animals, which were already created on the first day but had not yet come out (Rashi on Genesis 1:21 and 24). In both cases, as in that of man, *beri'ah* ("creation") is necessary, as opposed to production pure and simple or excretion by the waters. This can be understood easily if the expression *Nefesh Ḥayah* is read not as a repetition but as a paradoxical being, which I propose to translate "living rest," where *Nefesh* ("rest") has a connotation of stability and permanency, and "living" has a connotation of ceaseless change by destruction and renewal. As such, the paradoxical coexistence of the two opposite features needs the special act of creation (*beri'ah*), which (by definition?) is required to put together in the same world light and darkness, good and evil, as appears in the verse "Forms light and creates darkness, makes peace and creates evil" (Isaiah 45:7). See Atlan 1979, chap. 13.

different capabilities, responsible for growth, movement, and thought. In both positions it is not difficult to recognize the old tradition of vegetative, animal, and intellectual souls, and it is clear that what is at stake is the living character of the soul, more precisely, the function of the soul in making a body alive, life being described functionally by growth in vegetables, movement in animals, and thought in man. However, this does not solve the exegetic problem concerning the relative position of *Nefesh* and *Neshamah* and the relationship between them and vitality. The key to solving this problem, as Nahmanides already mentions, is first to be found in the Aramaic translations of Onkelos and of Yonathan, and then in the kabbalistic literature on the subject.

As Rashi notices, the expression *Nefesh Ḥayah* is peculiar, since it is used for all animals before being used for man, as produced by the waters and the earth as well as by the Creator. As we shall see, it combines two properties or two entities, *Nefesh* and *Ḥayah*, which can be found far apart, at the two extremities of the structure of the overall human soul in kabbalistic and Hassidic literature. An accurate translation of *Nefesh Ḥayah* would probably be "living rest,"[2] where *Nefesh* is translated by "rest" (as in *Nofesh* and *Vaynafash*). This expression is a kind of paradox, which defines living beings by two *opposite* properties: *Ḥay Vekayam*, as was noted by Rabbi A. I. Hacohen Kook (Kook 1969: 2), for whom *Kayam* meant permanency, as in "rest," and *Ḥay*, movement, and change, unrest. Now in the Aramaic translations of the Bible, while this expression is transposed with no change of meaning as *Nafsha Ḥaya* when it is applied to animals, it is translated as *Ruaḥ Memalelah*, that is, "speaking wind" (or breath) when it is applied to man.

This is a real change from the literal meaning, which needs some explanation. It seems to me that this change can be understood as follows. The new element specific to man is added by *Neshamah*, and the functioning result is what is characteristic in man: namely, as Rashi says, knowledge and speech. Now the word *Neshamah* by itself, as well as the context "He blew in his nostrils," means "breath," as in *Neshimah*, and also "blowing." Thus one can understand how the consequence of blowing a *Neshamah* would be to make him a speaking wind. However, the verse itself does not say "speaking wind" but rather *Nefesh Ḥayah*, and we must understand the link between the literal meaning and the exegesis, that is, the relationship between the meaning in the exegetic translation *Ruaḥ Memalelah* and that of *Nefesh Ḥayah*. That is

---

2. See the preceding note.

where the naturalistic context of the kabbalistic explanations is very helpful. They distinguish several levels in the overall *Neshamah* from the bottom upward: *Nefesh, Ruaḥ, Neshamah, Ḥayah, Yeḥidah*. We shall leave the last one, *Yeḥidah*, for the moment, since the term is not mentioned in Genesis (it appears only twice in the Psalms). Very schematically: *Nefesh* (which we suggest translating "rest") is understood as the vegetative soul, closer to the body (although the term *Nefesh* is not used explicitly for vegetables: Nahmanides dismisses this point, and we do not have time to get into it here). *Ruaḥ* ("Wind, spirit, or breath") is understood as the animal soul, in the sense that it is responsible for movement and sensation.

Neshamah in its particular aspect as a part of the overall *Neshamah* is understood as the intellect, more precisely *Binah*, that is, analytic intelligence.

Ḥayah is understood as the source of *Binah*, more precisely, *Hokhmah*, that is, synthetic intelligence, or wisdom, and more generally the basic matter of which the whole world is made. Whereas the first three are within the body and contained by it, *Ḥayah* (and also *Yeḥidah*) are considered as enveloping the body, because they cannot be contained within its limits. Thus, the higher qualities of the human structure and self are designated by the same word, *Ḥayah*, which is used to designate life in general.

The second important aspect of these explanations is connected with the particular role of language. We have seen that the *Ruaḥ Memalelah* ("speaking wind") links the different parts. We must consider the particular status of language in the philosophy of science. It works in a recursive way, being the vehicle to describe the different levels of reality and is itself structured into different levels of letters, words, sentences, and enunciations, each with a syntactic and a semantic level.

It thus seems to me that the exegetical difficulties we have encountered can be clarified by this kind of modern, naturalistic approach, based on the analysis of systems organized on different but interconnected levels, as we find examples in modern biology and in research in artificial intelligence and computer languages. A recent book on this subject by D. R. Hofstadter (1979) ends with the question of the articulation between self, creation of meaning, and dealing with chaos, which happen to be the three features of human language most difficult to formalize, because they appear in "tangled hierarchies," where different levels of organization are tangled without ceasing to be different.

EIGHTEEN

Israel in Question

(1975)

A People, Its Story, Its Culture

The story of the Jewish people and of their relationship with the Torah begins with their exodus from Egypt, three or four millennia ago.

The fact that we know the early millennium or millennia of this story only through the Torah does not dramatically change this historical reality. Even if we consider the biblical narrative to be more mythical than historical, it is the Jewish people's myth of origin and plays, at the very least, the same role in determining its identity as the origin myth of any people, tribe, or family. Moreover, the relationship between these myths (the complete Torah, both written and oral, if they are considered as such) and the historical Jewish people has a particular nature.

They are not only old narratives piously transmitted by elders to initiates and used by sorcerers (priests, later rabbis) to establish their power and

---

NOTE: This text dates to 1975. It was reprinted in Atlan 1979.

organize their society. They also concern the content of an entire teaching (*Torah* means "teaching") that has for centuries been the object of study and research, at once both popular and very intellectually refined, while also fashioning, producing, and founding the Jewish people itself. This teaching is *popular* because it was one of the first obligatory systems of instruction and general education to be established, approximately two thousand years ago, and it has been maintained more or less until today. It is *intellectually refined* for almost the same reasons: from the moment the study of the Torah, not only in its recitative mode but also in its analytical and critical mode, was established as the fundamental institutional pillar around which dispersed Jewish communities organized themselves, these communities were transformed, ipso facto, into factories and reservoirs of intelligence.

These factories and reservoirs emerged here and there thanks to conditions favorable to intellectual flowering in communities whose infrastructure was placed at the service of the school, of teaching and of research. This constituted the Talmud Torah in all of its forms, with its many levels of students and teachers, from the least gifted to the genius (*gaon*), who was recognized by the entire people as his works and reputation spread across the network of these community schools scattered throughout the world.

The story of the Jewish people, after the destruction of the Temple of Jerusalem, which two thousand years ago was the symbol of a national territorial founding, has been the story of their schools, divided into periods that marked the evolution of teaching by its content as much as by its institutions. There was the period of the "Tanaïm," the "Teachers" of the Mishnah, followed by that of the "Amoraïm," the "Speakers" of the Jerusalem and Babylonian Talmuds, followed by that of the "Gaonim," the dispersed "Geniuses," and then that of the "Poskim, richonim, and aharonim," the ancient "first and last" legal decision makers.

This system, set up at the time of the return from the first exile in Babylon and before the Second Temple, replaced the monarchal and sacerdotal national system after the destruction of the Second Temple not just for the dispersed Jews but also the communities that continued to live as best they could in the land of Israel, then under Roman rule and later under Byzantine, Arab, and finally Turkish rule.

In return, the masters (*rabbi* or *rav* means "master") were charged with the organization not only of the school but also of law and of the social and individual lives of these communities.

Indeed, the goal of their intellectual productions, in the end, even after the most abstract detours, was to enunciate the law, that is, practical rules for communities, a law and rules that needed always to be discovered and revealed, even though they were given in prototypical form, once and for all, "to Moses from Sinai."

This law covered and still covers all social and personal aspects of individuals' lives. A particularly important place was and still is reserved for the pursuit of the development of sexual and familial life, chiefly responsible for the generation of future students and their integration into the educational system.

In other words, these reciprocal relationships between this people's nature (meaning the type of humans and organizations constituting these people in their environment) and its culture, relationships recognized today for all people and all cultures, were consciously lived there through the initiating and generating role that the Torah played for Jewish society, in the eyes of those who at once studied it, renewed it, pronounced it, and applied it to their community.

Desert, Land, and Incest

For this people deprived of a land in which to put down roots, more than for any other people, the Torah and its culture took on the character of the source of identity and collective reference.

Each culture and each people has a similar character, but it is generally shared through geography, through a nurturing land, the motherland of all modern states, on and from which the group became a people while developing a culture, which completed the development of the people into a nation, or a state.

For the Jewish people, landless for two thousand years, the Torah encapsulated the generative role of culture and of land. Furthermore, and herein lies a key point, the privileged relationships that the Jewish people succeeded in establishing in their history with a particular land, the land of Israel, however fundamental and constitutive, are nonetheless absent from the myth of origin and, it follows, from the initial identity.

The Jewish people, in their own self-conception, were not born in the land of Israel. They were born in Egypt, more precisely, at the time of the exodus from Egypt, which the tradition compares to the expulsion and deliverance of a birth.[1]

The land of Israel is not, for the Jewish people, the motherland (mother-father as in the French *mère-patrie*), progenitor and nurturer of the people. The role of progenitor was played by ancient Egypt, which the tradition compared to a uterus and viewed as the birthplace of the whole of Mediterranean civilization. As for the role of nurturing mother, it was played by the Torah itself, with the desert as the place of its initial unveiling.

Instead, the land of Israel plays the role of the wife, the fiancée, the future wife who needs to be impregnated, the object of desire, the medium of accomplishment and overcoming once the mother, as woman, has been proscribed.

The Torah's proscription of a return to Egypt for the Jewish people is significant from this point of view, as are the biblical stories about the frightening fascination that the land of Israel exerted on the desert Hebrews, subject at the same time to constant but repressed longings for a "return to Egypt." Hence this return takes the form of an incest not to be committed. There is another incest, however, that is to be committed, and is consummated by the entire Jewish people over the centuries, with the Torah as at once nursemaid and wife. All of this takes place as though the incestuous relationship that each people has with its land, from which it is born, which it desires, and to which it returns, was violently sundered by the break between these two lands; on the one hand, Egypt the progenitor and, on the other hand, the land of Canaan, the desired. In the desert, nonland between these two, the Torah is discovered, filling the place of one and of the other. Israel in exile will have the strength to survive without land because the land to which it aspires is not its mother but its wife. Having overcome the wound of the proscription of their mother, the Jewish people could live, orphaned of a land (because *they always were*, even during their national existence), married to the land of Israel. Without any land, they have been like men without wives, condemned to oscillate between sublimation and decay, but they could carry on in exile, unlike the majority of peoples, for whom exile is a tearing from both mother and wife, like an obligation not only to live without a wife but to exist without ever having been born.

---

1. See Maharal, *Gvurot Hachem*, chap. 3; Midrash on Deuteronomy 4:34.

The Contemporary Jewish People: "Origin Myth," "Program," or Flight into the Unspeakable

We see how the reciprocal generic relationship of the Torah with its people, a special case of reciprocal relationships between culture and nature, takes on a particular intensity in the case of historical Jews because of the kind of relationships they have with the natural world—lands. Unless it is the inverse: that this division, this separation of the progenitorial land from the desired land is the *result*—and not the cause—of the Jewish people's special relationship with its cultural world.

Whatever the case, whether one considers this culture to be story, myth, law, philosophy, science, or religion—all of which it is at once, as with any culture—its constitutive effect for the Jewish people is sufficiently rooted to make it indispensable to invoke this culture beyond (and well within) considerations belonging to the order of Western historical science.

In other words, whether the birth of the Jewish people during the exodus from Egypt, narrated, taught, and renewed over centuries and millennia by the Torah, is historical or mythical matters little once it takes its understood place in the Torah, the generating culture of the historic Jewish people, which at the same time it portrays.

This is why the initial Jewish identity, the image that the Jewish people have held in mind since their beginnings and in which they recognize their real legitimacy, is provided by their exodus out of Egypt, as it was and continues to be experienced in stories, metaphors, and innumerable traditional teachings, continuously amplified and elaborated over the course of centuries.

From this perspective, it is remarkable that a nonreligious Israeli Jew, a Zionist from the very beginning, could have reacted to the uneasiness caused by a very militaristic and stately independence celebration in Israel in 1973—too Western and too "non-Jewish"—by declaring that, for someone who had participated in all of the struggles of national liberation and the achievement of independence in 1948, the real national celebration was not the day of the creation of the State of Israel but rather Passover, which marks the anniversary of the exodus out of Egypt.

In fact, more than the anniversary of a national celebration, it concerns an act of birth and at the same time a program. *It is in reflecting upon the meaning of the central role occupied by the exodus from Egypt in the birth of the Jewish people and the orientation of their history that we can find the connecting thread of responses to a series of questions about the identity of this people.*

These questions address, first of all, non-Jews, insofar as the Jewish people is not easily classifiable among other peoples by a simple taxonomic comparison, and Jews appear to non-Jews either as a historical scandal, as remnants that will not agree to disappear, or as mysterious depositaries to whom all philo- and anti-Semitic fantasies can be attached. But they are also posed, perhaps even especially, to the millions of Jews in the diaspora and in Israel for whom the conscious link with their culture was severed one, two, or even three generations ago. This acculturation to the benefit of the dominant Western Christian culture is the same as that of colonial countries and has produced the same type of alienation, with the same nomadic searches for identity, the same types of readymade pseudo-solutions, which are suggested by the environment—territorial or nonterritorial nationalism, religious "orthodox" tensions, assimilation by identification with the surrounding Western nation, or even frantic flights into internationalist messianisms, supposedly erasing through them all contradictions born of historical and cultural differences—and, finally, neuroses. At the same time, both the recent and the more ancient past remain stubbornly present, embodied in the ideological remnants of Zionism, in the reality of the existence of Israel, which prevents the forgetting of the enigma, and, in traditional orthodox communities, outwardly imperturbable witnesses to a Jewish existence that was authentic and unacculturated until approximately two hundred years ago.

Moreover, this presence strengthens and renews itself under the influence of various avatars of anti-Semitism, from the right and the left, the mechanisms (Sartre 1954) and the discourses (Faye 1974) of which have been extensively studied.

Jewish realities—Israel, more or less traditional orthodox communities, even the laicized communities based on nationality—preclude the disappearance of identity questions by keeping more or less alive the (unidentified) person of this people. But at the same time, the answers provided by these realities—nationalisms, religion, and assimilation—do not eliminate the neurosis of alienation. On the contrary, insofar as these realities are inspired by foreign systems of thought, suggested by imitative transpositions of the dominant, unidimensional Western culture, they are themselves the result of a loss of identity and of acculturation.

In Jewish culture itself, and primarily in the Torah in its entirety (the Bible, Talmud, Midrash, Kabbalah, and philosophy), we find a response to these questions, in other words, in what we can call Jewish myth, also inspired by definitions suggested by Western epistemology. By "myth," we do not mean the illusion of fable, but rather the most authentic expression of the specific organizing, identifying function that characterizes a society, even if the level of reality and signification is not yet very clear in the analysis and discourse of the Western humanities (Smith 1974: 714–30). And so the traits are both meaningful and absurd, real and illusory, open and closed to what in the Western consciousness (as embodied in universities, research institutes, the *mass media*) constitutes a myth in ancient civilizations, in its own and in those, Greco-Roman and Jewish, in which it recognizes its roots. It is in this sense that today the Torah can, from the external perspective of both non-Jews and acculturated Jews, play the role of Jewish myth and thereby justify our search within it for answers to the question "Who are the Jews?" asking for their birth certificate, their name, how they see and name themselves in relation to themselves and others, ever since their beginnings and throughout the twists and turns of their experience, up to and including the present period, which is marked by this loss of identity.

Still, we cannot allow any reduction of the Torah, even hidden, to a religious homiletic phenomenon, as we see in an unfortunate number of contemporary Jewish master thinkers. Is it not strange that "master thinker" has become pejorative, while "master" has simply disappeared from the vocabulary? Their sometimes brilliant, sometimes vulgar discourses always return in the end to the mystery of Israel, which is in some sense a privileged mystery, a kind of super-mystery if not a mirage, in some sense the ever-sought-after proof of God's existence, himself posited as the mystery of mysteries. To which, afterward, this mysterious function is proposed as an "explanation," an explicative answer to questions pertaining to our identity that by definition only believers can, in a pinch, accept. This at once permanently deprives all unbelieving Jews (why do certain individuals believe a priori and accept the God-mystery as an explanation while others do not? yet another mystery) of any hope of receiving an answer to the question "Who am I?"

During periods of crisis, when these questions are not only theoretical (of the sort that can be forgotten in daily life) but impose themselves through political or social events that affect us, a mortal danger of paranoia threatens those who accept these answers.

I mean to speak, for example, of the latest events that have shaken the Jews of the entire world, specifically, of the Yom Kippur War and of its repercussions.

People who think in this way perceive this war, which has revealed the relative weakness of Israel, to be a divine ordeal, while the diplomatic isolation of Israel and its unilateral condemnation by the UN stir up a resurgence of the quasi-archetypical images of the ancestral hatred of the Jews, images of a fundamental and essential anti-Judaism, a constant of history, that assumes different discourses and forms depending on context and circumstance. The enigmatic character of this anti-Judaism is often underlined—as though it were not obvious enough and needed to be emphasized—not in an attempt to circumscribe it and solve the enigma but rather to return us to an unfathomable mystery. Therein lies the danger of paranoia. The hatred and anti-Judaism that surround us is portrayed as the result of our particular relationship with a Creator who does this in an obscure way and for even more obscure reasons. Why do we not see that this attitude is symmetrical to the proclamations of the Inquisition into the Catholic and Spanish character of God, which are echoed today by Gaddafi, who proclaims the ontological superiority of Islam and the universal redemptive mission given to him by the All-Mighty, with its corollary of a liberating holy war? I say symmetrical and not analogous because one side attempts to "found" in the mystery of transcendence an election of persecution, while the other also attempts to "found" in the mystery of transcendence an election of power that must be imposed and cannot bear contradiction.

It is the same paranoia when an exceptional situation, either suffered or desired, is "founded" on the mysteries of transcendence—a paranoia because it is about closed discourses and attitudes, where nothing is founded; we do not understand our situation any better than before, when we associated it with the unfathomable mystery of the Creator, and we have no chance of taking hold of it. We remain at a standstill in our situation. As for our discourse, it remains incomprehensible to those on the outside and serves only to shroud us in words that entrap us.

And yet, we see that the temptation to be charmed and caught up in these paranoid discourses is great, as Jean Daniel has suggestively observed, saying that "there is in the heart of every Jew a place for Begin, and in the heart of every Arab, a place for Gaddafi."[2]

---

2. *Le Nouvel Observateur*, October 1973. The balance was upset by Begin's rise to power and his attempt to negotiate peace with Sadat, who, guided by Jimmy Carter, caused another aspect of the religious, very different and extremely interesting, to appear (*Le Nouvel Observateur*, no. 667, August 1977).

The paradox is at its most intense when this attitude of pseudo-explanation and of real flight into the mystery invokes the tradition of the Torah. At this point, we find all manner of ramblings, which culminate in the magical manipulations of numbers that use prophetic predictions and the ancient technique of *gematria*, taken out of its traditional context to "found" messianic speculations about the end of time. These, as it happens, reveal themselves to be particularly false, apart from a few resounding exceptions, which suffice to live down the others and authorize all a posteriori retouching (*Nouveaux Cahiers*, no. 39, 1974–75).

This is indeed a paradox, since we do not really need a Torah that presents itself as wisdom, teaching, and illumination if it does not illuminate, to some extent, the mechanisms and processes responsible for the situations and permanent features of our history. It is evident that, from the perspective of the Torah as illuminative knowledge and taught research, any analysis or hypothesis on these mechanisms, as partial or contestable as it may be, is worth more than an appeal "to the impenetrable ways of the Lord." If we really do not want to settle for partial and provisionary analyses and explanatory hypotheses, then we must criticize and demolish them in the name of a greater necessity or rigor by leaving open questions that are then posed again at a higher level.

In this essay, an analysis of the Jewish myth and of its central episode, the exodus from Egypt, will serve as the connecting thread to understand these mechanisms. In particular, the common ground of all these forms of anti-Judaism (Faye 1974 analyzes how their discourses have both stayed the same and been modified from Roman antiquity to the Christian anti-Semitism of the Middle Ages, the humanism of the century of Enlightenment and of the first socialists, and, finally, the national and Nazi anti-Semitisms of the nineteenth and twentieth centuries), a common ground that we find only at the heart of those who are the permanent object of these discourses, even though subjects and circumstances change, will clearly appear to us as one of the consequences of the birth and constitution of the Jewish people from a liberating movement experienced to its most extreme limits. This ground is the exodus from Egypt, in which a people and nation were born in the movement of a liberation of slaves that aims to be complete, definitive, and absolute liberation.

Exodus as Liberation-Program and Program of Liberation

The exodus from Egypt, the liberation, does not throw us upon the mystery of God. Rather, it defines the God of Israel, its inner motor, its active person, and its system of "self-organization." "I am the being of time, your God, who brought you out of Egypt, out of the house of slavery"—this is the leitmotif with which the God of Israel presents, names, and defines himself. And this cannot be a matter of one of many potential liberations, between two possible alienations, for a people born in this way can have no other program.

Every liberation from oppression carries with it the hope of a total, universal, and definitive liberation, just like the messianic movements found in colonized peoples, or even in Marxism. But once obtained, liberation always opens onto the establishment of a new society, where it is no longer a question of universal liberation, since, by definition one might say, this liberation has already taken place.

The Jewish myth narrates the errors and failings of the Jews themselves in relation to their project of liberation. Because of these stories, the liberation achieved by the exodus out of Egypt and the ensuing national existence cannot exhaust the hope or the demand for universal liberation.

The promised land beyond the desert is not a goal in itself, but rather a means of realizing the program expressed in the alliance between Israel and its God, that is, the equivalence between the concrete society of the people of Israel and their image, both the image they have of themselves and of their consciousness of time, of the project that is within them.

As we will see, one of the terms of this alliance is the relationship to the land. The other is the relationship to the Torah, which must guarantee the execution of the program and which must be understood, once again, in its ancient and mythical context, not least because this alliance continues to express itself: "I will be for you as a God, and you shall be for me as a people" (Leviticus 26:12, Exodus 6:7), as if the customary concepts of Gods and people were not befitting and we were dealing with a case of "everything happens as if," which we use to describe a situation by drawing on the only images available, even while knowing them to be false.[3]

What matters is to shape this people of former slaves into the movement of a liberation meant to be complete. The tensions between Moses and the elders of the people arise precisely from this issue: what could, for the ancients, be only liberation from the slavery of Egypt is, for Moses, the beginning of a liberation that must continue, a liberation from all alienation, all limitation, all particularism, opening onto the infinite and its desert, indispensable sites of the discovery of Wisdom and the Law, which are straightaway also presented as infinite, albeit intelligible. Wisdom and the Law

---

3. All translations from the Hebrew in this article are my own.

renew themselves incessantly and without end, and accompany the also incessant movement of life itself, of individuals, of the people, of societies, and of the world.

In the conflicts that oppose Moses to the traditional leaders of the people (the leaders of the tribes, an aristocracy made up of descendents of Hebrew ancestors for whom, quite naturally, this liberation took on a national form), Moses relies on a mass of non-Hebrew slaves that has joined the movement and that the Bible calls the *erev rav*, the "numerous crowd," or the "great mix" (Exodus 12:38; Rachi on Exodus 32:7).

Curiously, while the Hebrews properly so called are linked to the "as a God" for whom they are "as a people," this crowd is, following Midrash, called the people of Moses in the Bible (Exodus 12:38; Rachi on Exodus 32:7).

Of course, amid these tensions, Moses does not oppose the desire for national territory, but he tries to channel it toward the always more. "You will be for me a kingdom of priests and a consecrated people" (Exodus 19:6; again to be understood in its original context and not in the religious sense that has become the norm), such is the program, beyond simply realizing a nation, that Moses assigns to the people.

To avoid the risk of falling into new alienations after this realization, a general exhortation is incessantly repeated from generation to generation: "You will remember that you were slaves in Egypt: . . . you will tell your son" (Exodus 13:8; Deuteronomy 6:20); "I am the being of time, your God who led you out of Egypt, out of the house of slavery" (Exodus 6:6; 7:7; 20:2, Leviticus 19:14, Deuteronomy 10:19); "You will not oppress. . . . You will love your neighbor, for you have been strangers in the land of Egypt . . . for you were slaves in the land of Egypt" (Deuteronomy 5:15, 15:15).

In other words, now that the ancient land is no longer there to impose its limitations and the new is not yet there to risk replacing it, the alliance between Israel and its God imposed by Moses in the desert is the identification of a people made up of former slaves with a desire, an image, a program of universal liberation, a knowledge that, as A. I. H. Kook notes, coincides with a desire for knowledge.[4]

---

4. "The secret of Israel's existence, . . . obscure reasoning . . . where the paths of wandering are grades of ascent . . ., the darkening, a property of illumination. . . . While [attaining a knowledge of some kind] we believe we are following some kind of program, our thought and mind immediately alert us that we are only *desiring* an ideal of knowledge . . . [it is this desire, *chekika*] that, in itself, is characteristic of the knowledge of Being and

Idolatries are battled not for their own sake but because they pose a danger of standstill and detour. They are rejected as idolatries, as gods of death, because they consist in a certain kind of satisfaction of desires, where the desire itself risks being swallowed up. By contrast, to satisfy the desire of a work bearing life and liberty is to handle its source carefully, to resist mixing the fulfillment of a past desire that is exhausted with a future desire that must remain alive.

This is expressed in these commandments, which at first glance appear strange: "You shall not boil a kid in its mother's milk" (Exodus 23:19, 34:26); "When you happen to find a bird's nest, take the eggs and the young, but the mother, you shall dismiss her" (Deuteronomy 22:67).

These interdictions extend and generalize the interdiction of incest, which takes on, as a result, a new signification. The mother is the source of desire and of its fulfillment in that she is the source of life itself—the possession of a woman is fulfillment itself. Their fusion denotes this mixture (mortal because it destroys desire) of the fulfillment in which past desire is exhausted and the source of all future desires. The separation of nurturing motherland and impregnated land-woman of which we spoke earlier proceeds from the same movement.

It is in the desert, between these two lands, that the possibility of a total, universal, and definitive liberation is glimpsed, since it can in this place recharge and renew itself. "You will not act as they do in the land of Egypt, where you once lived, nor as they do in the land of Canaan, where I am bringing you" (Leviticus 18:3). Midrash reinforces this by underlining the already suspect and miserable nature, bearer of death, of all *residence* (Babylonian Talmud *Sanhédrin*, 106 [a] and Midrash Raba, *Noah*, chap. 38).

It is from this point, once the contract has been sealed, that the Torah begins to be taught, that the story will start—only the story, but in its entirety—of the insoluble but vital contradictions between the aim toward a total and universal liberation and the reality, which we always hope is temporary, of limitations and particularisms, including and especially our own.

The mortal danger of deleting the first term of the contradiction (and thus the contradiction itself—to establish oneself definitively, to stop and reduce oneself to fit the dimensions of an alienating particularism) was previously

---

of its service, where we situate the basis of our vision of the world . . ., incommensurable with any knowledge, whether intellectual or moral . . ., mechanical reasonings, which, in the end, take hold of strange trains of thoughts" (Kook 1964: 557–58).

averted. This danger presents itself as a false relation to the land, either as the return to Egypt out of fear of the new or as a relationship to the new outside of the alliance, as if it were the old one (Numbers 19).

It was averted after the program, the image of the liberated people, was written down. It took forty years of waiting for the old generation to disappear in the desert and the new generation to mature, having as its only determination the non-land of the desert, non-limitation, non-definition, and the Torah, that culture having "fallen from the sky."

From the Liberating Opening to the Conformism of a Social Organization

In the new nation thus created, which has as its mother the desert and the Torah, not Egypt, the program of definitive liberation that goes beyond this or that slavery without being satisfied is definitively inscribed. A people is born, not from the adaptations and accommodations of an existence at the limits and within the constraints of a determined land, but from an experience of travel outside these limitations, of a collective "trip" of forty years that definitively shapes the image that this people have of themselves: as living on earth and not in heaven (since the desert is still on earth), but on a journey in a "permanent elsewhere," an initiatory journey whose stages do not risk being reduced to this or that figure of a determined landscape.

Thus, from that point, that is, almost from the beginning, the movement turns around and engenders, among other "subproducts," a hatred that can disappear only with the end of the movement itself. In fact, universal liberation from all alienation also means the subversion of existing alienations, thus not only the rejection *of* but also rejection *by* every form of organized society. Yet, and this is more serious, this rejection implies and indeed provokes a folding back of these messengers of freedom onto themselves; in other words, it gives rise to . . . a new organized society. A group of marginal individuals initially marginalized by the demands of radical change is soon confronted with the problem of duration: whether because there is to be no change, or, because alibis for change stifle the hopes of the majority, the marginalized, faced with the perspective of a long wait, must reintegrate with the core of the central groups "through which history passes" or organize their own duration by waiting for the great night or the messiah—all of which contradicts at many points their initial marginalization by creating a new conformity and new alienations.

The vitality of the new social group thus constituted will depend upon the depth of the trace of the initial demands in its social organization, which in any case contradicts them: these demands stem from imaginary negativity and from utopia (the messianic change), while the organization chafes against the internal and external constraints that it adapts as organizing functions and to which it is adapted as organized state. And yet the first justify the second, at least from the perspective of individuals in the group. These initial demands generate the minimal personal enthusiasm necessary for the constitution and cohesion of the new social group. The organization is more or less assured of duration to the extent that it has more or less succeeded in integrating, without denaturing, the factors of negativity and utopia, which at once contradict it and serve as its guarantor.

In the case of the Hebrews, the newly organized society, "not like the residents of the land from where you come, nor like those of the land to which you are going," conserves the mark of the illumination and of the initial project. But as a social organization, it also has its slaves. For better or for worse, it is in the particular status enjoyed by slaves that the initial project of liberation from slavery succeeds or fails.

Liberation from slavery in Egypt is followed by the great ecstasy in Sinai, where the people discover their God as an inner driving force, usually concealed, with whom the program of alliance has been sealed. For them, that opens immediately onto specific laws of social organization (Rachi on Exodus 21:1), of which the first, the archetypical, concerns the status of two kinds of slaves (Exodus 21:1–6, 21:31, Leviticus 25:42, 25:44). On the one hand are the Hebrew slaves, witnesses to the sustenance of the constraints and alienations internal to the group. It would serve no purpose to content oneself with a formal declaration abolishing slavery when there exist conditions under which members of the group are in states of total economic dependency that eliminate their personal liberty and are in evident contradiction with the project inscribed in the birth of the group: "You will remember that you were slaves in Egypt." The organizational law, in instituting the obligatory liberation of the Hebrew slaves, thus tries hard to render transitory and reversible this state of alienation and to transform it into a source of new liberation. At the same time, it cannot do otherwise than institutionalize slavery, even if under the form of a necessary evil that must be abolished. Yet alongside the Hebrew slaves, foreign slaves witness the survival of exterior constraints that sometimes explode in wars (Deuteronomy 21:10, 23:16), which produce defeated individuals reduced to slavery, a situation that is also in contradiction with the initial project: "You will remember that you were strangers in the country of Egypt"—and that the organizational law cannot do otherwise than institutionalize.

As long as the pretense to universality is not itself universal, it comes up against the reality of particularist differences and rejections and finds itself inevitably thrown back onto its own particularism.

There, the pretense to universality is all the more easily mocked and ejected for having been forced to hide itself under the garment of a particular people, nation, land, and state. Thus the life of this people depends on a kind of contract between the garment and the body, the spokesperson and the speech. So long as Jewish particularism remains a means at the service of diverse expressions of the initial project, even if at first they are not understood as such, except by Jews themselves, it maintains enough force to resist the mockery and hatred that, both despite and because of this, it incites. At the same time, it keeps enough objective force to impose its existence among other particularisms, other peoples, nations, and states.

But when Jewish particularism is lived as an end in itself, then the justification that "one has a right to exist as much as others do" is no longer sufficient to sustain an interior force allowing it to resist the derision and assaults of others.

The Historical and Ideological Back and Forth

The existence of this particularism "like others" is not only contradictory but devoid of interest with respect to the initial marginal group. It leads, in the end, to the breakup of the group itself. Thus, exile and dispersion among the nations are accepted as both the consequence of these assaults by others and as the occasion for reconciliation with the initial project of universality. Moreover, this reconciliation prevents the exile from being an exile "like others," and, finally, allows it to last.

But then this exile is not like others. It enables a dispersed existence to last "longer" than is normal and engenders the hatred of nations organized against these ferments of subversion. These ferments are then perceived as bearing the scars of history's curses, whether in the form of a sinful people condemned by God, in the eyes of certain Christians, or in the form of history's venomous disease, for a current of Islam (Poliakov et al. 1972), or else in the form of a people-class (a parasite without legitimacy) for Marx and Léon.

This hatred and this rejection by organized societies—and by organizational and classificatory ideologies, even when revolutionary—of every bearer of hope of a universal liberation cannot cease as long so these societies and these ideologies are not liberated from the particular alienations on which they are founded. The closer a social system or an ideology approaches this liberation, the more its failure engenders hatred of the hope of liberation with respect to which the failure is denounced.

Among the diverse forms of anti-Judaism, anti-Semitism of Christian and Muslim inspiration, and, more recently, Marxist anti-Semitism, with its various Soviet and leftist variants, appear to be relatively clear examples of this mechanism.

As to Nazi anti-Semitism, its theorization by Hitler is sufficiently explicit for us to see there, as well, another sort of competition in the pretense to universality. The "true enemy of the German people" was the Jew, who was the first to have had a vocation, at least in the eyes of Hitler and of the Nazis, to "rule" the universe.

Thus, very naturally, the circle closes itself, or rather, the spiral loops onward: the dispersed who hold onto this hope tend once more to fold into themselves in order to escape this hatred and to allow their hope for radical and universal liberation to develop and be incarnated (again paradoxically), sheltered by their newfound particularism of people, nation, land, and state. And so on until the next exile, if it so happens, once again, that the dimensions of the particularist garment are decidedly too narrow for the initial project.

This particular, specific existence remains paradoxical and antinomic in relation to the initial project of universality, of which Zionism itself, in its redemptive socialist hope, was an avatar.

Indeed, the Zionist enterprise did not escape this paradox, constitutive of the history of the Jewish people: the desire for a Jewish state could only be born out of the conjunction of anti-Semitic rejection and the assumption once more of the initial universalist project.

Meanwhile, this project, in its traditional Jewish formulations, was reduced, at least in its public expression, to the dimensions of a battered religion, in the image of dispersed communities organized under conditions of persecution, hatred, disdain, and then acculturation.

The universalist project had, however, penetratred, albeit with considerable disfiguration, the dominant ideals of Christian morality and messianism, which were at the origin of prophetic visions à la Tolstoy and of demands for universal justice operating in diverse socialist ideologies, Marxism included.

Thus, in those prophetic and socialist languages, the first Zionists found a new avatar of the initial Jewish project around which to regroup, if they were not to resign themselves to their disappearance pure and simple by assimilation into the surrounding peoples and nations, which would have promised an end to persecution through an acculturating dispersion.

However, these languages were at the same time bearers of anti-Semitism, following the mechanism indicated above, insofar as they uttered organizing ideologies of universal liberation in relation to which the existence of particular, dispersed Jews, claiming to incarnate this universality, was a laughable scandal. This probably explains the refusal to assimilate to those verbally universalist ideologies, though the temptation was great for many Jews to see, at last universalized, their own pretense to universality.

This temptation has been at the root of the great debates between socialists of Jewish origin, on the one hand, who saw in socialism, first Internationalist, then Soviet, a "solution to the Jewish problem" that would accompany universal liberation, and, on the other hand, socialist Jews, Zionist or not, whose distrust of the claim that Jewish particularism would disappear into an internationalism that kept in place the specificities and powers of other nations had been acquired in the course of centuries of forced liberation, of Christian salvation and love imposed by fire.

The same debate is found today, and it divides young Jewish Westerners in their relation to leftism and to diverse internationalist revolutionary ideologies, "Maoist" and Trotskyist.

Some see in these ideologies a promise of liberation, where Jewish life experiences will blossom, while national particularism will disappear, whereas others cannot help but find suspect the desire that Jewish nationalism be the *first* to disappear.

We can thus only remain suspended, once again, between the two poles of this invigorating contradiction; always with the risk of a new exile, if, once again, the aspiration to cessation, to repose, to residency, to "have a king like the other nations" overcomes the initial project: "I will be for you as a God, and you shall be for me as a people."

This very going back and forth constitutes the alliance of Israel with its God and guarantees the strange durability of this people: the collective identification of Jews (as families and particular human groups) with this universal project, which inevitably overwhelms them and thus, in some respects, negates them, but without which they do not have any reason to be particular.

Such is the alliance, the identification with this God, no doubt "jealous," since he asks Jews always to go beyond themselves in his service and, above all, never to stop at an approximation, at an appearance, of liberation and generality, an idol all the more hateful in that, in truth, it nears a goal that is always receding. Given the makeup of organized human societies, including Jewish ones, this identification implies both the durability of this people and its rejection by others. Because it is both mechanistic and automatic, this process reveals itself to be an intelligible alliance and not a "mystery."

Durability is assured by this motion of back and forth between the universal and the particular, between a narrow nationalism and a dissolving internationalism.

At the same time, this motion keeps in place the rejection responses of other societies, whose own particular and alienating existences are incessantly denounced and which feel themselves threatened by the mere oscillating and multiform existence of Jewish society.

This contradictory superposition of durability and rejection traditionally constitutes the alliance, since an alliance in Hebrew "cuts" itself to "seal" itself. There is an alliance only between "cut," between contraries. The alliance itself is the contradiction in the union of what is fundamentally dissociated. The alliance of Israel and its God is the contradiction between human pettiness and finitude, accepted and proclaimed in familial, tribal, and social institutions, and the infinite of the imaginary, of thought and discourse, sensed in its experiences of liberation and of the desert, sought and found in the experiences of ritual, discourse, and thought.

The alliance is the contradiction that guarantees the durability of a limited existence that never ceases to die by encompassing it in a movement of back and forth that negates and kills it, but at the same time prevents it from dying by saving for it the opening and the minimal impulsion that allow it to begin again.

It is thus important to understand the terms and mechanisms of the alliance that runs through us, always essentially contradictory, for, at different moments of our history, they imply different aims and attitudes.

It is also important to understand that, for us who are always rootless, the reduction to one or the other term of the contradiction results in death, whether in the illusion of an always false and imperialist universalism or in the narrowing of national life.

At the same time, reflection on ourselves and comprehension of what is happening to us in terms of reciprocal interaction between our project and our surroundings—since these are the categories in which we think today—can save us from a paranoia suggested by a certain mysticism, where our being would found itself only in a transcendence that would be hunted everywhere save in a "Jewish destiny," and whose mysterious and unintelligible nature is capable, in fact, only of founding our nonbeing.

The Double Criterion

This situation of being suspended, yet in movement, this going back and forth that is the real life of the people in its entirety (beyond its national, community, religious, political, and humanist poles, etc.), must carry over to how we judge ourselves according to a double criterion: an external criterion, defined through comparison with other peoples, and an internal criterion, defined in relation to the project, to the hidden motor, that is the only truly efficacious source of internal force and vitality. If we are to have a good conscience and be "sure of ourselves" with respect to the external, only a bad conscience and a demand constantly renewed in relation to our own criteria can really assure, internally, our future.

As a people "like other peoples," engaged in its particular existence as a society settled on its land, as a nation and state, the Jewish people has rights and duties that must be perceived according to the same criteria as those of all other peoples, nations, and states. In particular, the right to self-determination, including national and political self-determination, accorded to all human groups, cannot, under any pretext whatsoever, be denied to Jewish societies, based on the more or less fantastical vision that other peoples have of Jewish actuality. In particular, the behavior of a Jewish state that has succeeded in constituting itself by methods no more and no less legitimate than other states, from whichever perspective we look at it—historical, from the right of peoples to position themselves, from the Brechtian right according to which the land belongs to whoever cultivates it, or from a victorious defense against another nation that has contradictory interests and contests its rights—cannot be judged according to criteria different from those of other states: that is, respect for human life; the justice of the society therein constructed; respect for the rights and sovereignty of other peoples; in sum, the contribution to the development of conditions enabling the blossoming of human potentialities in what they have in common with all of mankind, and this, taking into account the existence of historical, geographical, linguistic, cultural, and traditional differences that characterize each people and society. From this viewpoint, neither the Zionist movement nor the State of Israel that issued from it has "shown itself unworthy" or could see its legitimacy contested, neither from the viewpoint of the type of society and the interhuman relations that have been developed, nor from the viewpoint of specific contributions to the common search for new avenues of development, despite the difficulties and contradictions created by the opposition of the Arab nation in general and of its Palestinian branch in particular.

Even with respect to the war, to the fate of the Palestinian Arabs, and to the national aspiration that ensued (a hobbyhorse for all those who contest the legitimacy of the Jewish state), the behavior of the Jewish nation can only be judged "innocent," and even then, of an innocence almost angelic, if we compare it to that of other nations placed in analogous situations of conflicts of interest and struggles for survival.

But as it happens, this is not enough to keep us alive. For reasons having to do with our history and our project, it is not sufficient that we be right with respect to the criteria of others (and thus, that others be in no way "justified" in an enterprise of destroying the Jewish state) to find in ourselves the strength to fight. It is as though this were not interesting enough, as though it were not worth it. How many Jews are ready to fight and risk their lives for a national existence that is provincial and limited, where the only project is being "like others"? It is, moreover, a strange problem to have to define one's identity by a project of being or not being "like others." Of what people is the question of being or not being "like others" asked, and what national existence could be founded—and, above all, sustain itself—on this purely negative aspiration to the anonymity of "like others," the result of tragic exceptions and of persecutions?

Our real legitimacy is that which founds us, not only in law but in actuality, if we want not only to be right but also to continue to exist.

We can judge this legitimacy not only in relation to international law but also in relation to the internal criterion provided by our historical and cultural conditioning, our myth of origin and our inner motor. Judging by this criterion, the actual achievements of the Zionist movement and of the State of Israel are tragically insufficient: we have a society that is barely more just than others, even if it were to resolve the problems of the inequality of its communities, that is altogether closed in on itself, founded on the notion of fatherland and on the experience of a provincial and drab life apart from the heroism of the plow and the gun,[5] where the intellectual qualities that have defined traditional Jewish societies are contemptuously treated as "*galutic* mind" (i.e., shaped by the acculturating experiences of exile), while the spirit that is increasingly welcomed and favored—outside of scientific research—is but a pale imitation of the music hall and the "distraction" of Western Mediterranean societies, this famous Israeli *bidour*.

Of course, once again, all of this is perfectly comprehensible and justifiable from the viewpoint of Israeli society's conditions of existence and could not, much to the contrary, in any way justify a denial of legitimacy by whatever censor. Yet we must see that internally, in relation to the claims and aspirations, even unformulated, that the Jews have for themselves, this state of things cannot continue very long.

It is, moreover, remarkable that these demands and aspirations for a society more connected to the universal, both structurally and functionally, are found not only among the world's most active young Jews but also among young Israelis. All the research and opinion polls conducted among Israeli youth slightly prior to the Yom Kippur War and for a short time afterward show deep dissatisfaction with the official ideals of the pioneers' Zionism and the patriotism and cult of the state. It is as though the Jewish spirit of openness to the universal, decried as cosmopolitan and corruptive by certain anti-Semites, is applied also from the inside at the expense of Jewish particularism. Only if the two components, the universal and the particular, can recognize each other will the Jewish people, with the tension between these components as its normal state, survive.

---

5. The Jewish heroism or courage (*gvoura*) that, in the morning blessings, we say "surrounds Israel," is, according to Rav Kook, not the courage that conquers and dominates others but that of conquering oneself and of the life of the spirit (Kook 1969: 75). A contemporary Israeli author, Rachel Rosenzweig, projects these ideas into an analysis of the present situation by developing traditional themes such as: "Who is a hero? He who makes a friend of an enemy" (*Avot* of Rabbi Nathan, p. 75, in the Cherter edition); or "If you have overcome your evil instinct so as to make your enemy your friend, I promise you that I myself will make your enemy your friend" (*Mehlita* of Rabbi Shimon Bar Yohaï, p. 215). Her study, impressive in its depth and its erudition, was published in *Chdemot*, the journal of a kibbutznik group. She underlines the dangers of death contained in the heroism à la Bar Kochba that has imposed itself until now, in Israeli models, in place of the heroism of the spirit and of oneself from which the Jewish people has always drawn its strength. The title of this article, which aims to mobilize Israel toward a true "conquest" of Arab friendship, is "Zionism's Mission Today: Conquering Associates."

In conclusion, in these times when, once again, the legitimacy of a national Jewish sovereignty in Israel is contested, when numerous friends and enemies envisage the eventuality of the disappearance of the State of Israel, sacrificed to the policy of imperialisms rampant throughout the world, it is perhaps not futile to add that this essay does not in any way aim to prepare or justify a priori this eventuality. Much to the contrary, in the actual state of the relationships between nations, the disappearance of the State of Israel, which would not come to pass uncontested, would be a mortal blow to the entire Jewish people. It is also unthinkable from the viewpoint of other peoples, as an intolerable amputation of the image that men have forged of themselves. Just as the creation of this state gave a dignity, a new dimension, and an élan to the consciousness that Jews (whether Zionist or not) have of themselves and of their relationship to others, likewise such a catastrophe, happening less than two generations after the Hitlerian genocide, would transform Jews everywhere, in their own eyes and in the eyes of other nations, into eternal scapegoats, wandering from one catastrophe to the next, able to hope at best for a miserable survival (or a gilded one) at the price of a depersonalization and alienation that no people in the world today is ready to accept.

I have attempted to analyze the contradictory currents that "animate" (in the true sense of "make live"), from the inside, the Jewish people in their relationship to themselves and to other nations. Armed with a knowledge of these currents and forces that surpasses slogans and daily politics, we should be able, among other things, to master our history better, in order to avoid, perhaps, these kinds of catastrophes.

Translated by Danielle Dubois

NINETEEN

The Self, the Person, the "I"

(2004)

Introduction: Unconscious Self and Consciousness of the Self

Does such a thing as an unconscious self exist? Yes, of course: the biological self. The self, like consciousness, is not a reality that exists as all or nothing, as if certain things or certain beings had a self, while others did not. There exist different kinds or different degrees of self—as with consciousness—and from this point of view the name of the current publication, *The Self in All Its States*, seems adequate to me. When we speak about consciousness, the question to be posed concerns the eventual integration of its different aspects and the related properties that make it up. And as for consciousness, this integration isn't a simple matter of addition: its different components constantly modify themselves in relation to each other, and their composition is different from when their separate, isolated properties are simply grouped together. The unconscious self—or rather, the different unconscious selves—is modified by the existence of the conscious self, and obviously the latter cannot be considered in isolation, as if disembodied and unrelated to the unconscious body it also is.

One of the most spectacular expressions of the biological self is the distinction the immune system makes between self and non-self in our body's molecular structures and cellular activity. We can also conceive that the unconscious activity of our brain produces and expresses an unconscious self.

Both the immunological self and the cerebral unconscious self are not given once and for all: they constitute themselves through the activity of the immune and central nervous systems—in relation to one another as well as in relation to the rest of the organism—notably the endocrine system, from which it is increasingly difficult to set them apart. For both, what determines the structure and evolution of the self is stored memory of the events that make up their history, their development, and, to a certain extent, random encounters with other individuals, other selves, and other elements in their environment. We also encounter the role of memory and the history of the individual in the constitution of the self when we attempt to discern the nature of the conscious self. What are we conscious of when we are conscious of ourselves? In other words, how is the conscious self different from the unconscious self? One can answer: in that we *are* conscious of it. But what are we conscious of, given that we are not aware of the activity of the lymphocytes that make up our immune system or of the activity of the neurons in our brains? It seems that, in order to conceive what we are conscious of, we must make a detour through the question of consciousness. What is consciousness in relation to the cellular activity of our bodies? We seem unable to evade the classic mind-body problem: the emergence of consciousness and the individual mind's activities in relation to the development of the body and the brain in particular.

Obviously I am not going to attempt to solve this problem here. But we can approach it sideways, if one may say so, knowing that the position that seems to me the most fruitful, though neither the easiest nor simplest, is that of a Spinozist monism—neither idealistic nor materialistic—that affirms psychophysical unity in a radical way. We should make one more detour, however, this time through the question of the person. This isn't only an individual question, but one with a social nature, because the self that we are conscious of speaks to us, verbally or nonverbally; it expresses its *self*. It is the self that says "I" when speaking about him- or herself. This self constitutes *itself* in our dealings with others, meaning with other speaking beings, that is, humans with whom we coexist and constitute our society, directly or indirectly, in the social group or groups into which we are born, in which we grow up, and within which we have become human beings gifted with speech. That is why, perhaps, the most adequate fashion of researching the conscious self, which seems to vanish the moment we think we have laid hands on it, is to reflect upon the nature, also so present and fleeting at once, that expresses itself when somebody speaks and says "I." One can eventually say "I" to one's self. This person then takes on the role of a "you" in an internal dialogue, while continuing to fulfill its role as a "me." In effect, the "I" shouldn't be confused with the me or the *ego*, as it designates a certain conscious reality, though maybe one that is *too* conscious and therefore too enclosed and solidified in a defined self-image—objectified, if one may say so—from which the subject of speech, the one who says "I" has disappeared. The first person singular that says "I" thus assumes all these three persons at once, while remaining one, namely, the first.

Before arriving at this, let us see what can be said, very schematically, about a question that is a source of controversy: What is a person?

The Human Person

A NOTION OF PERSONHOOD IS ABSENT IN THE NATURAL SCIENCES

For a long time, the attributes of the person were considered to be related to the intrinsic properties of the human individual and the role of natural science and philosophy was to discover and circumscribe these properties. Philosophers like Kant and his successors were supported by the vitalist philosophy of their time in their bid to make natural science, physicochemical and mechanical, coexist with the study of a rational ethics and a law founded upon the observation of apparent purposes in nature. A beautiful metaphysical construction was believed capable of reconciling physicochemical mechanisms with a rational ethics founded upon free will and human dignity as a natural fact. This has allowed Western humanism to exist during the past two centuries.

But this reconciliation could hold sway only so long as finalist and vitalist theories in biology were deemed probable. Nowadays, we must admit that one corollary of the success of physicochemical biology is the progressive disappearance of the notion of the person—and thus of liberty, responsibility, and so on—from the scientific representation that we can draw from nature.

The ambition of modern biology is to account for animal behavior as well as our own, not by invoking our soul and our freedom but by showing the determinism of physicochemical interactions between the molecules that make up our bodies. Whenever this goal is realized, it excludes in advance the freedom of a person understood in the Kantian sense of a suprasensible domain of freedom and expressed in free will and free choices based upon reason. Yet social life cannot do without the law, and because of this dominant philosophical tradition, that seems to imply, in one way or another, the reality of this free will.

There is an opposition here, and it is helpful and interesting to return to Spinoza and to an idea of freedom different from that of free will, one that already posits the conditions for an ethics that implies, among other things, respect for persons and respect by persons for the laws of society—even though the idea of a universal mechanical determinism was renewed with reference to the ancient Stoics and certain currents in Jewish tradition. The necessity for an ethics of freedom through knowledge of determinism imposed itself upon Spinoza in the seventeenth century, at the birth of modern science, and he held that free will was merely an illusion created by ignorance of the causes of the body's affections.

The opposition between an increasingly depersonalized science of the living and the necessity for an ethics and law based upon the reality of the person poses concrete problems that call for legal and administrative decisions. Confronted by these problems, we must first renounce the idea of defining the human person as a reality grounded in facts, facts that could only be taught as natural science—biology and psychology included. On the contrary, one must accept the idea that this is a reality of the *law*.

Indeed, what can biology tell us nowadays about the person? Actually, not a lot, even though it can tell us a lot about the individual and the process of individualization. But this is not the same thing. We know nowadays that biological individuality within the human species is pushed very far. That becomes evident, for example, when we look at the unique character of each individual's immune system after only a couple of years of development. This explains why tissue transplants from one person to another are usually rejected. But even this individuality is not a given, as it results from the maturation of the immune system, which commences in the embryo and continues during the first years of life.

The only really unique characteristic proper to each individual from the moment of fertilization is its genetic map, which results from what is referred to as the genetic lottery of sexual reproduction. This is not unique to human species, and thus it cannot serve to support a biological definition of the human.

Moreover, the uniqueness of the genome is not sufficient to determine the biological uniqueness of the individual, because epigenetic mechanisms contribute to its maturation and development—as we have seen in the case of the immune system. The brain develops in the same way. That is why two genetically identical individuals, like two "identical" twins or two cloned individuals, are already biologically different.

THE WESTERN NOTION OF THE PERSON AS ROOTED IN ROMAN LAW

In fact, the initial idea of the person appears not to have originated in scientific or even philosophical thinking but rather in the domain of the law, mainly Roman law. As a first principle, in it men are considered persons as beings whose legal status is different from that of things and animals. To them pertain responsibility, legal obligations, and rights, but above all they are accorded a dignity and respect that is central to the principles on which human rights in the West are founded. At the same time, the individual cannot exist as a pure abstraction without any physical support, namely, the human body. From that is derived the second principle, also a legal one: after the distinction between persons and things comes the indivisibility of body and mind.

These ideas—the distinction between person and thing and the indivisibility of body and mind—are strongly determined by the cultural and linguistic context in which they were developed, namely, the Western Christian world, heir to the Greco-Judaic world and also that of Rome—especially in matters concerning the law. It is obvious that other civilizations rate the notion of the human person differently: some, for example, attribute an important role to reincarnation; in some, an animal such as a cow is at least as sacred as a man, and, in general, there are cultural contexts in which the notion of soul or spirit is not specific to humans and can be extended to animals, plants, or other things.

HEBREW NOTIONS OF PERSONHOOD

Let's see what the concept of person is in Hebrew civilization and the Talmudic law derived from it. This law differs from Roman law in more than one way.

Hebrew terminology, from the narrative of Genesis 1–2. First of all, it is remarkable that *biblical* Hebrew contains no equivalent to the Latin *persona*, no word that exactly translates the notion of "human person." According to context, we find *adam* or *enoch* ("man" in general), *nefesh* (a term designating an aspect of the soul), *guf* ("body"), *ich* (another word for "man"), or *panim* ("face"). It would be easy to show that none of these terms exactly signifies "person." *Nefesh*, also translated as "soul," designates the activity of a living body and in this sense approaches the idea of a person, in which the body and mind unite in an indivisible fashion. Nevertheless, the expression *nefesh ḥayah*, which, when applied to a human person, can be translated "living person," is employed in the biblical narrative for animals as well (see Genesis 1:20, 21, 24; 2:7).

The term that most closely approaches the Latin idea of a "human person" might be a derivative of the noun ʾ*ish*, namely, the term ʾ*ishi*, an adjective that in modern Hebrew has acquired the sense of the personal, while the nouns ʾ*ish* and ʾ*ishah* simply designate, respectively, "man" and "woman." A noun derived from the same root, *ishiʾut*, approaches the Latin term a bit more closely, though it in fact designates someone's personality rather than the concept of a person. Meanwhile, another noun, *ishʾout*, also derived from the same root, designates the whole domain of relations between man and wife, meaning, more or less, sexuality.

Thus, a relationship between the Hebrew root ʾ*ish* or ʾ*ishah* and the Roman notion of a person appears to exist, but this relation is only expressed—at a pinch—in the adjective ʾ*ishi* ("personal"), not in a noun. The biblical narrative can help us catch a glimpse of the possible nature of this connection. Let's take a look at the curious context in which the word ʾ*ish* first appears, the narrative that treats the separation of man and woman from the androgenic Adam of Genesis. After the woman has been separated, Adam sees her, meets her, and names her ʾ*ishah*, the feminine of ʾ*ish*, because, as the text reads, she was taken from ʾ*ish* (Genesis 2:23). But the narrative is enigmatic, because she was "taken" from Adam. And it is Adam who names the animals and the woman, while naming himself ʾ*ish*—in the first instance of the word in the biblical narrative. From that moment onward, only the word ʾ*ishah* is used to designate a woman, while the two words ʾ*adam* and ʾ*ish* continue to be employed, according to context, to designate a man. From this we can learn several lessons.

First, a *transformation* from ʾ*adam* to ʾ*ish* is described: ʾ*adam*, masculine of ʾ*adamah*, the soil from which he is derived, is transformed into ʾ*ish*—the masculine of "woman." More precisely, the telluric couple *'adam-'adamah* (i.e., the couple of the initial human androgyne with mother earth) is transformed into a personalized couple: *'ish-'ishah*, man-woman. Furthermore, this transformation appears as a process, still ongoing, of personalization, or even becoming human—of humanization—that is earlier and more accomplished for the feminine part of *adam* than for the masculine part. That is why no noun could designate this always-ongoing process, neither *ich* nor *icha*, because each one would be too static, too limited, too enclosed within an *ego* of a particular man or a particular woman.

In other words, the human person is established as such not in an(y) individual—which is why there is no unique word to designate it—but in the meeting with and recognition of another's gaze. And not just any other: not that of an animal, too different, and not one of the same sex, too similar, but an other of a different sex.

Everything thus happens as if humanity was at the beginning nothing more than a collective property, that of the Adamic species, just like the animality of another species, and possessed nothing personal to the species' individuals at this stage. For this property of the person to become that of living individuals, and something different from living animals, the individual needs to acquire this humanity during his or her development, as if, in a certain way of speaking, he or she not only becomes *human* but also becomes *humane* by being *humanized*. The sign that humanity in this sense of humaneness is accomplished within the individual, and which permits him or her to be designated not only by the generic name of *adam* but as one personalized by *'ish* or *'ishah*, is the possibility of recognizing oneself in the encounter between a man and a woman.

Obviously, a question remains: What does a man recognize in a woman and what does a woman recognize in a man that is different from what both of them recognize in an animal? In other words, what is given in *'adam* that permits him to humanize himself and to be recognized as *'ish*? Or, once again, what in *'adam* is potentially *'ish* or *'ishah* or, as we perhaps say nowadays, a "potential human person"?[1]

In this context, one can show that what is immediately recognized is simply the shape of the human body, what the Bible refers to as *tselem*, image or

---

1. This problematic expression was invented by the French National Bioethics Committee, as a compromise between looking at the human embryo as a real person starting from fertilization and looking at it as simply a cluster of cells, i.e., a thing.

icon, when it is written that "Adam was created male and female in the image of Elohim" (see Genesis 1:27). I cannot enter here into detail but would like to remark only that this does not refer to a certain image of God as we would know it from elsewhere. On the contrary, it is this human form that is, in a certain manner, deified, by abstracting and generalizing its structure in a world in which divinities are multiple—as indicated by the plurality of the name "Elohim."

Ethical and legal applications. Don't believe for a moment, despite appearances, that all this is only about poetic metaphors and/or myths designed for religious edification. In fact, it has immediate ethical and legal consequences, which directly concern the practical problem of the definitions and limits that are associated with the idea of human personhood.

One radical way of posing these problems in the present consists in asking oneself who is the "human being" that is the subject of human rights. How can we recognize him or her? Isn't it, in the end, through the shape of his or her body? The most visible sign of the humanity of an individual, the most obvious one, is the human body. The body of *Homo sapiens*, with its face, is immediately recognizable as such. Needing no scholarly investigation, it cannot be confused with that of another species or with a robot—even an intelligent one. The humanity of the species—that is, of all of us—is concentrated in each human body, in its physiological reality and in the appearance of its shape, in which the social reality of our existence is also expressed. This empirical definition of humanity can serve as a regulating principle to resolve difficult ethical problems posed by biomedical interventions at the beginning and at the end of life: for example, the famous question "At what moment does an embryo become a person?" The biological criteria, whatever they may be—fertilization, the blastocyst differentiation at five days, development of the neural crest at fourteen days, and so on—are always arbitrary, even if they can be of some utility.

Instead of looking for a factual criterion with a scientific origin, which would always be fleeting and open to contestation, one might as well accept a criterion based upon our sensibility and our immediate perception. In other words, we will take the capacity of a living individual to perceive the shape of the human body and, in particular, its face—meaning our proper shape—as the criterion through which we recognize a person, not through a long scientific and philosophic analysis, but immediately through a process of projection and identification that starts from our own proper form.

The idea of this criterion of recognizing a person based upon the immediate and exterior perception of his or her shape, the *tselem*, through projection and identification gives sense to the seemingly bizarre considerations of the Talmud and of Maimonides on the subject of abortion and the status of the embryo before forty-one days.

We know that Talmudic law considers forty-one days to be a threshold; before that an abortion isn't considered an abortion, because what is removed isn't yet recognized as a child. The reason given is that "everything that has no human form is not a child" (Babylonian Talmud, *Nidda*, 21a). In other words, an individual is only considered to be a child, with a status different from a thing or an animal, when we perceive a *tselem* in it—meaning a human form, especially its face. Maimonides specifies in his treatise the characteristics that are proper to the human face. As long as the eyes are on the sides of the head, the embryo is still considered an animal form, which resembles, as he puts it, a fly. One must wait until they are near each other in order to perceive a human face, from which, among other things, stems this delay of forty-one days.[2]

For us nowadays, even if the threshold can be modified to a lesser or greater extent according to the problems under consideration, the principle, implying a pragmatic instead of a substantialist attitude, can remain the same. In other words, it is our acknowledgment of a human form and face that causes us to attribute the status of a person, not research into intrinsic properties that would bear witness to a human nature. This position has the great advantage of providing us with an efficient criterion for solving problems concerning limits. In the case of the embryo, it leads us to accept, in some ways, the idea of a pre-embryo that is a shapeless cluster of cells and rely on our immediate sensibility, which does not recognize a human person here. This obviously concerns its present, excluding any project of manufacturing a future person that could possibly result from its development.

This criterion is also efficient with regard to the respect paid to a corpse. It is the still-remaining human form to which the dignity of a person is attributed.

Furthermore, the same criterion easily extends toward a problem that is being more and more frequently posed, namely, animal experimentation.

---

2. One can trace back the origin of this idea at least to Aristotle. Probably under Aristotelian influence, Islam also adopted this criterion, and Christianity as well, at least in its Thomist conception of delayed animation, which the nineteenth-century Catholic Church replaced with that of early animation, i.e., at fertilization.

There too, when one is obliged to experiment upon animals, our sensibility is determined by identifying to a greater or lesser degree with the different possible objects of experimentation. The biologist is thought to be more "humane" when working with cultured cells, even though they might be of human origin, than when working on animals. Why? Because these animals bear a much greater outward resemblance—from the viewpoint of our immediate perception rather than a scholarly one—to man than do cells. It involves the human researcher's relative identification with the object of experimentation. The problem is all the more severe because an experiment carries a tremendous risk that the objects of experimentation will be destroyed or—in the case of animals—"sacrificed," as the specialized literature refers to it. Once more, this identification isn't very scientific: it is an immediate sympathetic identification. One identifies easier with a monkey than with a mouse, more with a mouse than with a frog, more with a frog than with a fly, and more with a fly than with cultured cells—even if those cells are of human origin.

Some want to find a criterion of distinction in the existence of consciousness, which would permit us, for example, to differentiate primates from other animals. However, this criterion is as much open to discussion as any other in that the apparition of consciousness and even of sensitivity is a progressive phenomenon, in phylogenesis as well as in ontogeny. In fact, in the end we must fall back on our immediate sensibility rather than seek illusionary essentialist definitions of the nature of a human being, whether we say it is potential or real.

PROVISIONAL CONCLUSION: A "DIALOGUE" BETWEEN SCIENCE AND ETHICS

One can see that the scientific truth concerning the nature of nature is not always and necessarily the most pertinent for questions concerning law and personhood. That doesn't mean that scientific truth can be ignored or neglected. One does not want to return to the confused obscurantism of the middle ages or to drape the situation in the virtues of religion under the pretext of defending morality. One cannot ignore the merits and techniques of science under the pretext that it has nothing to say about "how to live." Monsieur de la Palisse would doubtless say that we cannot pose problems without knowing how they arise. There is an incessant coming and going between objective scientific research, on the one hand, and the concern for justice and dignity, on the other, to both of which we must adhere—even if that is more difficult today than it used to be, now that the illusion of a built-in unity of truth and justice, the true and the good, has dissipated.

The "I"

We have so far considered the person or the *tselem* only in its generality. What about its individuality, the uniqueness of each person, his or her singularity?

THE "I" IN THE SYMBOLISM OF A GOD OF THE PERSON

Here we encounter the same temptation to seek an objective definition. And it is still toward genetics that we are supposed to turn in order to find the concrete signature of this unity. But here there exists a real danger, because genetic individuality, even if not open to discussion, is not, as we have seen, characteristic of the human. Individual genetic outlay is not necessarily what supports a human person: it implies the possibility of impersonal individuals. The use of genetic imprints should not mislead us. They are, precisely, imprints: that is, traces. Like digital imprints, they mark an individual without determining it—without even acting as its support.

Here, too, it would be better to look elsewhere than toward objective science, for example, to grammar and the symbolism of mythic narrative, for the reality of every human being and his or her singularity, that which each person can consider his or her own, so long as one is conscious and designates that by saying "I."

Such symbolism is suggested by the idea of a "god of the person," which should not be understood as a personal God, who would be a person himself, being one or three, but rather the somehow polytheistic idea of a god of persons in the same way we speak about a god of the oceans, of the forest, of fire, of storms and wind, or of other dynamic realities whose effects we perceive as forces of nature. As we have seen from the outset, the question of the self does not impose itself only at the level of our unconscious cellular activity but, after the acquisition of language, as a "who" or "what" that is capable of saying "I." It is in the tension between these two terms, *who* or *what*, that the reality of the human individual is constantly played out, either as person or as thing. And when we try to identify what is also being said when we say in Hebrew *ani* ("I"), we cannot but be pulled up short by a curious expression that frequently recurs in the Bible, one that requires us to conceive the notion of a god of the person.

This expression is, in Hebrew, *ani* ("I"), followed by the tetragram YHWH (Genesis 28:13, Numbers 28:13, etc.). This is usually translated as

"I am God," "I am the Eternal," or "I am the Lord," depending upon how one wishes to translate the tetragram. But all of this is rather enigmatic. What, for a staunch monotheist, might this discourse of God that finishes by saying "I am God" actually mean? It is either self-evident, from the moment the discourse begins with "God speaks" or "God says" and doesn't need to be repeated as its conclusion, or it isn't obvious, and the fact that it is being repeated at the end doesn't render the divine origin of the discourse any clearer.

Indeed, this expression can be read another way, once we recognize that the verb "to be," in Hebrew, is only implied. In consequence, for the "I am [the tetragrammaton]," one can read "I *is* [the tetragrammaton]" or, more precisely, "an 'I' [is] [the tetragrammaton]." As this is a combination of the letters of the verb "to be" in its three tenses, a literal reading of this expression could be as follows: "an 'I' [is] *was-is-will be*." Therefore, instead of "I am God," or even "I am the being," expression of an unmoving tautology, this formula might be trying to transmit to us, in a schematic fashion: "an 'I' is being in becoming, which designates being in its three tenses." Here, we learn an additional something concerning the Bible's principal character, precisely that which is designated by this name, namely, that it is through the experience of the human "I," the *ani*, that being in becoming is experienced.

In other words, we have here a reversal, in which human personal experience is perceived as an opening to a kind of divinity. The experience of a personal existence, which is what humans do when they think and say "I," maintains its strength, notwithstanding its problematic nature. Being an experience of a natural dynamics, it can be perceived as that which human persons make of the tutelary angels, guardian angels, "spirits," or individual gods that the ancient Romans attached to each individual—as many different ways of perceiving oneself as a matter of history or destiny. The unification of these experiences in an abstraction of a "god of the person" that would designate the *was-is-will be* of the tetragrammaton would thus express the possibility that persons can come together in a human collective person, or at least the aspiration to such a coming together.

THE PARADOX OF THE "I"

All this becomes much clearer when we leave behind the confines of theology for those of psychology. We must realize that what is at stake in this discourse concerning a god of the person is not the existence of the God of theology and the belief in dogmas and articles of faith but the existence of human persons themselves. Thus, the most universal word to denote the divinity of each and every one is the word *I*, with which, in all languages, he or she who speaks designates oneself as a singular subject, the *first person singular*. Human nature, the result of the perceptions of one's own proper body, memory, imagination, and linguistic relationship to others—human and nonhuman—is such that everyone feels at one with the consciousness of his or her own proper personal existence. Everyone perceives him- or herself differently from that which is not this consciousness, given that everyone's existence depends on it, in space and in duration. It is this consciousness that makes somebody say "I," especially when remembering and making plans. But this "I" cannot be localized in space or in time, and it can otherwise be localized only in a trivial way, in the body that speaks: the "I" vanishes like a phantom, a nothingness without concrete existence, each time at the exact moment one stops and tries to denote it.

This is what some kabbalistic interpretations convey when they concentrate upon the fact that the word *ani* ("I") is made up of the same letters as the word *ain* ("nothingness"). For this tradition, the universe is described as an infinite chain of worlds, each one made up of ten categories or spheres in such a way that the tenth in one world constitutes the first in the next. The word *ani* designates the tenth category, whereas *ain* designates the first one. Due to this sequence of scores of tens, one after the other, it is in the same category as zero, a nothingness that is not nothing but that expresses and accomplishes itself at the end of the tenth. Thus the same reality can be seen either as the first initial zero, a nothingness opened to the possible, or as the termination of an accomplished entity that calls itself "I."

The difference—and the tension between the "I," as "myself," an *ego* closed in upon itself, and the "I" as an opening, as pointing toward an infinite that traverses it from side to side—is expressed in a formula that is assigned to Hillel in the *Pirkei Avot* or *Chapters of the Fathers*: "'Im 'ein 'ani li, mi li? Ve-'im 'ani le'atsmi ma 'ani? [If I (am) not for me, who (will be) for me? And if I (am) for myself, what (am) I ?]" (*Pirkei Avot* 1, 14).[3] A simple way of

---

3. Hillel is the author of a negative transformation of the famous biblical precept "Thou shall love thy neighbor as thyself" (Leviticus 19:18) into the Talmudic precept "What you wouldn't want to be done to thyself, do not do to others." This injunction purports to summarize the whole Torah, when supplemented by "Go and study, the rest is interpretation" (Babylonian Talmud, *Shabbat*, p. 31a).

putting it would be somewhat like this: "If I don't care for myself—for my salvation, most notably by my own work—who will do it for me? But if I am reduced to myself and to my own strength, what am I?" But this sentence may also be read in a different fashion: "If the 'I' is a nothingness to me, then a who is for me [meaning, a questioning concerning the personal mode "Who?" is for me]. But if the 'I' is for myself, closed in upon me, then 'I' is a what [a thing]."

Here one can hear an echo of the voluntarily paradoxical statements, in the fashion of Buddhist koans, that break up the appearance of the principles of identity and noncontradiction, like the following statement by Yumen Wenyan, a ninth-century Zen master: "I am I because I am not."[4]

THE "I" ACCORDING TO A MODEL OF PHYSICAL INTENTIONALITY

In the monist context that I announced at the outset, the following question naturally imposes itself: How can this capacity to say "I," with all the connotations that we have briefly glimpsed, be produced by, or associated with, the structure of the human body and especially its brain? Without being able to reply to this question exhaustively, as we can contemplate it only on a completely different level from the plane on which we have situated ourselves so far, I would just like to mention a model that we might qualify as a one of physical intentionality, that is, a model of self-organized networks of automata that simulate neurons, each one obviously nonintentional, functioning in a purely mechanical fashion in such a way that their level of activity taken as a whole causes new, unprogrammed projects to emerge and that resemble decisions resulting from voluntary actions. These ensue (or do not) in execution, and the model simulates the behavior of conscious subjects, whose special mental states referred to as "intentional" seem to be the cause of the brain activity and the resulting bodily movements.

These networks simulate voluntary behavior that is still primitive, similar to what we can observe when looking at great apes capable of using objects as new tools. The movie *2001: A Space Odyssey*, directed by Stanley Kubrick, shows great apes playing with bones up until the moment one of them deals a blow to another with the aid of a bone. He then discovers that these bones can be employed as weapons. He then repeats the movements he accidentally carried out the first time, though this time with the objective of dealing a

---

4. See the article by Victor Sogen Hori in the same issue of *Theologiques* (Hori 2004).

blow. A causal sequence of movements is transformed into a directed procedure, as if the project of dealing blows with a stick had emerged from this
transformation.

This primitive, but by no means trivial, form of intentionality can be simulated on a computer by a self-organizing network coupled to an associative
memory.[5] Very schematically, the model functions as follows: a causal
sequence of states of the network that produced some effect a first time,
possibly by chance (like, e.g., administering a blow) is stored in memory. It
may then happen that the last state of the sequence corresponding to this
effect is recalled on the occasion of other events, in other encounters, by
mechanisms of associative memory. Then, what is recalled and repeated is
not only the last state of the sequence but the whole sequence that produced
this effect the first time. As with a voluntary action, there is an apparent
inversion of time, in which the last state—or a neighboring associated state—
causes the repetition of the sequence that produced it. In other words, a
causal sequence is automatically transformed into a procedure that is apparently directed toward a goal.

What this model of the emergence of a project suggests is that consciousness of the self and of things, although sometimes seemingly directed toward
the future, is in fact a *memory* in this process of constructing procedures, a
reminder of states that are found to be the most frequent because of properties of both the environment and the network. One step further toward
human intentionality would imply another level of memorization. This
would permit us to extend the model of constructing projects toward our
experience of a general *capacity* to fabricate projects, independently of this
or that specific project. In other words, this would permit us to conceive a
mechanical model of our consciousness as consciousness of intentionality,
even though still in the limited and restricted sense that we have
contemplated.

The idea that consciousness is fundamentally memory of past events, not
directed or orientated toward the future, doesn't mean that no novelty whatsoever is to be expected of the future. But this novelty is first unconscious.[6]

---

5. [For more details, see Chapter 12 in this volume and also Louzoun and Atlan
2007.—Eds.]

6. In a reflection upon the relative character of time, defined by the dynamics of different systems that produce their own time, we have suggested replacing "past" and
"future" with "known" and "unknown" (see Atlan 1982: 24–35).

Obviously, the novelty isn't known as such before it occurs, and isn't conscious. When it suddenly occurs, in an unplanned fashion when one is paying no attention, and thus, at least in appearance, randomly—otherwise it wouldn't be novel—consciousness takes possession of it after the fact. Consciousness of the new is thus the result of a transformation of indetermination and randomness into structure and meaning by the dynamics of *self-organized memory*.

This is one of the reasons we conceive our self in a confused way when we seek to locate in it an absolute origin, endowed with free will in that it would be capable, by conscious will, of triggering causal series out of nowhere, without any other cause but final causes, that is to say, giving itself projects in a nondetermined fashion.

In this light, it is interesting to recall the results obtained by Benjamin Libet, who measured the time elapsed between the electrical activity of the brain associated with a conscious decision to make a voluntary movement and its electrical activity corresponding to the movement's initiation. In an apparently paradoxical way, he consistently found that the initiation of the movement in the brain *preceded* by about 300 microseconds, and did not follow, consciousness of having decided to make the move. This apparently paradoxical result, which affronts common sense, is the product of a systematic antedating of conscious cerebral events. It is a kind of temporal illusion constructed by the brain, analogous to well-known optical spatial illusions in which the brain constructs images that are different from what is objectively given. This observation, since repeated, has given rise to many disputes and misunderstandings. However, the latter diminish substantially if one understands, as Spinoza has said and as is suggested by our model of intentionality and the emergence of projects, that "will and understanding are one and the same thing" (*Ethics*, pt. 2, prop. 49, corollary; 2002a: 273); that is to say, voluntary decision and the affirmation of a state of things are one and the same thing, seen from two different temporal perspectives: apparently oriented toward the future, in one case, and toward repetition of the past, in the other.

These models of memorized self-organization show us how, contrary to the idealist conception of consciousness aimed at, oriented toward, a nonexistent future or simply at what is exterior to it, consciousness is fundamentally a memory of states and past processes. Yet creative intentionality, inventor of projects and of new meanings, is the property of self-organizing phenomena that are at first unconscious, produced by the emergence of new structures and functions, as in our automata networks. These new structures and functions appear as such in a secondary activity, when consciousness as memory observes their emergence and memorizes them in turn. As we have suggested elsewhere,[7] our voluntary consciousness is the result of memorizing self-organized phenomena that produce, in an unconscious fashion, the novelty of the future. It is this *association* of an unconscious wish with consciousness-memory that establishes the frail temporal unity of a self. An intentional consciousness or a conscious intention thus does not appear as a primary phenomenon, as a kind of foundation for the self, but as secondary, produced by and through this association.

Translated by Jaron Paktor

---

7. [See Atlan 1979. For a more detailed presentation of the concepts that are presented here, see Atlan 2003, chaps. 5 and 7.—Eds.]

PART FIVE

Fabricating the Living

TWENTY

Golems

(1999)

To compare present-day operational notions with ancient philosophical ideas that are still permeated by myth is not a meaningless game of association. It is useful because such comparisons make it possible for us to step back from the current situation and can help us enunciate the philosophical and sometimes mythological context, which would otherwise remain implicit and unanalyzed, in which these new biological concepts are employed.

By means of the many myths it has inspired, "spermatic knowledge," as found in certain kabbalistic texts, similar to but also different from the seminal reason of the Stoics, can help us define an anthropological context for analyzing the theoretical and ethical questions raised by technologies that manufacture life. In essence: the main difference between the roles played by semen in the myth of the sparks of randomness and in the Stoic notion has to do with the part assigned to chance in making this knowledge efficacious, which significantly alters their common experience of the absolute determinism of nature.

Elsewhere we have discussed the ethical problems posed by present and future applications of biotechnology.[1] We must deal with these problems case by case, paying special attention to the consequences of a particular technology for our biological, emotional, social, and cultural life, to the extent that they can be foreseen. The exercise becomes even more difficult—but also more necessary—because the same technology may produce different effects, some of them desirable, others far from it, depending on the conditions under which it is applied.

Here we propose, instead, to analyze origin myths, whose anachronistic juxtaposition with the new situations created by these technologies can yield new insights. They make it possible to project diverse (and perhaps contradictory) meanings onto the moral and legal positions we are tempted to take in a given situation. This does not make our decisions about what is and what is not desirable any easier. But whatever the decision, and whatever the criteria applied—utilitarian, intentional, religious, or secular—these decisions acquire added anthropological scope and weight from the new meanings conferred on them by the ancestral myth into which they are set.

This is why, rather than describing yet again all of the possibilities cooked up by the untrammeled imagination of Frankenstein-like biologists, we shall begin by examining the myth of the fabrication of a man or woman, as recounted in the many narratives about the creation of a golem by wise and righteous rabbis who, as Rashi reports, employed the teachings of *Sefer Yetzirah*, the Book of Creation, from which "they learned to combine the letters of a [divine] name."[2]

Gershom Scholem devoted a long chapter to this question (Scholem 1965, chap. 5), and now we have a comprehensive study (more than three hundred pages) by Moshe Idel (1990), who reviews the many accounts of golems from the talmudic era through the nineteenth century. These two authors eliminate several misunderstandings based on preconceptions: for example, the most famous golem of all, that of Rabbi Judah Loew, known as the Maharal of Prague (ca. 1525–1609), is certainly apocryphal; the Maharal himself never refers to it in his copious and detailed writings, although, like many other authors, he discusses the theoretical question of the manufacture of an artificial humanoid. The legend of the golem of Prague first appeared long after

---

1. Atlan and Bousquet 1994. See also Atlan 1987b: 209–26, 1992a: 52–63, 1994b: 27–36, and 1997: 5–18.

2. Commentary on B *Sanhedrin* 65b, s.v. "created a man."

his death. What need was this posthumous attribution of a golem to the Maharal meant to fill?[3]

For many, these accounts are merely so many fantastical superstitions or popular legends, like the tales of Faust and Frankenstein. This may be, but the evocative power of myth is undeniably present, as indicated by its resonance with questions that the development of modern science has made increasingly weighty. In 1982, a U.S. presidential commission referred to the

---

3. See Atlan 1992c. One answer to this question may lie in the prescientific magical activity characteristic of the Renaissance and in the Maharal's genuine and documented interest in what would prove to be the earliest accomplishments of the scientific revolution. André Neher (1986), who has studied the relations between the Maharal and his scientific milieu, has proposed an analysis of the golem myth focused entirely on this legendary attribution (see Neher 1987). Drawing a striking parallel with the legend of Doctor Faustus, created a century earlier, he analyzes what he calls "the profound identity between the golemic structure and Faustian soul of modern and postmodern man" (1987: 198), where "Faust is the myth of the modern man and the [Maharal's] golem is the myth of postmodern man" (ibid. 9). Despite the evident similarities between these two myths, Neher underlines the differences, notably the element of demonic possession and the pact with the devil that is the core of Faust's career, quite unlike the case of the golem. The latter's adventure merely reflects the irreducible ambiguity of human nature, both determined and determining, created and creative, bearer of life and death. The subsequent centuries of the machine age have merely highlighted this ambiguity and brought it to the breaking point. It is in this way that the golem, a human machine in every sense of the term, is the creation of postmodern man, as was suggested by the father of cybernetics, Norbert Wiener (1964). As for why later legend credited the Maharal with the manufacture of a golem, we may offer the following hypothesis. In retrospect, the sixteenth and seventeenth centuries were seen as a turning point when modernity struck deep roots, thanks to the scientific and technological revolution in Europe. How the alchemical, magical, and astrological traditions inspired by hermeticism and Kabbalah influenced the origins of this revolution is well known. In this context, as Idel's book clearly shows, fabrication of a golem represented a zenith, evidence of the profundity and divine truth of the knowledge that makes it possible to build one. Better yet, to borrow from the sixteenth-century kabbalist Meir ben Ezekiel Ibn Gabbai, the rabbis' ability to make a golem was, for those who asserted it, an indication of the superiority of Hebrew science and Kabbalah to pagan lore and magic. The Maharal of Prague was a major figure, perhaps the most important, in the intellectual milieu at the intersection of ancient Kabbalah and modern astronomy. It was thus natural that legend attributed to him the knowledge that leads to the ultimate *imitatio Dei*—the creation of a human being—despite the absence of a specific account in his writings. In this way, popular imagination preserved and re-created the memory of the preeminence of the sage of Prague over the other sages of the era that saw the beginnings of modern science, when, later, the formidable practical efficacy of the sciences had become the evidence of that modernity.

legend of the golem in its discussion of the possible benefits and dangers of genetic engineering as applied to human beings: "each of these tales [the Maharal's golem, Dr. Frankenstein's monster] conveys a painful irony: in seeking to extend their control over the world, people may lessen it. The artifices they create to do their bidding may rebound destructively against them—the slave may become the master."[4]

Now we can appreciate the extent to which the traditional literature about the golem, which is much more extensive than the legend of the Maharal, comes into contact, sometimes in quite unexpected details, with the ambiguity of the quest for power through knowledge. When human beings discover that they can deploy their knowledge and technology to create a human being who is "artificial" but who is in every (or almost every) way identical to themselves, the essence of the human condition is revealed in its fullest complexity. Nowhere in the golem literature (at least originally) do we find a negative judgment about human knowledge and creative activity performed "in the image of God"; in this it differs starkly from the Faust legend.[5] Quite the contrary: it is in and through creative activity that human beings achieve their fullest humanity, in an *imitatio Dei* that associates them with God in an ongoing and ever-more-perfect process of creation. But like a two-edged sword, this activity—the essential humanity of human beings—jeopardizes the continuation of the process for which it is, nevertheless, indispensable.

The legend of the golem—a male or female robot, built by wise men—is no marginal or anecdotal theme, but rather a constant presence in traditional Jewish literature. It occupies a quasi-canonical place there because it has the sanction of the Talmud itself (B Sanhedrin 65b). In a discussion of the various types of sorcery condemned by the Torah, we are told, by way of contrast to these practices—which, though impure, are nevertheless considered to be efficacious—that the righteous have the power to create worlds and that only their transgressions deprive them of this ability. To illustrate this statement, we are told that the fourth-century sage Rava "created a man." The talmudic account continues: "He sent him to R. Zeira. R. Zeira spoke to him, but received no answer"—because, Rashi explains in his commentary, the man in question was not endowed with the power of speech. Whereupon Rabbi Zeira told him: "You are a creature of the rabbis [thus Rashi, glossing the

4. President's Commission for the Study of Ethical Problems in Medicine and Biomedical and Behavioral Research 1982: 58.

5. See Neher 1987.

Aramaic *havrayya*; others understand the word to mean "sorcerers, Magi"]. Return to your dust." To drive the point home, the text continues: "R. Hanina and R. Oshaia spent every Sabbath eve [i.e., Friday] studying the *Book of Creation*, by means of which they created a fat calf and ate it."

Given this narrative, reported by the Talmud and vouched for by Rashi as literal truth, the possibility of making a golem was an incontrovertible part of rabbinic thought. Some, such as Maimonides, did treat it as symbolic. But other leading rabbis, such as Isaiah Horowitz (1565?–1630),[6] one of the leading halakhic authorities of his time, read it as a tradition about an efficacious and authentic lore in which the ancient sages were proficient, thanks to their great wisdom and holiness. Moshe Idel, in his book on the golem tradition, lists many kabbalistic accounts of the various techniques used to create one. The question, at least in some cases, is whether these authors believed in the material reality of the being created in this fashion. For many, the golem was a purely mystical phenomenon produced by meditation on the letters of the divine Name, it being understood that combinations of letters are insufficient to produce a material being.[7] The thirteenth-century itinerant kabbalist Abraham Abulafia, himself a great adept of meditation on the letters, mocked

---

6. Often referred to as the *ShLaH*, from the acronym of the title of his major work, *Shenei Luḥot ha-Berit* (*The Two Tablets of the Law*), first published in Amsterdam in 1649, after his death in Tiberias, following a long rabbinic career in Poland, Frankfurt am Main, and Prague. The book's reference to the ancient traditions about the creation, by application of *Sefer Yetzirah* (pt. 3, *Vayeshev*, p. 103), of artificial human beings and animals is particularly fascinating, given its wide circulation in the Orthodox rabbinic Judaism of the period. Horowitz refers to the tradition in an excursus from his commentary on the biblical text, offering a possible solution to an exegetical problem. Commenting on Genesis 37:2, which states that Joseph slandered his brothers to their father Jacob, he quotes the midrash that fills in the details of this report: Joseph asserted that the brothers were having illicit relations with local women and were eating meat that had been cut from living animals—a gross violation of the Noahide laws that were incumbent on them. Horowitz propounds a dilemma: either they were really guilty of these misdeeds, which is hard to accept of Jacob's sons, the ancestors of the people of Israel; or the accusation was false, in which case Joseph was a liar. As a way out, Horowitz cites the tradition that *Sefer Yetzirah* was written by Abraham and handed down to Isaac and then to Jacob and his sons. As the talmudic account affirms, this would have made it possible for them to create artificial animals to which the prohibition of unslaughtered meat would not apply; similarly, by following its instructions they could have created artificial women . . . But Joseph, not knowing this, believed that what he had seen were women and animals with two natural parents.

7. Thus the eighteenth-century kabbalist Hayyim Joseph David Azoulai, quoted in Scholem 1965: 188–89.

those who took the talmudic account literally and believed it possible to create a calf by studying *Sefer Yetzirah*: "Those who do so are themselves
calves."[8]

Whatever the case, the creation of a golem, far from being sacrilegious,
was seen as the fulfillment, by means of the secrets of the Torah, of a long
and difficult ascent toward wisdom and holiness. This is, no doubt, a perilous
path; but, like the Torah itself, it can be "a potion of life or a potion of
death,"[9] depending on the student's orientation and intentions. Of course,
the golem may be more or less perfect, depending on the standing of its
creator. We have seen that Rava's golem was still quite imperfect because it
lacked the power of speech; since it was not really a man—perhaps only a
robot—R. Zeira had no scruples about destroying it. But we also hear of
other golems whose makers were evidently on a higher level and able to
endow them with the power of speech and sexuality.

In texts of the thirteenth-century German Pietists, cited by Idel, we read
that the prophet Jeremiah created a golem. This was not an act of rebellion
against God but the culmination of a long path ascending toward holiness
and knowledge, which are necessarily coupled in any *imitatio Dei*. Indeed,
how can we know that an initiate has truly deciphered and understood the
principles of creation other than by seeing that his knowledge enables him,
too, to create a world? The only way we can be sure that his knowledge of
human nature is correct is that it permits him to create a human being. The
truth criterion of the sage's knowledge—like the criterion of scientific truth
today—is empirical.

Unlike Rava's golem, Jeremiah's artificial man could speak and soon
rebuked the prophet: Was he aware of the confusion he might introduce into
the world? From that day forth, whenever people encountered a man or a
woman, they would no longer know whether it was God's creature or Jeremiah's. The prophet, who does not seem to have thought of this before, asked
the golem how he could repair the situation. The artificial man replied that,
just as Jeremiah had made him, he could unmake him: "Reverse the combinations of the letters." The lesson would seem to be that we need not renounce

---

8. Ibid. 188. See the Appendix to this essay.

9. B Yoma 72b. It is "a potion of life" (*sam ḥayyim*) for one who learns with "fear of
heaven" or is a craftsman of the Torah (*'uman*) and rejoices in his learning. But it is a
potion of death, and even of double death, for those who do not. As Rashi explains, such a
person inherits hell twice: he loses this world by renouncing the common pleasures of life,
but also loses the world to come because his learning is sterile.

the attempt to attain the total knowledge that would make us able to create a human being; once we have achieved it, however, we must abstain from applying it. This is a great lesson, which merits serious reflection.

First, we must wonder about this confusion that the prophet Jeremiah, despite his holiness and wisdom, had not foreseen, so that his own creature had to enlighten him. If the man he made was as perfect as a man conceived and born in the normal fashion, what was Jeremiah's mistake? What was wrong with what he did? The fault lies in the ambiguous nature of this project of creative knowledge, which is simultaneously good and bad.

Jeremiah was not aware of this problem because, as in the Promethean projects of modern science and technology, ambiguity vanishes from the horizon of the learned man, even the sage, the moment that his creative activity carries him to the heights where evil no longer exists and, consequently, there is neither good nor evil. The success of the knowledge-driven ascent toward holiness and beatitude carries him, as it were, into a world in which the Tree of Knowledge fuses with the Tree of Life, where Adam's transgression can be repaired. It is a world out of time, a world of eternity, "the world to come" (*'olam ha-ba'*) rather than simply a "future world." With regard to our existence, which is in time, this flaw, inherent in the perfection of eternity, is expressed in part by the relationships with time, death, and imperfection that are intrinsic to the golem narratives: one does not *kill* a golem, because, in the abstract eternity of the world in which it is fabricated, the golem's time is reversible. Instead, it is *unmade* by means of the same formulas and operations used to fashion it, carried out in reverse order, or by removing the letter *alef* from the word *'emet* ("truth"), engraved on its forehead, leaving the word *met* ("dead"). In other words, golem time, unlike the time of human life, is reversible, just like the fully deterministic time of mathematical physics. By contrast, when randomness is introduced into a machine, when the determinism is no longer totally controlled, time becomes irreversible. That is the kind of time we experience in our lives.

Translated by Lenn Schramm

Appendix: The Word Golem

The word *golem* has several meanings, which support diverse interpretations concerning the nature of this artificial being. In the talmudic narrative, the being created by Rava is a *gavra* ("man"), not a golem. The first use of the word with the sense of an artificial human being is much later. In its only occurrence in the Bible (Psalms 139:16), *golem* is variously rendered as "embryo," destiny," etc. and taken to apply to Adam, the "natural" man. In Midrash and the Talmud, *golem* designates an intermediate stage in the creation of Adam from the dust of the earth, before God breathed a living soul into him (Genesis 2:7, Naḥmanides on this verse, Gen. Rabbah 14:8), when he was still a shapeless mass (B Sanhedrin 38b) or, in other versions, a lump of clay with human shape but not yet alive (Leviticus Rabbah 29:1). By extension, a women who has never had sexual relations with a man is called a *golem*, in the sense of the raw material used by a potter for an unfinished vessel. According to Maimonides (commentary on M *Avot* 5:7: "Seven traits [are characteristic] of a *golem* and seven of a wise man"), following the lead of M Kelim 12:6 and Rashi's commentary thereon ("unfinished metal vessels"), this sense of something crude and unfinished is the most general meaning of the word. The same root appears in the modern Hebrew *ḥomer gelem* ("raw material"). In the introductory section of a liturgical poem that recounts the ritual performed by the High Priest in the Temple on Yom Kippur (part of the Additional Service for the Day of Atonement), the word refers to the raw material from which the first man was created.

Thus from the outset the word *golem* expressed the ambiguity found in all its later occurrences: both an "inanimate human body," like a statue or a corpse, which is in some sense the raw material without the imposed "form," in the Scholastic sense of that which provides the essence and life, and an "animate human form"—Idel uses the term "anthropoid"—that is, a copy of a real man or women. For the Spanish kabbalist Joseph Gikatilla (1248–ca. 1325), a golem is a human body animated only by the *nefesh*, which is the part of the soul that is closest to the body and common to human beings and animals (a sort of vegetative soul that provides the capacity for movement), but which still lacks the affective soul and the intelligent soul (*Ginnat 'Egoz* [Hanau 1650; Jerusalem 1989], p. 208). Following this line, the question is whether Rava's creation was a "real" man. As we have seen, the answer was obvious as soon as R. Zeira discovered that the creature lacked the capacity for speech and recognized it as an artificial being. But this is not always the case, as we shall see; there remains the theoretical possibility that one human being could fabricate a "real" human being who in every respect resembles a person with a father and mother. Thus the manufacture of an artificial human being raises the question of the nature or essence of man. It is in this context that the word *golem* designates a human being made by another human being, by means of the lore derived from a natural science that is held to be "true" because it is identical with that employed by the Creator to create the world. Most often, nevertheless, the synthetic being is an unintelligent brute, a caricature of a man or women. This is the sense in which Avot contrasts the golem with the wise man or sage.

But we cannot exclude the possibility that, thanks to its maker's knowledge, virtue, and purity of intention, a particularly successful golem might be endowed with speech and intelligence. What is more, because we all know that brutes devoid of rational intelligence—golems in that sense—can be produced in the normal fashion by a father and mother, the question of the human or nonhuman status of an artificial anthropoid remains open. We are left with the "form" of the human body and face—the *tselem* that the golem reproduces no less than the authentic man Adam—as the only criterion for the humanity of a living being.

This permits the assimilation of golem and *tselem*, of the matter of the body being formed with the form of the human body. Thus the "man" produced by Rava can be interpreted not as a golem in the sense of an artificial man, but as a symbolic representation of the world in the shape of a human being. This is the theme of the first man in Genesis, whose dimensions, according to Midrash, were those of the entire universe—what Idel calls a *macranthropos*. This theme was later taken up by the Lurianic Kabbalah as the primordial man, *Adam Qadmon*, both created and creator, the telos of creation as well as the formative intermediary between the Infinite (the *Ein Sof*) and the infinity of worlds and formed beings, to the point that he himself is sometimes referred to as the *Ein Sof*.

Even some kabbalists who hewed to a literal interpretation of the talmudic text saw the "man" created by means of combinations of letters from *Sefer Yetzirah* as no more than a robot—endowed, perhaps, with animality, but not with true humanity, as indicated by its inability to speak. This seems to have been the case, in particular, of the philosopher-kabbalists Solomon Ibn Gabirol and Moses Cordovero. It is interesting to compare this attitude with what can be inferred from the stories (with their interpretations) about Albertus Magnus, theologian, philosopher, alchemist, teacher of Thomas Aquinas, and the inspiration for Meister Eckhart and medieval Neoplatonism. On his deathbed, Albertus Magnus defended himself against charges of having practiced black magic, but stated his desire to "verify several elements of this art" in order to know "whether it contained even the tiniest grain of truth," in his effort "to study thoroughly everything that the human mind has attempted" (see de Libera 1990: 12). Albertus, too, was said by legend to have created an "android" during thirty years of labor "with the favor of heaven and the planets, a veritable speaking statue, a Venus of Ille [a reference to the story by Mérimée] or a golem," which his pupil, Thomas Aquinas, is supposed to have discovered and, horrified, shattered to bits. Contrary to this legend, Albertus was interested only in the theoretical aspects of the operation of an automaton, in connection with the wooden goddess that Daedalus was supposed to have made (ibid. 13–14).

Finally, as Yehuda Liebes remarks, in a review of Idel's *Golem* (*Ha'aretz*, August 2, 1996, p. D2), it is not always easy to distinguish between the material sense of a golem, whether a "true" artificial man or a robot that is more or less androidal, and the allegoric or symbolic sense in which "making a man" may mean healing the sick or reviving those who seem to be dead, as in the story of the son of the Shunnamite matron, restored to life by the prophet Elisha (2 Kings 4:34). Although the *Zohar* does not explicitly mention the fabrication of a golem, it does explain that the prophet effected this miracle by writing the divine name of 216 letters on the child, as in certain techniques for making a golem (*Zohar*, Introduction, p. 7a–b). "Making a man" may also mean teaching someone the road to salvation, as in the traditional understanding of Genesis 12:5, which refers to the persons whom Abraham and Sarah had "made" in Haran. This allegorical interpretation with a pedagogical sense may remind us of the myth of Pygmalion and its twentieth-century literary avatar. In Ovid's *Metamorphoses*, Galatea is an ivory statue; the sculptor falls in love with his handiwork and begs Venus to bring it to life. In George Bernard Shaw's *Pygmalion*, she is a flower girl whose horizons are expanded by a professor of philology. Finally, in the context of the kabbalists' mystical experiences of the supernatural, the metaphorical sense can be assimilated to the intrinsic sense if one accepts that sages operate in a world perceived to be different from that of our normal experiences. What we consider to be symbolic they take to be real, like alchemists, whose experiments in fact focused on themselves through the symbolism of their project of transmutation.

In the kabbalistic tradition, the manufacture of a human being by a human being, whether taken literally or metaphorically, is part of the sage's imitation of the divine creation; study of the words of the Torah (perhaps by means of the letters of *Sefer Yetzirah*) associates this imitation with the unfinished work of ongoing creation. This is stated explicitly in the *Zohar*, shortly before the interpretation quoted above about the child restored to life:

> It is written, "And I have placed My words in your mouth, and hidden you in the shadow of My hand, to plant a heaven and to lay the foundations of an earth" (Isaiah 51:16). It does not say "the heaven" but "a heaven" [implying that the reference is to a new heaven and not the old one]. . . . The verse continues: "And to say to Zion, 'You are my people (*'ammi*).'" Do not read *'ammi* ("my people") but rather *'immi* ("with me")—i.e., be My collaborator. Just as I made heaven and earth by a word, as it says: "By the word of the Lord the heavens were made" (Psalms 33:6), so should you do [by devoting yourself to the study of the Torah]. (*Zohar* 4b–5a)

In the ellipsis above, the *Zohar* explains that the "shadow" refers to the fact that God covers over the words of the Torah when they are placed in the mouth of Moses and the righteous, to protect them from the "jealousy of the angels"; the angels, envious of the material nature of the human body, try to prove that the body makes men weak and unable to receive the divine words, which will necessarily incinerate them. Whence the need for God to hide both human beings and the words until "a new heaven and a new earth" have been established.

TWENTY-ONE

The Mother Machine

(2005)

As early as 1985, U.S.-based researcher Gena Corea foresaw that the medicalization of procreation would result in a continuous progression from artificial insemination and in vitro fertilization to an "artificial uterus" and cloning. In her startling and well-documented *The Mother Machine: Reproductive Technologies from Artificial Insemination to Artificial Wombs* (Corea 1985a), Corea criticizes the age-old exploitation of women as "mother machines" bound to the reproduction of the species or society *for men*, in accordance with the classic patriarchal schema. Corea's principal target, however, is biomedical power itself (made up of individuals she terms "pharmacrats"), insofar as its techniques of procreation enable a radical twentieth-century renewal of this long-standing exploitation of women.

Several studies have insisted upon the pathogenic—one could even say iatrogenic—role played by hypermedicalized procreation: for example, a 1983 work by Geneviève Delaisi de Parseval and Alan Janaud. Corea's study, however, appeared as part of an ongoing exposition of masculine domination and the dispossession of women's reproductive agency, phenomena that the technical skill of the pharmacrats intensifies. This work was not Corea's début; she had already published, under the provocative title *The Hidden Malpractice: How American Medicine Mistreats Women* (Corea, 1985b),[1] an investigation revealing that the negative consequences of medical practices affect women more frequently and more systematically than men.

In general terms, Corea's work supported various feminist movements, which in turn supplied her with abundant source material. Her emphasis, however, is on showing how reproductive technologies offer new means for the masculine appropriation of women's reproductive capacities. Herein lie the interest and originality of her book, as well as its limits—due to the very generality of its central claim, and despite the precision of the examples she marshals in its support. We should immediately acknowledge that her thesis is convincing on more than one point. It reminds us that the medicalization of birth and delivery, following the shift in authority on these matters from midwives to obstetricians, was accompanied by the slaughter of countless newly delivered mothers, who died of postpartum fevers as a result of infections transmitted by the doctors performing the deliveries, infections originating in corpses the same doctors had previously been dissecting. This was in the nineteenth century, before the role of microorganisms had been recognized and before hygienic practices—at a minimum the washing of hands—had generally been adopted. But this is only the beginning of a long history of medical interventions, paved with good intentions, to be sure, but in which women's bodies have been unjustifiably sacrificed on the altar of reproduction. Even in the twentieth century, the Distilbene scandal provides another example (well known and well documented today). Intended to prevent miscarriages, Distilbene instead caused vaginal and ovarian cancer in girls whose mothers had taken the drug.

Faced with the accumulation of (at times lesser-known) examples compiled by Corea, we cannot content ourselves with invoking the inevitability of medico-surgical accidents, not least because it is difficult to find equivalent examples in which men's bodies provided the victims. Furthermore, the observations doctors offered to justify their practices—given in good faith and also, perfectly naturally, taken up by female doctors—often prove quite revealing about the ideological aspect of medical intervention in reproduction, a dimension of medical practice all the more powerful because it is

---

1. [This is an updated version of Gena Corea, *The Hidden Malpractice: How American Medicine Treats Women as Patients and Professionals* (New York: Morrow, 1977).— Trans.]

largely unconscious. The treatment of infertility—including male infertility, long hidden from public view—serves to justify practices that in other branches of medicine would be difficult to accept or would even be strictly forbidden on ethical grounds. For the author of *The Mother Machine*, the colonization of human reproduction by reproductive technology is simply an extension of the mastery of animal reproduction achieved by the "cattle industry." Deprived of its status as the representation of a primitive mother goddess, the cow has become a mother machine, an object of exploitation for its milk and for industrial reproductive experimentation. The "woman industry" has followed its lead. This recognition of the cow raised for meat or milk as the first instance of desacralization produced by technology recalls, curiously, the remarks of J. B. S. Haldane, though here in the context of a polemic antagonistic to Haldane's thinking.[2] Whatever we may think, it is noteworthy that recently developed techniques of in vitro fertilization, in which pathological and at times immature spermatozoa are locally injected into eggs, have been carried out on women without prior animal experimentation, in violation of every accepted rule of experimental practice on humans. It would appear that, in the field of reproductive technology, experimentation on women can dispense with the usual rules of sound medical practice. This frankly astonishing observation joins many others accumulated by Corea in support of her thesis of the medical mistreatment of women via the practices of reproductive technology.

It is in this context that the concept of the artificial womb, along with cloning, was presented, beginning in 1985, as an ineluctable technique of human reproduction.

Yet Corea's thesis is flawed in its unilateral and overly systematic character. Nothing is said of the interests, masculine or feminine, that ectogenesis would serve. Human reproductive cloning could have been presented as the eradication of men and their sperm from the process of procreation. In defense of ectogenesis, the desire of single women or lesbian couples to have *their own* biological children is often presented as legitimate justification for medical procedures. In her description of the reproductive technologies

---

2. [John B. S. Haldane, a geneticist, coined the term *ectogenesis* in 1923. He is discussed at length in Atlan 2005, chap. 1. His remark here refers to Atlan's earlier discussion of Haldane's view of this "desacralization." According to Atlan, for Haldane biological technologies first appear to be a perversion; then they come to be accepted and are even transformed into social ritual. (For Haldane, this seems to be a good thing, whereas for Corea it is an instance of relationship gone awry.)—Trans.]

already in use in the United States—artificial insemination, embryonic transplantation, the increasing commercial availability of in vitro fertilization, surrogate mothers (for infertile couples or single men), sex selection, and so on—Corea pays little attention to the liberating effects of some of these techniques. The initial practice of artificial insemination by donor (AID) presents, in her opinion, an exception that confirms the rule: the procedure was welcomed by women who saw it as a way to have children by themselves, or at least without a known male partner. But medical power in the service of patriarchy was quick to try to reappropriate the technique by limiting its application to traditional family contexts. Actually, it turns out that France is one of very few countries in which AID is effectively prohibited outside of the "parental project" of stable heterosexual couples; in most developed countries, including the United States, the major site of Corea's observations, AID can be obtained by single women or lesbians without need of further justification. In a similarly one-sided move, Corea remains silent about pills—birth control pills, abortion pills, and the morning-after pill—although their industrial production is the result of technical achievements in reproductive biology that has played a significant role in feminist struggles for the right to contraception and abortion.

In other words, as convincing as Corea's account is, it is marred by its dichotomous character. The feminist movement is presented as if it had no other goal than the reappropriation of reproductive capacities, which have been monopolized by men. By contrast, current and future reproductive technologies count as nothing more than means at the service of patriarchy to effect an even more radical dispossession of women's reproductive agency. We could cite evidence for this line of argument in literature from antiquity to Freud on the menacing character of the "dark continent" of the female sex and of the hidden machine where the embryo takes shape, out of sight, beyond control. To this literature we can now add a flowering of more recent and more precise declarations by reproductive biologists, arguments sometimes taken up by "bioethicists," whom Corea is also quick to cite, regarding the future "benefits" of artificial wombs: prevention of abortions, prevention of embryonic damage resulting from mothers who are alcoholics, drug addicts, smokers, or who simply don't eat a healthy diet—or, more generally, a more thorough understanding and control of the various stages of embryonic and fetal development, finally accessible to medical observation and perhaps ultimately to intervention, therapeutic or simply "ameliorative," genetic or otherwise. Ultrasound, with which we are now familiar, prenatal exams, and the current state of neonatal medicine and surgery will then look like a prehistory, and a primitive one at that, compared to the ubiquitous medical regulation that ectogenesis will permit. Control over human reproduction will have passed definitively into the hands of the "pharmacrats" in the service of patriarchy. Even if female doctors and biologists take part in this process, even if women "volunteer"—while actually being conditioned to sign supposedly "enlightened" statements of consent—for the necessary experiments, they will only reinforce the masculine domination of the realm of human reproduction and contribute to rendering it irrevocable.

The rigorous argumentation of *The Mother Machine* and its accumulation of testimony regarding the explicit aims of reproductive technology, as revealed by participants—sometimes eminent ones—in the revolution now under way, make this book a valuable document. Twenty years after its publication, the debates foregrounded by Corea retain their currency.

In fact, the work's major defect, the one-sidedness of its argument, can be seen as a strength if it is taken as an incentive to debate, an injunction to future philosophical and political reflection about how to prevent the potentially deleterious, even barbaric and inhuman, consequences of these technologies. Furthermore, the author seems aware of the dangers of restricting resistance to the "pharmacrats" to a feminism exclusively focused on the reappropriation of procreative power and a return to the era of the primitive mother goddess, in which men were excluded from the drama of reproduction. She calls for the development of a "philosophy of birth" that (male) philosophers have thus far neglected. Marx, Freud, and the philosophers of existence have pushed the limits of thinking about how vital necessities organize our conscious and unconscious representations in terms, respectively, of subsistence, sexuality, and death. Reproductive consciousness alone, with its different modalities—the feminine continuous and the masculine discontinuous—that accompany birth and the raising of children, has remained a fallow field, perhaps because it has been contested terrain in the everlasting battle between the sexes. Taking up the analysis of Mary O'Brien, theorist of this battle and of male uterus envy, Corea nonetheless distances herself from a theory of reproductive agency that would limit itself to this dimension alone. She draws, therefore, on other feminist authors, such as Mary Daly, who caution against theories of uterus envy that lead to a fetishistic fixation on women's reproductive organs—whether to glorify or to disparage them—at the expense of "a female creative energy in *all* of its dimensions" (Daly

1978:60; cited in Corea 1985a: 299n). Corea also recalls the remarks of Elizabeth Fisher, author of *Woman's Creation*, about celebrations of women as mothers or nourishers that in fact reduce their role to slavery, the cult of their "natural powers" threatening the recognition of their humanity.

Yet in Corea's book reservations about a philosophy of birth overly fixated on uterus envy are confined to one footnote. By contrast, the work's main argument plows on ever deeper in a single direction: that of combating the dispossession of women's creative power by means of technology. In this context, Corea introduces the "artificial womb," with the reproductive cloning it will enable, and the "reproductive brothels" that will re-create the delirium of the Nazi *Lebensborn* project (but with greater efficiency, thanks to technical advances) as a hitherto unattained summit of masculine domination and the exploitation of women's bodies.

Things are not so simple, however: in this domain as in others, we cannot just assume the worst. By trying too hard to drive home its argument, *The Mother Machine* ignores everything that might contradict it, whether such contradictions concern biomedical practices or feminist claims. With the single exception noted above, when it comes to the search for a multidimensional philosophy of birth, Corea seems to want to ignore the various—even opposed—orientations of different currents of feminism. She also doesn't seem to realize the fundamental ambivalence of technologies, including those of reproduction, insofar as their social effects are concerned. The ambivalence of technology here is effaced and covered over by condemnation pure and simple, with no room for appeal.

And yet it is important to keep in mind precisely this multiplicity of possible effects if we want to free ourselves from the risky game of extrapolation concerning future biotechnological achievements and their anthropological consequences. From this point of view, consideration of the sociocultural consequences of future ectogenesis constitutes a double challenge. On the one hand, in addition to the properly therapeutic advantages we could imagine, we cannot ignore the potential effects on the conditions of birth and child rearing, or the multiplication of these effects through ectogenesis's facilitation of other practices, such as reproductive cloning and all sorts of conditioning, genetic or otherwise. This aspect of the question has already been raised.[3] On the other hand, this situation brings to light the contradictions in various feminist claims.

---

3. [See Atlan 2005, chaps. 5 and 7.—Trans.]

Modern feminism developed in the context of the individualism of the rights of man, which imposed itself as the dominant ideology of Western democracies. Following the foundational assertion of Simone de Beauvoir, "one is not born, but rather one becomes, a woman," completed today by Élisabeth Badinter, who affirms that "one is not born, but rather one becomes, a man," the *identitarian* current of feminism asserts the equal rights of women and men on the basis of their essential identity, holding the differences between the genders to be socially constructed. But a *differentialist* current has also developed in a reactionary manner,[4] on the basis of differences both biological—where maternity, actual or potential, plays a determining role—and psychological, which constitute properly feminine moral characteristics. Taken to the extreme, these differentialist models confer a sort of "natural" moral superiority on women, resulting from their more immediate relationship to nature and to life, experiences that arise in the maternal relationship—above all, that between mother and daughter, potentially passed on by lesbian relationships. This ideology culminates in a feminist form of "deep ecology," in which the human domination of nature is denounced as the domination by men of everything that is not men, that is to say, women just the same as animals and trees. Thus arises a form of feminism that renews in modernity the ancient battle between the sexes. Across the oppositions of these different currents, however, there exists a common denominator of modern feminist struggle, rooted in women's right to control their own bodies. For Nancy Huston, this deeply seated right plays the role of a

> moral conviction which, though not always explicit, is not for this reason any less capable of being expressed. This conviction can be formulated more or less as follows: every girl and every woman, whatever her national, cultural, or religious background, must be respected in her integrity and considered as the sole proprietor of her body. This body cannot and must not, on the basis of its belonging to the female sex, be sold, exchanged, beaten, mutilated, penetrated by force, or obligated to bear a child." (Huston 1986)

While the struggle for equality, especially economic and political equality, is far from having attained its goal, this widely shared conviction constitutes the principal force of feminism, of whatever current. This is why, once its technical achievement is possible, it will be very difficult, if not impossible, to hinder the diffusion of ectogenesis if it is considered, like contraception

---

4. ["Identitarian" here corresponds to Atlan's *identitaire*; the neologism "differentialist" follows Atlan, who coins the term *différencialiste*.—Trans.]

and abortion, to be part of the claim to this right by those who, for whatever reason, wish to employ it. Yet it is easy to see how such an argument will conflict with the struggle for the maternal-uterine control of procreation *against* the power of men developed by the "pharmacrats."

It is therefore probable that many women will not wish to renounce this control, be it real or imaginary. Without even entering into the dynamic of the battle between the sexes, and still in the name of the individual woman's right to control her own body, many women are not prepared today to give up bearing children and giving birth to them. Far from seeing pregnancy and delivery only as servitude to be avoided, these women see them as an extraordinary blessing and happiness, a privilege missed by men and by women who do not experience them. Yet if tomorrow's societies remain individualist and democratic on the model of contemporary Western democracies—which is far from certain, incidentally—we can bet that things will evolve as they did with respect to contraception and abortion, and nursing as well: women will claim and will obtain the freedom to have children as they choose, mobilizing their bodies to this end for nine months or employing "their" artificial uterus for this labor. We should note that, in the second case, their bodies will still be put to use, for less time, certainly, but long enough for the removal of an egg by a slight surgical intervention. Equality between the sexes would thus remain imperfect, as the removal of sperm from men spares them such intervention—unless, of course, between now and then technology advances even further, making the removal of ovules less and less invasive, or even dispensing with this step via the laboratory cultivation of stocks of ovules identical to those each woman carries in her ovaries.

Whatever may happen, the feminist declaration of the second half of the twentieth century, "a child if I wish, when I wish," will be completed by "*as* I wish," and technology will have to follow, for better or for worse. As we have seen,[5] the availability of and the demand for new reproductive technologies reinforce each other. "A child when I wish" seemed to follow as a matter of course from "if I wish." But the technology that facilitated contraception, even rendered it a banality, was not and still is not equally effective for allowing women to bear a child at any age. Whence follows the explosion of medically assisted reproduction, allowing women who could have had children at

---

5. [Atlan 2005, chap. 3.—Trans.]

a younger age without medical assistance to have children later in life, potentially after having assured their professional careers.

When this evolution reaches a new level with the availability of the artificial uterus, new feminine and masculine identities will necessarily emerge. New relationships, never before seen, will be established between those whom we will continue—perhaps—to call men and women.

What will comprise the masculine and feminine genders and their articulations in a world where the asymmetry of the sexes in reproduction will have disappeared? The roles of progenitors, paternal and maternal, will be practically the same, reduced to the provision of cells for fertilization or other modes of procreation derived from future reproductive technologies. The range of social and familial structures invented by human society is already expansive, as ethnological research has abundantly demonstrated. But the constant of sexual reproduction and maternal asymmetry has constituted a foundational constraint, around which structural variations have formed. Tomorrow, this constant will perhaps cease to exist. So, in an apparent paradox, the hypermedicalization of procreation will end up de-biologizing the relationship of parents to each other and to their children. Continuing and perhaps completing a transformation already in progress, kinship will be increasingly a social function, and less a matter of biology. This is at least a possibility for future generations. But nothing is certain as yet; it is very difficult, if not impossible, to predict what will make up the future of a society in which the nature of the genders and their relationships will have been shaken up to such an extent.

If we want to venture out onto this uncertain terrain, we will need to think in terms of imaginary *gender prospectives* to replace the contemporary sociological and historical concepts of *gender studies.*[6]

Translated by Cara Weber

---

6. [In English in the original.—Trans.]

TWENTY-TWO

Human Cloning: Biological Possibilities, Social Impossibilities

(1999)

Human cloning: what reasons can we give for a recommendation that it be completely prohibited or authorized only in certain cases? Contrary to what is generally assumed, it is not easy to answer this question in a clear and straightforward manner. On the contrary, the more we wish to present clearly and convincingly the arguments that are likely to support our position, the more we come to discover how arguments we thought were sound can in fact be questioned, can be turned back against us, and are difficult to stabilize. That is what I want to show, by examining the main reactions to the prospect of human cloning, which has now become conceivable from the viewpoint of biological technology. I will only initiate this discussion, for this whole book [Atlan, Augé, et al. 1999] is dedicated to analyzing the difficulties we encounter when we try to grasp the stakes and consequences associated with cloning. Let me add that these first considerations reflect, at least partly, the evolution of my own personal attitude. While trying to provide a rationale for my early positions, I came to acknowledge that they were not supported by strong enough arguments. By searching for indisputable arguments, I have realized that these questions cannot be addressed in the way I first conceived.

Two Types of Cloning: Reproductive and Nonreproductive

Before we begin, I want to define what we are talking about and therefore to recall, from a biological standpoint, what is meant by "cloning." Today this term generally refers to interventions that are scientifically distinct and that it is important not to confuse. To recall this distinction is important not only from a technical point of view but also from the viewpoint of our use of language. First of all, it amounts to distinguishing between two types of human cloning that can be called, for want of better terms, "reproductive," and cell-lineage "nonreproductive" cloning.

The first type of cloning, reproductive cloning, consists in having a new individual be born by way of a nucleus-transfer technique. It aims to reproduce genetically identical organisms by transferring the nucleus of a somatic cell into an enucleated ovum, that is, an ovum deprived of its nucleus.

In discussing the terminology of cloning, one may wonder why these techniques, which result in the birth of a new individual, have been called "cloning." The usage is quite new. Formerly, when researchers carried out nucleus-transfer experiments—there had been a whole series of these, in which nuclei were transferred into enucleated ova in order to produce embryos that would then develop—nobody had the idea of calling this technique "cloning."

The nonreproductive cloning of organisms consists in using either the same technique of nucleus transfer or other techniques of cellular cloning proper, that is, of production of genetically identical colonies of cells by way of successive divisions from a single cell. These techniques can apply to both embryonic and nonembryonic cells and can lead to the production of cell lines or tissues. They can also produce cells endowed with all the potentialities of an embryo, but the technique is not designed to bring growth to its completion, and therefore to result in the birth of a child. We will return later to the medical perspectives offered by the ongoing development of these techniques and the quite varied ethical problems they raise.

Embryo splitting is yet another technique, which starts with an embryo produced in the usual way, either in vivo or in vitro, and which is designed to achieve artificially what nature does when it produces twins. It amounts to taking the fertilized embryo when it is still a cell, letting it divide into two cells, and then separating these two cells in such a way that each of them comes to produce an embryo. That is the way identical twins are produced in nature. But this can also be done artificially in vitro. This technique is widely used on bovine and ovine embryos.

Reproductive cloning by way of nucleus transfer is, therefore, fundamentally different from embryo splitting, which enables one to get several identical twins from a single fertilized ovum by separating the embryonic cells produced by early cellular divisions. Embryo splitting takes as its starting point an embryo produced by way of ordinary sexual reproduction, that is, through the fusion of a spermatozoid and an ovum, which associates chromosomes, recombines parental genes, and results in a unique genome, different from both that of the father and that of the mother. Thereafter, twins produced by embryo splitting obviously have the same genome, as real, naturally produced twins do, but there can be very many of these "twins." An attempt to apply this technique to human embryos resulted in dozens of twin embryos. However, for ethical reasons, these embryos never were implanted in utero, and there was no attempt to bring their growth up to the term of pregnancy. The embryos used for splitting were chromosomically abnormal (triploid), so the possibility that they might develop normally was excluded from the outset.

One now sees how different reproductive cloning is—for instance, as it has been carried out on the ewe Dolly. In this case, the embryo is produced by way of asexual reproduction: its nucleus, with the totality of its genome, comes from the cell of an adult organism, without any new fusion of gametes or recombination of parental genes. This nucleus is then transferred into the nucleus of an ovum that has been deprived of its original nucleus. If the ovum were removed from the same adult female from which the nucleus to be transferred has been removed, then the result would be close to a parthenogenesis. In Dolly's case, the female from which the nucleus was taken and the female from whom the ovum was taken were different individuals. If the organism giving the nucleus had been a male, this would have produced a genetically identical male organism, almost a real twin of its "father," just as Dolly is almost a real twin of her genetic "mother."

Of course, nothing stops us from imagining the association of these two techniques. The cloning of an individual by way of nucleus transfer and the splitting of the embryo thus produced, repeated a certain number of times, would allow for the birth—if the embryos are implanted in several female uteruses and if pregnancies come to term—of several individuals genetically identical to one another and to the individual from whom the cellular nucleus was removed.

Finally, not every nucleus transfer is cloning: recently, a baby was born to a woman whose ova were affected by a disease of the cytoplasm and could not develop after fertilization. An in vitro fertilization of one of her ova by one of her husband's spermatozoa could be carried out, and the nucleus thus produced was transferred to the enucleated ovum of another woman. The embryo could then begin to develop normally and was thereafter reimplanted into the uterus of the first woman. It is easy to see why, in that case, we are not dealing with cloning: there had been fecundation by a spermatozoon, fusion of gametes, and constitution of a new genome, identical to no other, as in any instance of ordinary sexual reproduction. In that case, nucleus transfer allowed for the replacement of the faulty cytoplasm of the fertilized ovum by that of another woman's ovum. Accordingly, the other woman, in a way, plays the role of a "surrogate" mother, not by lending her body for pregnancy but by giving an enucleated ovum.

The great scientific interest of the experiments on nucleus transfer in animals is that they allow to study the respective roles of the nucleus—which contains the embryo's genome—and the cytoplasm of the maternal ovum. Experiments had been carried out for decades, but to no avail. Despite the achievements of pioneers such as John Gurdon (who experimented on batrachians), researchers admitted that a systematic impossibility existed, since embryonic development was held to be determined by a "program" that was fully contained in the genome of the embryo. As cells divide and differentiate to produce the various organs of the adult, the activity of their genes is modified by this developmental "program." These modifications were thought to be irreversible: a liver cell, a kidney cell, a brain cell, or the cell of any other organ, is the result of the "commitment" of the embryonic cells that have produced it, via a path of differentiation in which only a part of the individual's genome is active and produces the cells of this organ rather than those of another.

The reproduction of Dolly brought about a revolution: it delivered experimental proof that differentiation is not irreversible and that it partly depends on nongenetic factors linked to the properties of the cytoplasm. Thanks to Dolly and to the successful application of the same technique to other animals, most notably and most recently to mice, it has been discovered that the genome of an adult cell behaves like that of the initial embryonic cell: it recovers all its potential to produce the various organs of the adult organism when it is implanted in an ovum's cytoplasm. It is now believed that the genome as it had earlier been differentiated, its activity becoming limited to the cell of a specific organ or tissue, is "reprogrammed" by the ovum's cytoplasm. Now, this cytoplasm does not contain genes (except for the DNA of mitochondria, whose role in development, as things now stand, is quite improbable). Cytoplasmic proteins play a decisive role in the activity of the developmental program, not only at the initial stage of the fertilized egg, but also throughout the process of embryonic differentiation. Accordingly, this developmental program is in fact temporally and spatially spread over the set of interactions between DNA and regulative proteins, and over the temporal sequence of the various states of activity produced by the dynamics of these interactions.

The genome can be compared to data stored in memory, rather than to a computer program. The metaphor of genetic memory is probably more appropriate than that of the genetic program. Contrary to what has long been claimed, the organism controls the activity of the genome at least as much as the genome controls the development and activity of the organism. That is why experimental reproductive cloning by nucleus transfer matters so much for understanding the dynamics of complex interactions between genetic and epigenetic determinations that take place during embryonic development.

Reproducing Human Beings Through Cloning?

The reproductive cloning of a human being would consist in the production of an embryo by way of nucleus transfer from a somatic or embryonic cell, then its full development up to the birth of a child. If the cloned cell were a somatic cell—removed from an adult or a child—the result would be a child whose genome would be identical to that of the original adult or child. If the cloned cell were that of an embryo, the result would be an induced quasi-twinship, slightly regressed in time and not restricted to two exemplars. If these techniques were to be applied, one or more children would be reproduced through asexual reproduction, like copies that would be chromosomically identical to one another and to the organism from which they originated. On the biological level, it would amount to producing individuals more or less numerous, as genetically identical as real twins are, but born in a way shifted both in time and with respect to generations. As we have seen, this *reproductive cloning*—leading to the full development and birth of a child—must be distinguished from nonreproductive cloning, which applies only to adult somatic or embryonic cells and does not produce an embryo that will reach the term of its development.

The cloning of human cells, that is, the reproduction through culture of a great number of genetically identical cells originating from a single initial cell, has long been used for numerous applications within biological research and medicine. In most cases, this technique does not raise any ethical problem when the cloned cell is an adult somatic cell. The cloning of embryonic cells can raise ethical problems linked to the conditions of experimentation or research on the embryo. The problems raised by the prospect of human reproductive cloning are very different, for they bear on the status of children who would already have been born and then of adults who might have been produced in such a way. These are major ethical and legal problems, which bear on the identity of the human person and the very definition of humanity. All ethics committees that have been consulted have recommended a complete ban on this practice, at both national and international levels, even if this means that the issue must be reconsidered in a few years.

Several reasons contribute to rendering the development of such a reproductive practice unacceptable—at our present stage. It risks leading to a serious moral regression in the history of mankind. But it is important to analyze in detail the reasons supporting this ban, since they do not all partake of the same kind of argument and do not all have the same weight, given the circumstances.

Would Human Reproductive Cloning Constitute a Crime Against Humanity?

Some have suggested that the practice of human cloning—when it consists in reproductive cloning—should be compared to a crime against humanity. This comparison of human cloning to a crime against humanity rests on an analogy with the forced matings that took place in the human stud farms set up by the Nazis, with the goal of creating human beings in accordance with their ideology. The crime would be even greater in the case of reproductive cloning, for the individuals thus created would no longer benefit from the unforeseeable aspects of sexual reproduction and the genetic uniqueness it guarantees.

The idea of reproducing something exactly unquestionably plays a role in the feeling of horror that, spontaneously and prior to any reflection, is often inspired by the prospect of human cloning. However, it is important to analyze in detail the relation between the uniqueness of the genome of every individual and the uniqueness of the person, to whom ideas of dignity and respect are linked. There is ample empirical evidence that, despite the similarity of their genome and the likeness of their physical appearance, identical twins are not the same in the constitutive elements of their individuality. Neither the structure of the nervous connections in their brains nor the structure of their immune systems is totally determined by their genes, for the history of their development includes epigenetic factors and partly random elements. This is even truer of their psychic personalities, whose history is influenced by biological, social, and cultural factors. Those personalities cannot be considered identical, pure and simple. In their persons, such twins are no less unique than any other human being. Reproductive cloning—producing individuals from the nucleus of a cell taken from a male or female donor and thereafter multiplied by embryo splitting—would result in a situation like that of identical twins. Indeed, the donor would probably be less like its clone or clones than identical twins are, although we do not yet know how and how far the likeness would manifest itself. Identical twins have in common not only the genome of the initial fecundated ovum from which they originate but also the cytoplasm, with its protein factors. By contrast, an individual produced by nucleus transfer from an already-existing organism has in common with the latter only the DNA of its chromosomes and the other elements constitutive of the nucleus.

That is why the argument that reproductive cloning would erase the uniqueness of a person does not hold. As the American National Bioethics Advisory Commission has argued, even for those absolutely opposed to this practice (in that instance, the Catholic Church), human cloning would be an offence to human dignity, but in no way would the dignity of the person resulting from cloning be diminished.

In fact, the reasons for banning human reproductive cloning are more social than strictly biological. Moreover, each of these reasons could be turned around, given particular circumstances that would allow for possible medical applications of reproductive cloning as a new technique of medically assisted procreation. However, unlike already-existing techniques of assisted procreation, reproductive cloning has not yet been applied to the human species. Furthermore, even when it is applied to animals, the technique is not yet safe, especially with regard to the normal development of the individual thus produced.

That is why the question whether it is lawful to transpose onto man and woman techniques aimed at producing a child through reproductive cloning has so far received negative answers. In its report to President of the French Republic Jacques Chirac (April 1997), the National Advisory Committee on the Ethics of the Life and Health Sciences "considers that there is cause to oppose, by all means available, the development of practices aiming to reproduce exactly a human being as well as research that may lead to that end." (Note that this recommendation does not concern only reproductive cloning but also embryo splitting.)

The National Bioethics Advisory Commission of the United States considers irresponsible, in the present state of our knowledge, any attempt at human reproductive cloning and recommends that a federal law banning it be passed, although that law could include a clause that would allow for the question to be reconsidered after a certain delay.

Finally, in most European countries legislation prohibits, in one way or another, human reproductive cloning.

In addition to the fact that cloning technology is not yet reliable when applied to animals, several reasons are put forward to justify these bans. Their common denominator is the offense to human dignity. This is briefly mentioned, for instance, in UNESCO's "Universal Declaration on the Human Genome and Human Rights." But this notion of an offense to human dignity can be viewed differently depending on the social consequences one can foresee if this practice were applied and on the motivations at the origin of these applications.

Chaotic Kinship?

Individuals produced by reproductive cloning would be genetically identical to the twin brothers or twin sisters of those from whom they were cloned, but their birth would be shifted in time so that they would belong to the generation of "children" or "grandchildren." Such a situation might totally disrupt all known human landmarks with respect to kinship. Although anthropologists have described systems of multiple filiations that are quite different from the one traditionally established in our societies, no system of kinship does without one of the two biological parents pure and simple, since they all rest on the universal experience of sexual reproduction. Asexual reproduction as carried out by reproductive cloning would disrupt any existing system of filiation and might eventually result in the abolition of filial relations. In addition, the coexistence within the same population of persons born of a father and a mother and of persons born by way of genetically asexual reproduction would create problems of civil identity difficult to solve, as well as social conditions of possible discrimination that would be morally unacceptable.

From the Risk of Instrumentalization to the Risk of a New Slavery

Reproductive cloning leading to the birth of human beings would necessarily obey aims external to the human beings produced in that way, for they would be the result of projects whose goal would be, by definition, the exact reproduction of a determined genome. The organism of an individual would be created to serve as a means of expression for a genome chosen by a third party. The genetic lottery would be abolished. Now, as we have seen, because of the role of cytoplasmic heredity and epigenetics, the biological identity of an individual cannot be reduced to its chromosomal genetic identity. That holds even more for the identity of a human person in his or her broader social and cultural dimensions.

Nevertheless, the production of a human person by reproductive cloning would result from an aim—explicit and planned—external to that person, an aim that would not be the autonomous and unpredictable flourishing of that person. It therefore follows that accepting such fabrication amounts to denying the possible autonomy of the person by alienating him or her from the outset via an instrumental project that would define, both in the literal and the figurative meanings of the term, his or her "identity card."

In fact, human reproductive cloning would totally disrupt the relations between genetic identity and the identity of the human person in all its dimensions. The uniqueness of each human being, on which human rights and the dignity of the person rest, is *visibly* expressed as the unique *appearance* of one's body and one's face, which results directly from the uniqueness of one's genome. Identical twins are the exception—relatively rare, for it is restricted to brothers and sisters born at the same time—that allows one to imagine roughly the social reality that would be created by making adult clones, in varying numbers, who might come into being out of phase with the usual course of generations. Although their genetic identity would not be equivalent to their identity as persons—and they would be human beings in their own right, individualized as persons—they would be *seen*, both in the literal and the figurative meaning of the term, as exact replicas of one another and of the ancestor from whom they had been cloned. The symbolic value of the human body and face, seen as bearers of a person's uniqueness, would tend to disappear.[1]

Unlike the ewe Dolly, who knows nothing about genetics and ignores being a clone, unlike the sheep around her, who equally ignore that they are not clones, *human clones would know that they are clones and would be known to be such by other humans*. These human clones could then be considered as different "races" or as subhuman or posthuman varieties of the human species. They would be produced according to an aim external to themselves. Their existence would tend to be instrumentalized and might be reduced to a new form of slavery, within which clones would be used as a means of expressing the alleged qualities of their genomes and would have been selected for that very reason. They might thus become the slaves of their genomes, as well as the slaves of the other human beings, who would have created them with that aim. Yet their peculiar human character would remain as irreducible to their genes as is the case for other human beings, and one could imagine that their very humanity would probably lead them to rebel against that state of affairs. But their fabrication, far from being a progress, would be a social and moral regression that would lead to recreating the conditions of a new kind of slavery.

---

1. A talmudic passage about the unique appearance of every human being is particularly relevant to our considerations. It first appears as a remark that could be called factual: when a man imprints several shapes with a single seal, the shapes all look alike. Although the Creator has imprinted the shape of each man (and fixed his nature) with the same seal that he used for the first man, nobody looks alike. That is why everybody must say to himself: "It is for me that the world has been created." But the text does not stop there. It goes on to ask a rhetorical question: "Why is it so? Why are the faces not all the same?" To which it answers: "So that if somebody sees a beautiful house or a beautiful woman, he does not say: 'She is mine'" (Babylonian Talmud, *Sanhedrin*, 37a and 38a). In its particular style, this text starts by underlining both the universal and the particular character of humanity in each individual. But the "reason" put forward to explain why it has to be so is not metaphysical; it does not refer, for example, to an essence of the human person that would disappear if faces looked alike. The reason is simply social: one would no longer know who is who, and there would be chaos in both family relations and the organization of property.

What Reasons Might Support Possible Authorization?

Some of the witnesses before the American commission envisaged reproductive cloning only in an individual dimension, restricted to certain particular situations, rather than within a social context in which it would be used as a somewhat common mode of reproduction. According to this conception, cloning would be only a new way of satisfying the individual desire of having a child when no other technique of assisted procreation could be applied. In these cases, the possibility of medical applications of cloning was raised, and that led to an analysis of the motivations that could justify the possible use of this technique.

It is hardly worth mentioning the fabrication of clones, whether adults or children, that would serve as stocks of transplantable organs. This would properly constitute a revival of ancient practices of human sacrifice.

Other applications have been evoked for a candidate's own cloning or that of his or her relatives. These requests always involve a phantasmagorical aspect, which seems to have its roots in ancient myths of reincarnation and immortality, reinterpreted in pseudo-biological terms. The resonance of these myths in the individual and collective imagination must lead us to a heightened vigilance concerning the prospect of using this technique for pseudo-medical reasons.

Parents, for example, might want to reproduce by cloning a child they had lost. A child produced under these conditions would in fact be a new person, but she would also be, in her parents' eyes, the dead child revived, because of physical resemblance and the obviously false idea that genetic identity is equivalent to complete identity. A further step has been taken by those who have expressed a desire to have a dead spouse, or any other dead relative, cloned. In the representations underlying these desires, it is as if the genome of an individual were endowed with the attributes of the soul according to ancient traditions. The idea of the immortality of the soul, seemingly incarnate in the permanence of the molecular structure of genes, naturally leads to the idea of a reincarnation falsely grounded on a mythical conception of genetics.

Similarly, those who proclaim themselves candidates for their own cloning often seem to be motivated by the same muddled view of an immortality produced by the preservation of their genome's structure in an individual who would start a new existence while being supposedly the same.

In all these cases, the representations involved make light of the personal dignity of future individuals, who would be produced as a planned means of fulfilling these fantastical desires. In no way could biomedical technology serve such confusions without perverting its scientific and ethical nature. Medicine would then be serving a magical, pseudo-scientific thinking, at the origin of projects that would scorn the dignity of persons to come.

One occasionally envisaged medical application of this technique consists in compensation for male or female infertility resulting from a total absence of gamete production. The reproductive cloning of an adult cell of a man or a woman in this situation, perhaps using the cytoplasm of the female partner's ovum, would give him or her a twin who would be a substitute for a child. In some cases, a couple might even put forward a parental project, as required by French law, and propose to use one ovum of the woman in order to receive and activate the nucleus of one of the man's cells. The genetic individuality of this child (a boy, in that case) would be the exact copy of that of his father, which would include the possible anomalies of the genome that are responsible for the man's sterility. In that case, too, the instrumental character of the production of such a child, that is, according to an aim external to the child, and the abolition of its genetic indeterminacy are obvious. It is difficult to see how the desire to have a child at any cost could justify this practice. Just as there are heroic treatments medicine must know to give up, it seems that here we go beyond the threshold of procreative acceptability when we replace impossible procreation with asexual reproduction.

Yet, a few biologists and philosophers and even some Protestant, Jewish, and Muslim religious leaders, consider that we are dealing here with a case in which, contrary to the general rule, human reproductive cloning could be morally justified. This leads us to contemplate, by way of concluding, the possibility of other exceptional cases in which reproductive cloning might appear to be an acceptable therapeutic practice.

Acceptable Cases?

One very particular case is often presented as a first-rate example of a possible medical justification for reproductive cloning. It is that of a child suffering from leukemia who could be saved by a bone-marrow transplant, but for whom there is no compatible donor. The parents could then ask for the production of a clone of that child, from whom bone marrow could be taken without threatening her life. She would obviously be raised like any other child, and perhaps even more lovingly, since she would have helped save the life of her sister or brother. This possibility is considered by some to be morally justified. Others reject it with horror, insisting that the child would not be produced for herself but in order to serve as a means to cure the sick child. The Kantian principle according to which one must not use humans as a means is sometimes invoked, forgetting that the child may be desired *both* for herself and as a means to cure her brother or sister. Accordingly, she would not necessarily be used merely as a means. In fact, the desire to have a child is, most of the time, somewhat ambiguous: it has as its object as much the satisfaction of the conscious or unconscious desires of the parents as the flourishing of the child to come—considered as an autonomous individual independent of her parents' projects. Finally, it has already occurred that the parents of a child suffering from leukemia have conceived a child naturally so as to be able to use her bone marrow to cure the sick child. However, contrary to the case of a clone, there was no certainty that the bone marrow of this child would be compatible with the sick child. Here, one sees that we are faced with borderline cases, in which moral judgment is particularly difficult and for which it would probably be necessary to examine the details of every particular situation.

So, with respect to the uses of human reproductive cloning in the present state of technology and knowledge, the existence of such borderline cases must lead us to ask the following question: Is the possibility, in these exceptional situations, of arguing for the moral justification of human reproductive cloning sufficient for society to authorize the development of the technique and its first applications to humans, with all the dangers of social disruption and moral regression we have briefly evoked? It seems that this question must receive a negative answer, as illustrated by the current legislation of many countries and by the recommendations of all the national and international ethics committees that have been consulted. One may hope that, during the time that would be necessary for the research and development of this technique, the progress of other biological and medical techniques—notably the *nonreproductive* cloning of human cells and tissues—will allow the particular cases for which human reproductive cloning today seems to be the last resort to be treated in other ways.

The Prospects of Nonreproductive Cloning of Human Cells

Using embryos resulting from abortions, or using surplus embryos resulting from in vitro fertilization, it is possible today to cultivate cells (also known as embryonic stem cells) that might develop as specialized cell lines, that is, as tissues, and maybe one day as organs. Research in that field is active in some countries (the United States, Great Britain) that either do not or only loosely restrict research on embryos in the days following fertilization.

Now, let us suppose that we remove from a patient in vital need of a transplant—for example, a child suffering from leukemia—the nucleus of one of her cells (e.g., the nucleus of one of the fibroblasts of her conjunctive tissue, located underneath her skin). Let us suppose that this nucleus is transferred into an enucleated ovum—coming, for example, from her mother, her sister, or any other woman ready to contribute to her recovery. Let us suppose that a totipotent cell is thereby artificially created—through the "reprogramming" of the nucleus—with the same properties as those of an embryo to the extent that it will produce embryonic stem cells by dividing itself. Finally, let us suppose that a technique capable of producing from stem cells cellular lines, tissues, or even the organ necessary to the patient can be perfected. The patient would then benefit from a perfect transplant, since the cells, tissues, or organs, being genetically identical to her own cells, would pose practically no rejection problems.

It is sometimes claimed that this amounts to instrumentalization. But is that really so? Undoubtedly, cellular elements—the nucleus, the enucleated ovum—and cells created and cultivated from these elements are instrumentalized for a therapeutic purpose. We have long been instrumentalizing nonembryonic human cells for numerous biomedical applications without that raising any specific ethical problem. So, we would be dealing here with a cell artificially created through nucleus transfer, without any fertilization or fusion of gametes, which we hold to be an embryo only because of the possibility that a clone of the individual from whom the nucleus has been removed might develop—on condition, of course, that it be placed in the indispensable favorable environment, namely, a female uterus.

Let us recall that a human embryo is considered to be a real person by the Catholic Church as soon as fertilization takes place, even if this has not always been the case, since the thesis of late animation by the soul—which has been accepted, under various forms, by other religions, notably Judaism and Islam—had long been accepted by the Church. Let us also recall that the notion of the potential human person was invented to guard against misuses of human embryos after fertilization. Following this, in France fertilization is held to define the moment at which a cell is considered to be an embryo, by contrast to the United States and the United Kingdom, where the notion of a "pre-embryo" (lasting until fourteen days after fertilization) is allowed.

In the case of nonreproductive cloning we have just imagined, the totipotent cell is produced *without fertilization*, from the nucleus of an adult cell and a nonfertilized, enucleated ovum. In fact, the latter might even come from another species, for instance, from a cow—as has already been carried out—which eliminates the need to retrieve an ovum from a woman. However that may be, *the totipotent cell produced by way of nucleus transfer is not an embryo with regard to the way it has been produced, although, in certain conditions, it might have the properties of an embryo to the extent of possibly developing into an adult individual*. One now clearly sees how the essentialist definitions that ascribe to one single cell properties that in fact result from numerous interactions between the elements of that cell and its successive environments come to disappear. All that can be argued is that there exist, within this cell, the potentialities of a clone's embryo, just as they exist in an ovum or a spermatozoon, although in another way; these are the potentialities of potentiality of a human person, which is quite different from a potential person and even more different from a real person. It is therefore perfectly consistent, on the one hand, to consider that an individual born through cloning would be an individual in his or her own right and to ban this practice for that reason and, on the other hand, not to consider the therapeutic instrumentalization of a totipotent cell produced through nucleus transfer as an instrumentalization of an embryo.[2]

Translated by Vincent Guillin

---

2. [For a more recent statement on this question, see Atlan 2007.—Eds.]

TWENTY-THREE

Possibility in Development

(1991)

Negation, Totality, and the Possible in the Development of the Individual

Today, in the aftermath of two scientific revolutions, the former during the seventeenth century, the latter now taking place, at the heart of the question of education lies the status of truth. But what kind of truth are we talking about?

Given that the purpose of education is to form a human person, it is clear that the transmission of knowledge is not restricted to the transmission of objective and disinterested knowledge. Accordingly, the kind of truth modeled after logical and scientific truths is not the only one we can encounter. To realize this we must examine negation, totality, and the possible, not as words or concepts, categories or pseudo-concepts playing a particular role at the periphery of logical discourse, but as dynamic processes that accompany one another in human individualization and socialization.

Within this context, negation, in its most archaic form (i.e., understood as the process of separation and individuation of the child from her mother), consists in breaking the continuity between mother and infant by establishing an intermediate zone, which no longer belongs to the mother and is not yet the reality of the child, the originary negativity of the "real-unreal" (Winnicott), which will later be expressed, as soon as the basics of language have been acquired, by the *jouissance* of the "No," through which the individual to come will assert herself by her opposition.

It is in education's first circle—that of the family, whatever its structure may be—that patterns of event-perception and behavior start to be transmitted. There for each individual the core of habits and mores characteristic of a given society are constituted. Mores determine to a large extent the modalities of parental transmission, but the main feature of parental transmission is the fact that the workings of the unconscious make it very difficult to know what is transmitted. The child picks up from her parents or familial environment not only behavioral habits but also a set of stimuli that shapes her sensitivity without herself or her parents being aware of how this takes place and what it leads to. In other words, the results of this formation are quite unpredictable, even if psychoanalysis tries to untangle it. And the role of *negation* is the most important factor of indeterminacy in this process: what is transmitted by parents to children is both their image and its negative, according to the well-known effects of reaction and opposition. Both the command "Do this, don't do that" and its opposite, which immediately inverts it by ordering "Don't do this, do that," are transmitted. It is quite difficult to determine which traits—whether of character or behavior—are transmitted as such and which reversed. Parents are neither systematically imitated nor systematically opposed and rejected. The mixture of these two impulses, which constitutes the underlying strata of the child's personality—associated with even more ambiguous genetic determinations—creates something unique that it is not possible to predict, even less to control or to program, as they now say.

Superimposed on this familial formation are the effects of the social environment as they are immediately perceived in daily life, beyond the parental circle. But these effects—originating in the tribe, the village, the town, the region, or the country—are not just superimposed on the familial environment, for they also influence the intricacy and combination of the positive and negative transmissions originating from it. Although psychosociology endeavors to discover laws that would allow us to understand its mechanisms, the transmission of knowledge that takes place at this level has become more

and more difficult to analyze. Here one can observe the effects of an undefined demand that is issued by these transmissions—transmissions of mores
and habits by the family, and transmissions of knowledge by the third vehicle
of education, namely, school. So long as access to the social environment is
restricted to the tribe, within which a seemingly immutable ritual (because it
modifies itself only very slowly, with the passing of generations) determines
mores, this demand elicits effects that are relatively predictable, at least with
respect to behavior, if not to sensibility. It reduces to a large extent the indeterminacy caused by the genetic and unconscious workings involved in familial transmission. In a way, at least on the surface, it tends to make individuals
"trivial."[1] But even the mores that are the most tightly regulated and ritualized can only repress—without eliminating—negative drives, which are
exerted not only on what comes from parents but also on what comes from
other sources of power within the social organization. These negative
impulses are at once deadly—to the extent that they tend to sever the individual from the group on which she depends for her life—and liberating, to the
extent that they prevent the stifling of her personality by the impersonal rules
of society. That is why submission to the law always involves a range of possibilities, ranging from complete submission to rebellion and including a dialectical mode of coexistence exemplified today by those called "dropouts."
Of course, the distribution of individuals across that range of possibilities
differs widely according to the kinds of societies one considers: an individual
who would be a dropout in a closed tribe or a modern totalitarian regime
might integrate well in a democratic and permissive society; by contrast, an
individual who is well integrated in the hierarchy of a traditional society or
a bureaucracy might be marginalized in an open society. Eventually, some
institutions allow the status of dropouts to be defined by submitting them to
particular rules, either valued or deprecated: those of the sacred, the artist,
the institutionalized lunatic, the offender, and so on.

But none of this matters in the end: the effects of the direct transmission
of mores and knowledge by the social group as a whole also display a degree
of indeterminacy, which is, of course lesser or greater according to the societies considered but which remains irreducible. Moreover, this same demand,
rooted in multiple factors, balanced in a different and unpredictable way for

---

1. [In the sense of Heinz von Foerster's famous distinction between trivial and nontrivial machines as a starting point for recognizing the complexity of cognitive
behavior.—Eds.]

each individual, constitutes her as a *whole*, as a singularity whose limits are fuzzy. At the same time, it introduces the individual to the *idea* of the whole, of the totality: the wholeness of the "world," of the "universe," which is grasped starting from *her* world, *her* universe.

The *no* and the *whole*, negation and totality, thus are bound up with the acquisition of language. The question of education is present throughout this process, since language both transforms the child into a *talking* individual and constitutes an essential element of her integration into the group, without which the language proper to that individual would not be spoken.

That is how the ground is prepared for school to intervene, for the third circle, in which the question of education rests at the level of the acquisition and development of knowledge—that is, of an explicit and conscious kind of knowledge, whether learning techniques or introduction to the laws of nature or society. One of the main stakes of education, at this stage, is access to the worlds of the *possible*.

Possibilities of Persons or Potentialities?

The possible and the potential play different roles in vernacular and scientific languages that are far from purely speculative or devoid of practical significance. This can be seen in their application to an old problem that new biotechnologies have brought back to center stage in recent years and that is at the heart of a public debate with significant moral, religious, and legal implications, namely, the problem of the relations between the human person and the biology of development and individuation. At what stage in the development of an embryo can one begin to speak of a human person? The more biotechnologies allow us to intervene in this development and the more they provide us with opportunities to use the products that result (the sampling of cells or embryonic tissues for therapeutic or research purposes), the more an answer to this question becomes necessary. On it depends whether or not one has a duty to grant this embryo the respect and dignity that in our societies we grant a human person or whether we have a right to use embryos as objects for aims to be determined and judged independently of them.

Different cultures, different religions, and different philosophies offer different answers to this question, but developmental biology is of no great help in the matter, for the notion of person, unlike that of individual, is far more sociocultural than biological.

From a pragmatic standpoint, however, many agree to grant the embryo an intermediate status, with respect to our rights and duties toward it, by considering it neither a simple object nor a person comparable to a human being who has already been born and has a name and a place in familial and social structures. This intermediate status is, of course, difficult to determine. This is why, given the temporal character of its development (and with an eye to a future when it might be possible for the product of these transformations to be recognized as a real person), it has been suggested that at present the embryo be defined as a potential person.[2] Our previous considerations show to what extent this definition remains ambiguous and why it is eventually of no use for solving the problem.

Taken literally, and with the connotation of biological potentiality we have already pointed out,[3] this notion of a potential person assumes that the whole of the person—the totality of her attributes—is already there, potentially, in the embryo and is ready to express itself, if all the obstacles acting to repress it were to be removed. Now, potentialities for development do exist in a fertilized egg; they also exist before fertilization, even if fertilization sets up a biological threshold, that of genetic individuality. But, above all, these potentialities are only those of a biological individual and how they play themselves out *varies with the course of development*: it is not the case that, in the fertilized egg, the individual is contained as a totality of properties already present in the form of repressed potentialities. What is lacking is what will result from interaction with the environment. This kind of development, which is called epigenetic, will result in both more and less than the outcome of the genetic determinations that are already present and that are not to be confused, contrary to what the metaphor suggests, with a deterministic computer program.

---

2. See Comité Consultatif 1988. In *Dire l'éthique*, Philippe Lucas offers a pertinent analysis of this text. He underlines the difference between the normative, nonfactual character of the notion of potential person and the factual but non-normative character of the potentiality (which is only one among several) to lead subsequently to the development of a person: "In other words, the embryo *is only* the potentiality of a human personhood, but *should* nevertheless be recognized as a potential human person. This ethical step marks the current moment in a discussion that, for the National Advisory Committee on the Ethics of the Life and Health Sciences [*Comité consultatif national d'éthique*] has not been concluded" (Lucas 1990: 194).

3. [Atlan is referring to an argument concerning biologically determined potential in an earlier section of Atlan 1991a, "The Question of Education."—Eds.]

A fortiori, with regard to the human person (and not only with regard to the biological individual), one may indeed claim that some potentialities of a person exist in the embryo. (The same may also be claimed for the unfertilized egg and the spermatozoid). But in it also exist numerous *other potentialities that could result in something other than a human person*. First, it might eventuate in an aborted development; as is well known, under "natural" conditions—that is, in the absence of any intervention—an early interruption in development and the death of the embryo at a stage before it has even been noticed is the "normal" fate of a great number of fertilized eggs. Moreover—as, for instance, in the case of an anencephalous child—it may be the outcome of an abnormal development that is not viable without a voluntary intervention and that results in a permanent and irreversible lack of the minimum of autonomy, not to speak of freedom. We acknowledge this in handicapped and impaired individuals, yet we nonetheless treat them as real persons. With regard to abnormal development, the technical possibility of creating "chimeras"—that is, half-human, half-animal individuals—by grafting embryonic cells from another species (as with the half-goat, half-sheep chimeras that have already been created in labs), is quite enlightening for our concern, independently, of course, of the truly monstrous and medico-legal aspects of the matter. Provided they are carried out early enough, such grafts are not rejected, although once the very first steps of development have begun, they quickly become impossible. This means that the potentialities contained in a fertilized egg genetically belonging to the human species are initially probably more numerous than those of a human individual: they are indeed potentialities of life, and even of individuation, but they are more numerous than those of a human individual and, a fortiori, of a person. In other words, embryonic development consists not only in expressing potentialities but also in *reducing* potentialities so as to limit them to those of a human.

It follows from this that nowhere does there exist a human person who is already there, just waiting to be expressed—no more in an embryo than it does in the project of having a child. Of course, nothing prevents us from considering an embryo or a fetus to be a *real* person, just as we do a child or a handicapped adult. In that respect, we pass a judgment that might eventually be justified from a moral, social, or even political standpoint, but that is totally arbitrary from the standpoint of objective knowledge. What we mean by a potential person is in fact a possible person, that is, an unreal person (who is therefore not the bearer of any of the attributes of a person and who can impose nothing on us from the perspective of her rights and our duties).

But we willfully deceive ourselves when we think we grant her the status of an objective reality that the notion of potential has acquired in physics and biology.

Accordingly, totality, negation, and the possible are, respectively, transformed into the countable, subtraction, and potential; just as intentionality is transformed into formal finality and the principle of the optimum. This does not mean, however, that they have no usage without this transformation. Of course, there is no question of reverting to their precritical ontological and theological usages, in which, starting with the representation of the totality as a thing, the idea of Being or of God served as an alleged foundation for rational theology. But their use in everyday language, with no other meaning than a pragmatic one, must not be ignored.[4]

Translated by Vincent Guillin

---

4. [See Chapter 6 of this volume.—Eds.]

PART SIX

Ethics

TWENTY-FOUR

Does Life Exist?

(1999)

Let us take a look at some considerations of a historical nature, which I think it is important to take into account.[1] Let me first dispel any misunderstandings: the evolution of biology that I am describing to you, moving from the dominant paradigm of "everything is genetic" toward something much more balanced, integrating the interaction between genetics and epigenetics, obviously does not entail a return to the vitalist representations predating the advent of molecular biology. On the contrary, in a certain way this new representation is much more mechanistic than that of classical molecular biology, since it tends to free itself from the notion of program, itself an end-seeking, teleological one. To questions about the origin of the genetic program, the classic response is always: "It's natural selection." Natural selection would thus play the role of a sort of programmer who writes genetic programs. Without wanting to enter here into a discussion of the legitimacy of

---

1. [This is the second part of *La fin du "tout génétique"?* being the continuation and conclusion of Chapter 5, above.—Eds.]

this comparison, I emphasize that the approach I am proposing allows me to offer a more mechanical representation of these phenomena.

The Essence of Life

Current conceptions have led to misunderstandings and new difficulties in the propagation of information concerning biology, with important consequences for debates relating to problems in biomedical ethics. These misunderstandings concern the new signification, or the absence of signification, now accorded to the notion of life. Already at the beginning of the twentieth century, Albert Szent-Györgyi, who discovered vitamin C, scandalously declared that life does not exist. Following upon this "good news" announced by Szent-Györgyi, my entire approach consists in putting a few more nails in the coffin—and in recognizing, in effect, that life does not exist as such, at least not as an object of scientific investigation, since its mechanisms can be entirely reduced to chemical interactions. These interactions sometimes present us with some particular characteristics, notably phenomena of memory—for instance, genetic memory—but that's all.

This conceptual transformation is even more visible if one goes back not just thirty years but one hundred, to the beginnings of genetics and to its founders (Hugo de Vries, Wilhelm Johannsen, August Weismann, William Bateson, etc.). De Vries, emulating Darwin, invented the word *pangene* in the context of Weismann's discovery of the separation of *germen* and *soma*. For all these biologists, genes were purely formal entities, theoretical entities, about whose material support nothing was known. These genes were considered to be what determined the development of organisms. By definition, genes were the elements transmitted in the moment of reproduction of organisms, responsible simultaneously for hereditary characteristics and for the development of organisms. This is why Johannsen called them "genes."

Genetic is a very old word, used by Scholastic philosophers (not in a biological context but in a purely logical one) to define something that produces something else: something genetic is that which is responsible for a genesis. One thus speaks of genetic definitions in opposition to other definitions that are not, since they do not produce the object defined: for example, the definition of a circle as a figure produced by a line segment rotated around a point is "genetic," while the definition of the same circle as a collection of all the points equidistant from a given point is not. The utilization of this term clearly translated the idea that there exist in living organisms elements that are responsible for their development, for their genesis. De Vries explicitly said that the gene was the smallest unit of living cytoplasm and refused to consider the idea that it could be a protein. At the time, proteins were the most complicated molecular structure that one could imagine (moreover, they still are). But in the eyes of de Vries, a protein was just a molecule and therefore could not be a gene because a gene necessarily had to be "living" in order to have genetic capacities, that is, the capacities to initiate and direct development. This is why he maintained that his "pangenes" had to be units of protoplasm, in the sense of units of living matter, and not chemical molecules, since, he said, "they [the pangenes] are much larger than these and are more correctly to be compared with the smallest known organisms" (de Vries 1910: 4).

The discovery of the actual structure of genes has revealed that they are molecules markedly simpler than proteins. To tell the truth, one has no reason whatsoever to consider these molecules of nucleic acid as living. However, just by calling these molecules genes, one is using them to explain the development of the living and a set of properties that are habitually attributed to life in general. There is in this a source of extraordinary misunderstandings. Moreover, in the last analysis, one should not call pieces of DNA "genes." If one does so, one is obliged to recognize that genetics (i.e., all of the genetic processes properly speaking) is not in the genes. Or inversely, one is obliged to consider a gene to be, at minimum, an ensemble of DNA together with proteins that is capable of performing a certain activity. The geneticist Richard Lewontin, who has been opposed to the attribution of extraordinary capacities for directing action to DNA, remarked that DNA molecules are among the most chemically inert molecules one can imagine:

> DNA is a dead molecule, among the most nonreactive, chemically inert molecules in the world. . . . DNA has no power to reproduce itself. Rather it is produced out of elementary materials by a complex cellular machinery of proteins. While it is often said that DNA produces proteins, in fact proteins (enzymes) produce DNA. The newly manufactured DNA is certainly a *copy* of the old, . . . but we do not describe the Eastman Kodak factory as a place of self-reproduction [of photographs]. . . . Not only is DNA incapable of making copies of itself . . . but it is incapable of "making" anything else. (Lewontin 1992)

By contrast to proteins, and even to RNA (which can have enzymatic properties), DNA molecules have only the least bit of chemical activity and no enzymatic activity whatsoever. As Lewontin emphasizes, these molecules are incapable of doing anything by themselves, not even replicating themselves. Any activity demands the presence of active molecules, at minimum proteins and RNA, in conjunction with DNA.

We currently find ourselves in an extremely delicate situation at the level of vocabulary, which makes communicating biological information difficult. In his introduction to my talk, Etienne Landais mentioned opinion 45 of the National Advisory Committee on the Ethics of the Life and Health Sciences (Atlan 1999a: 8). In this opinion, we tried to explain that the difficulties in communication are not necessarily the fault of bad research organizations or of malicious biologists. Rather, there is an intrinsic difficulty, arising from the evolution of biology itself. It has been classically considered to be a science of life, but today it is no longer any such thing. Rather, it has become a science of physicochemical systems, in which what one calls "the gene" is not genetic, and where one is concerned with a certain number of chemical reactions, integrated in processes of metabolism and development.

Here is an example of the confusion to which this situation gives rise, taken from the report of a British commission on the ethics of gene therapy, dating from 1992 (Clothier et al. 1992). This text begins with the following fantastic declaration: "Genes are the essence of life." Right away, therefore, the report is convinced that there is an essence of life and that this essence is to be found in the genes. It forgets, thanks notably to the utilization of computational metaphors and in particular that of the program, that genes are nothing but molecules of DNA. Having announced that genes are the essence of life, the report continues: "Quite apart from necessary questions about the safety of any new medical intervention, it is not at all surprising that there are public concerns about a procedure which could be used to change inherited human characteristics, especially if those changes might be passed on to future generations." Having thus put the accent, with good reason, on the possible dangers of transmissible modifications of the genome, the authors then add: "In addition, there are likely to be irrational fears which derive from misunderstandings in biology, and are compounded by the effects of popular creations of fiction, such as Frankenstein's monster."

Unfortunately, the same report then issues some reservations about somatic gene therapy, which are a little difficult to comprehend: "It is not merely a fear of the unknown that engenders caution, but also a recognition that the ability to modify the genetic endowment of human beings provides opportunities to influence life and health more fundamentally than could any treatment available hitherto."

The reason it is necessary to take particular precautions, even for somatic gene therapy, is that gene therapy—whether somatic or germinal—in touching the gene, therefore is touching the essence of life.

After all that we have seen, even if it is, of course, necessary to take particular precautions, it is certainly not because we are touching on some supposed essence of life: this language of "the essence of life" is totally inadequate if what we are talking about is molecules of DNA. It would have been adequate if we had discovered that de Vries was right after all—in other words, if we had discovered that the material support of genes were in fact those famous minimum units of living protoplasm. One could then, perhaps, have been able to say that there exists something that one could call the essence of life. But this is not the case, since what we call genes are nothing but simple molecules of DNA. If one still wants to speak of the essence of life—though it is not obvious that one still can—then perhaps one should locate this essence in the ensemble of dynamic systems that constitute the biochemical networks in which functional states maintain, transform, and transmit themselves. This is why, if one is preoccupied with questions in biomedical ethics and, as we are here, with the precautions to be taken in this domain, it is necessary to try to define these precautions in a pragmatic fashion, to analyze each situation in its particularity—that is, for each disease, each technique, and so on to analyze the specific potentially dangerous or undesirable effects, without launching into notions as woolly as "the essence of life."

The Temptation of Unique Causes and the Streetlight Effect

In closing, I would like to adopt a point of view more theoretical than historical, which should allow us to avoid a return to the old errors of preformationism. One recalls that the old quarrel between preformation and epigenesis, which lasted close to two centuries and traversed the history of embryology, was brought to an end by a kind of compromise, namely, the acceptance of the necessity of associating attenuated forms of these two opposed but ultimately complementary ideas. At the end of the nineteenth century, the necessity of eliminating the errors of these two extremisms was recognized—in other words, the errors to which each of these theories led when posed in an extreme form. The error of extreme preformationism was symbolized by the old theory of the homunculus, which conceived of the germ as a tiny, already completely formed adult and considered development to be nothing more than augmentation of the quantity of its material mass, without any changes in its form. Symmetrically, the error of extreme epigeneticism was to conceive the germ as completely unorganized, devoid of all structure, and to consider all the organization present in the adult as stemming entirely from development. These theories, defended under various forms by various people, had coexisted since at least the seventeenth century. At the end of the nineteenth century, a certain consensus was established, admitting that the germ certainly possessed some organization but that nevertheless something beyond this was contributed by epigenesis.

Molecular biology has always been located in the space in between two alternatives: on the one hand, it has confirmed that the germ is not without structure and that there is, therefore, a certain element of preformation (at minimum the structure of the genome); on the other, it has also brought to light that elements of epigenesis are necessary in order to explain development. I have tried to show that, when analyzing the nature of the mechanisms put in play here, one definitely finds this balance between preformation and epigenesis on the molecular level. Under the influence of the progress and massive diffusion of molecular genetics, however, an obvious asymmetry has installed itself between the two factors, not just in popular conceptions but equally in laboratories. This can be explained by what is called the "streetlight effect." In effect, scientific research often proceeds in the same manner as the fool in the famous story, who, in the middle of the night, looks for his keys under the streetlight, even though he knows that he has lost them somewhere else, because "that is where the light is." Likewise, the largest number of research programs are developed in areas where new, maximally exploitable techniques are available. Given the great developments in DNA technology, it is much easier to know the structure of genes than to know the structures of epigenetic mechanisms, which, as I have suggested, are much more complicated to untangle. What one knows how to do most easily today is to spot what are called "genetic determinants." Not one week goes by without someone identifying some new gene claimed to be responsible for some normal or pathological characteristic. Much could obviously be said about the lack of rigor that too often mars the manner in which such an identification is established. In general, such claims are based on approximate statistical correlations, without any demonstration of causal relation. But it is much easier to perform this kind of research than to analyze epigenetic determinants, which necessarily involve a multiplicity of mutually implicated causal factors. As a result, we are witnessing a return to the extreme form of preformationism in the guise of a new avatar, in which everything is contained in the genes. The metaphor of the genetic program obviously reinforces this new reductionism. From this we see the utility of applying alternative metaphors to epigenesis, such as that of self-organization, which, drawing from both chemical kinetics and the dynamics of complex systems, can help stop the scale from tipping to the side of an extreme preformationist interpretation.

If the "streetlight effect" is important, we can also wonder—at perhaps a more speculative level this time—about another reason reinforcing this preformationist tendency: namely, the eternal temptation to find a unique cause. What researcher has not dreamed of identifying the unique cause of the phenomenon he studies? In the 1950s, everything had thus to be brought down to a hormone. Today, it's to a gene. Apparently, there is nothing more satisfying than to discover the unique cause of something, especially if this cause contains the effect, to take up a somewhat Scholastic notion developed by Étienne Geoffroy Saint-Hilaire. Saint-Hilaire was a partisan of epigenesis who violently attacked preformationism (defended, in particular, by Cuvier), charging that it corresponded to magical or theological thinking, according to which the cause must contain the effect. The cause, in this case, was the germ, which, according to the theory of preformation, was supposed to contain already in itself its effect, that is to say, the adult.

Now, inferring an unknown cause from an effect supposed to be already contained within this cause is what the philosopher David Hume called a "cause not proportionate to the effect" (Hume 1999, §11). He pinpointed as a characteristic example of such a disproportionate cause the famous theological argument that claims to prove the existence of God from the organization of nature, "the argument from design." Leibniz, for one, used this argument, which consists in saying: "When one observes in nature a fantastic organization, one is obliged to recognize that it could not have produced itself, nor could it have been produced by chance, and to admit that an intelligent author—therefore God—is the origin of the cause of the organization of nature." Hume completely demolishes this type of reasoning, by showing that one is typically dealing here with a cause disproportionate to its effect. To illustrate his analysis, he presents the following example: for those who cling to the argument from design, the observer of nature is in the same situation as someone who finds a human footprint in the sand. When they see this footprint, they cannot say that this form produced itself, or that it coalesced by chance: they infer from the observation of the print that it had a cause and that this cause was the passage of a man. The same type of reasoning is apparently used in the argument from design. But in the case of the footprint in the sand, the cause is proportioned to the effect: the observer knows beforehand that there exist human beings and that they walk on their feet (whose form he also knows). Faced with the print in the sand, he can therefore justly establish a correlation and even a causal induction, which justifies his conclusion that one of these beings, *which he has observed elsewhere*, has passed by this spot. The cause is known and can contain its effect. The case of nature and its Creator is altogether different: we see nothing but nature—in other words, the supposed effect—and not the Creator. The Creator might exist, but if we do not have at our disposal (and this is the hypothesis here) any other means to know this Creator except by observing nature, then to infer from this single observation the existence and properties of the Creator is a leap of a magical order, producing a cause disproportionate to the effect.

One often imagines that only scientific thought investigates causes. This is not so at all. What characterizes magical thought is precisely that nothing, for it, is without a cause. Everything that happens has a cause, usually a unique and hidden one. The principal role of magical rites consists in identifying these causes, in order either to neutralize or to provoke their effects—even if this means inventing a cause disproportionate to the effect. Is it possible that some elements of magical thinking persist in how biologists operate?

The preformationist interpretation of molecular genetics is today more widespread than one might think, probably because, among other reasons, it responds to this magical need for explanation by causes disproportionate to their effects. This need leads one to attribute to the genome things that one *formerly* used to attribute to the germ. Thomas Huxley, for example, defined the germ in 1878 as "matter potentially alive, and having within itself the tendency to assume a definite living form" (Huxley 1884: 290). This was normal, since the vitalism dominant at the time did not permit one to imagine otherwise. But when today we interpret the role of the genome in this fashion, we obviously forget that the genome, reduced to molecules of DNA, is nothing but a piece of matter, structured, certainly, but not living. And it is thus that, in a magical fashion, and by an erroneous conception of causality in which causes are disproportionate to their effects, we attribute to the genome the mysterious properties that at another time we called Life.

Behind the metaphor of the program then appears "the essence of life," and this is quickly transformed into a "sanctuary" and an "endowment." The genome then becomes a fetish, generating as much fear as fascination. And like all fetishes, it presents itself as a source of non-negligible profits, to be reaped by those who play skillfully on these fears and fascinations. In other words, as wherever there is a fetish, the merchants of the temple are not far off.

Translated by John Duda

TWENTY-FIVE

The Knowledge of Ignorance

(1991)

We can now analyze the effects of the underdetermination of theories by facts. Our goal is to differentiate the domain of the sciences (where one supposes that the underdetermination will be reduced thanks to more and more empirical observations) from that of another knowledge, one condemned to confront the irreducible underdetermination of theories. We will propose that the domain of this knowledge is none other than that of philosophy in the shape of wisdom.

We say, therefore: "The more complex and singular a phenomenon is, the more underdetermined any theory giving some account of this phenomenon will be."

Complex: in that the phenomenon is constituted by many elements or interconnected processes.

Singular: in that it is difficult to observe a large number of relevant facts under conditions of experimental reproducibility that would accord with the rules of the scientific game. The conditions of reproducibility oppose the singularity of the phenomenon because they imply that one can include the phenomenon exactly in a class of definite equivalence by means of a property observed in a repetitive way. The singularity of the phenomenon would then disappear, since it would be shared with other phenomena considered identical from the standpoint of the equivalence relation that this common property would define.

Since the phenomenon is both complex and singular in these senses, any theory capable of giving some account of it will be underdetermined. But this does not mean that it will have no explanatory value. Its truth value will be weak, but only in relation to a criterion of absolute truth. Its predictive value may be significant enough, however, if the theory permits us to predict a large number of facts. The fact that other theories can do the same has no bearing on the issue. Each theory is capable of constructing a world that is the same (the one of the observable facts) but different with regard to what we don't observe, that is, what is only implied by what we observe, what is logically possible given what we observe.

In particular, if a norm must be erected, rightly or wrongly, on the basis of a theory (either by deduction and predictive projection of what must be on the basis of what is or merely by indicating the constraints that limit the possibilities), then each theory will permit us to erect very different norms. Thus, without sacrificing any rigor in predicting observable facts, we can choose among different theories the one (or the ones) favoring the norm that suits us (for reasons very different from those internal to theorizing itself). This choice of theory will be an exercise in controlled wishful thinking. This is, apparently, how the choice of theory works—without its being conscious, of course—in the development of ethical or political norms that proclaim themselves true because they are erected on scientific theories.

This activity, though dangerous because it is at the origin of modern ideologies, is not pointless if it manages to recognize itself for what it is, that is, a construction of a rational world, coherent with a certain explicit or implicit project, a project that expresses itself in the norm. It must be understood, however, that this norm does not come from our rational knowledge of reality, nor is it grounded in or founded by this reality. This knowledge does not necessarily emanate from transcendence, so to speak, but from concatenations of the imaginary and desires. Evidently, this knowledge can be modified by such or such findings abstracted from observable facts. But such modification will occur only with a considerable degree of underdetermination, as I have shown.

When it escapes the trap of ideology, this activity seems to me the essence of constructivist philosophy, drawn to the analysis of complex and singular situations. This philosophy is unlike science, since science can never, without destroying itself, give up epistemological realism or the ideal of being totally constrained by facts. The very essence of the original project of science consists in an indefinite pursuit of the greater determination of theory via the observation of more and more relevant facts.

Since the underdetermination of theories seems to be the rule rather than an exception in the case of theories of complex natural phenomena, it is up to constructivist philosophy to take care not to surrender this domain—that of the singularities and complexities in our lived experience—either to rationalizing ideology or to discourses of the unspeakable or ineffable that can never can stop speaking.

Today's science restricts itself to the enormous domain in which it is increasingly preoccupied with mastering artifacts born in laboratories for the sole purpose of being mastered and, so to speak, isolated in the real. Under these conditions, theorizing comes into its own—although in a local way, limited to the domain of the factual thus circumscribed—and permits the operational and much-sought-after mastery of prediction.

Thus, contrary to the ideal of the neo-positivist philosophies that tried to imitate the logical-mathematical shape of physics, the role of philosophy is to speak of what cannot be formalized, to use a natural language with its metaphors, analogies, and all the vagueness that comes with them, yet without giving up rationality. This means not giving up on distinguishing good analogies from bad ones, enriching metaphors from misleading ones, the vagueness [*le vague en moins*] that conceals what should be said from the vagueness [*du vague en plus*] that stands for the potential of creation. For the sake of this new ideal of philosophy, one must envision how this philosophy based on natural language—nonformalized, but rational and rigorous nonetheless—distinguishes itself from scientific language, and does so without confounding itself with either mythology or ideology.

This philosophical language aims at a discourse of wisdom, but unlike mythologies it takes as its starting point the ensemble of scientific knowledge at any given moment—that is, our knowledge of the design and performance of the natural and artificial machines that surround and constitute us. Though starting here, this new philosophical language overflows such knowledge and thereby radically differentiates itself.

More precisely, the role of philosophy is to recuperate our pseudo-concepts without, as science tends to do, having transformed them beforehand into concepts susceptible of becoming scientific because they are operational. We have seen how the concepts of totality, negation, and the possible are distorted and transformed into operational concepts. We have also seen how use of these pseudo-concepts is always guided by the idea of intentional finality that accompanies our experiences of projects and intentions. This idea cannot be disdainfully dismissed as illusory, because reflection on the good and evil in human behavior, eventually leading to the enunciation of norms and of the law, cannot happen without reflection on the goals that one sets oneself. If one pretends to root this in a knowledge of nature, and thereby to found a natural law, how can one not have recourse to a theory of the ends of nature, that is, a teleological conception of the universe? Yet our scientific representations have imposed on us a mechanical, nonteleological conception not only of the sky and the movements of the celestial bodies but also of life, up to and including its richest and most unexpected, most creative productions. This is why the complete victory of modern natural science in the form of a generalized mechanicism inaugurated *the crisis of natural law*. And this crisis has only deepened as the physico-chemical mechanicism has kept on spreading, contrary to the predictions of Kant, until it has encompassed our representations of the living.

The desire, the need to stem this crisis of natural law, while regaining the source of the sovereign good (which would also be scientific truth), probably explains the desperate pursuit of grand meta-theoretical syntheses that seek to unify the evolution of the species and the meaning of the universe, the theory of consciousness and the Big Bang.

If we reject these panoramic visions—which lack neither grandeur nor nobility, but in which the natural sciences serve only to supply a jargon that lends credibility to what, after all, is nothing but rationalization and ideology—it would seem that we cannot avoid either letting the irrational reign over human affairs or falling into the skeptical relativism in which anything goes. But that would be a cop out for philosophy and for ethical reflection, stemming from the fact that philosophy has long been exclusively dominated by the idea of the ends of the nature. If today this idea is no longer tenable, that does not mean that the determinism advocated by Pierre-Simon Laplace has replaced it. Certainly, in the place of natural finality, everything in our scientific representations impels us to conceive natural determinism and nothing but determinism. But this determinism is different from a Laplacian one in that it takes into account the underdetermination of theories by facts—that is, the local character of our knowledge—and our ignorance of the *total* order of nature.

Laplace's hypothesis suggests an all-or-nothing situation: either we know everything about determinations and the world appears to us as it is in itself, a universal mechanism without a place for anything unforeseen, or we don't know everything about these determinations but they exist *as if we knew them*, and our perception of the unforeseen is only an illusion due to our ignorance.

Today we find ourselves in a different situation: our ignorance of the *totality* of determinations is part of natural reality *as much as the determinations themselves* because this ignorance produces effects—our behavior—different from what they would be if we had a total knowledge of natural determinations.

Moreover, the absence of total knowledge doesn't imply total ignorance: we already know much more than Laplace, thereby validating his confidence in science, though in a way qualitatively different from the one he seems to have imagined: we have theories that are effective in predicting facts, yet the more general the theories, the more we have too many of them in relation to the number of facts we are trying to predict. (Of course, the fundamental equations of mathematical physics, where the work of theorization comes closer to the classical ideal of a unified theory explaining all facts, is an exception. But this is obtained at the expense of considering facts only in their fundamental aspect, that is, according to the nature of the forces responsible for the interaction among elementary particles.) Thus, our theoretical knowledge can be very rich and very powerful, but it remains underdetermined in the sense that several coexisting theories contain very different implications concerning a global vision of things, and if we obey the protocols of experimental method, we cannot gather enough observable facts to choose among them.

For all of these reasons, our ignorance of the totality of determinations is equivalent, *insofar as what matters to us*, to the *real* existence of indetermination in nature.

We have seen that it is easy enough to appreciate the significance of the underdetermination of theories by facts in relatively simple systems where the metaphysical questions of determinism and freedom, the relations between body and mind, do not arise, and in an ontological context that is perfectly determinist. This means that to recognize the underdetermination of theories by observed facts at the infinitely more complicated level of the psycho-social systems does not imply a dualist ontology in which freedom would slip between the cracks in determinism. No, in these systems too the underdetermination of theories can coexist with a total determinism of nature, but on a level that to us can only be unknown because of the intrinsic complexity of the level on which the phenomena in question would need to be observed (in other words, because of the number of coupled elementary processes that determine the phenomena).

Our subjective experience of voluntary acts—what Spinoza designates as what *we* call the decree of free will from the viewpoint of the attribute of thought—is no more illusory than our objective experience of the physiological mechanisms that we discover in these acts. It is a consequence of our ignorance of a part—a very large part—of their determinations. And we have no means *of acting as if* this ignorance did not exist; *we have no means of knowing what our behavior would be like if we had a total knowledge of the determinisms in nature.*

This is why our behavior can be directed only *both by what we know of our determinisms and by the fact that we know that we don't know everything.* In our knowledge of our ignorance lies our experience of our will. And indeed, our will is all the more alienated in that it would be truly free if it coincided (as it does in the case of Spinoza's God) with total knowledge. But it is this will, such as it is, whether determined by passion or by reason, and not the hypothetical finalities of nature, that supplies the foundations of the law.

Translated by Oleg Gelikman

TWENTY-SIX

The Fraternal Utopia

(2005)

Like other domains involving unprecedented futures, fiction and mythology often have a better chance of representing reality than does a simple extrapolation and amplification from the present. Polysemy is another advantage of myth: the same story can be interpreted in multiple ways, multiplying the meanings of the images—good or bad—that we can then project into the future. This multiplicity is evident in origin myths, which tend to involve extraordinary, more or less "monstrous," modes of reproduction. Athena's birth out of Zeus's brain, for example, is generally understood as an episode in the battle between the sexes for the exclusive control of procreation. But matters are not so simple: unlike Dionysus (who was born out of the divine thigh), Athena is female, yet has characteristics commonly attributed to males. Zeus gave birth to her after having swallowed the Ruse. Later, she appears as a figure of wisdom and intelligence, and she goes on to inspire and protect Odysseus, a king who is wise, intelligent, and artful at once. In other words, in this myth the female is nothing more than a passive receptacle for male seed, formless matter animated by the shape contained in the sperm, as Aristotle would affirm.

Similarly, the biblical myth of Adam and Eve is often understood as the installation of patriarchy (which indeed it is). But this installation is an unhappy one, since the expulsion from Eden signals a *malediction* on both sexes: to Eve, it brings the curse of giving birth in pain and being dominated by man; to Adam, the curse of having to work "in the sweat of your brow." In the Edenic state, by contrast, what must happen occurs without pain, including the encounter of Adam and Eve, who *recognize one another* after having been separated. In the creation of woman from Adam's rib, some see another expression of procreative rivalry between the sexes. Even though the woman, Eve, is then called the "mother of all living," in this story it is the man who gives birth to the woman, as Zeus gives birth to Athena. The legend of a primordial woman, Lilith—the nocturnal, demonic rival of Adam who would have preceded Eve before becoming the wife of Satan—gives some support to this interpretation.

But this interpretation is challenged when one realizes that the Hebrew word for "rib" can also be translated as "side." This translation coheres with the image of Adam as an androgynous being, "male and female at the origin," the image found in the first chapter of Genesis. Far from seeing the woman's creation in the second chapter of Genesis as an extraction of woman out of man, the kabbalist tradition treats it as a separation in an original bisexuality, in view of its future reunification: fused back to back in the initial androgyne, in the act of the creation of woman, the male and the female are separated, so as to be able to turn around and meet face-to-face. In kabbalist typology, the figures of the "father and the mother from above," whose union is so permanent that one cannot be separated from the other, are sexualized representations of two complementary aspects of a living knowledge: respectively, wisdom-father and intelligence-mother.

For some bioethicists (e.g., Abel Fletcher), the new biology would recover the artifice of birth without a mother and masculine mothers in the Garden of Eden evoked by Genesis. God the Father created Adam artificially from the dust, playing the role of the mother: "Then Adam became mother again, birthing Eve, while God acted as an obstetrician" (cited in Corea 1985a: 292). Another bioethicist, Janine Raymond, sees the birth of the son Jesus from God the Father without coupling as the "fundamental patriarchal myth" affirming the mono-parenthood of the father (ibid). One could object that the story of Mary, who was both a mother and a virgin, portrays mono-parenthood as being exclusively female; in this view, Jesus, when claiming to be son of God the Father, would be dispossessing her of motherhood and thus would be playing a role in reaffirming patriarchy.

Origin myths are undeniably full of images of archaic rivalry that patriarchal organizations resolved by reversing the power relations between the sexes. But in these stories one can also see nostalgia for or prefiguration of a friendly and brotherly peace released from the struggle for procreation, a win-win situation for both sexes. In an episode reflecting the origin of Greek marriage, the story of the Danaides shows that not only women but men too were losers in the perpetual war for the mastery of reproduction, even when seemingly suppressed by the institution of marriage. A talmudic legend presents this nostalgia—the prefiguration of a friendly equality between the sexes—in a parable of the diminishment of the Moon.

According to this narrative, a cosmic mistake, committed by—who else?—the Creator himself, preceded the one committed by Adam and Eve. Initially equal to the sun, the moon was reduced in size for no other reason than that she had understood that "two kings cannot be under the same crown." The biblical rite of a sacrifice of atonement each new moon recalls this diminution, an interpretation that attributes the fault to the Creator.

Looking at reproductive technologies associated with genetics, Lee Silver, a contemporary biologist, envisions the future of humanity as a reconstruction of Eden (as the title of his book *Remaking Eden* [2002] makes plain). He argues that the use of these technologies in a more or less distant future is unavoidable, especially in a society as dominated by individualism and the laws of the market as the United States. Silver has also tried to predict the effects of genetic biotechnologies in 2010, 2050, and 2350.

According to him, in 2350 humanity will be divided into two classes of individuals: the "natural" and the "genetically enhanced." Moreover, these two classes will appear to be on the path of separating into two different species, because mixed unions are less fertile than unions between members of the same class. Before pointing to the limits of the genetic reductionism that permeates this book, one can be skeptical about the technical possibility of "enriching the genome," producing some children on demand and programming them with all sorts of physical and intellectual properties. But in three hundred fifty years, we can imagine what seems technically impossible today. Once again, the use of myth facilitates imaginary representations of a future shaped by biotechnology. According to Silver, the artificial uterus is not going to be the main focus of the future, because he considers the obstacles to its technical realization more difficult to overcome than those to a baby delivered on demand and genetically programmed by parents who are able to afford that. Here, the mythical dimension is amplified by a fetishism of the gene. I do not share this opinion. Only the future will tell which of these techniques will be developed and disseminated first. We cannot eliminate the possibility that none of them will be both practical and socially acceptable.

After all, as we have known since Jeremiah, it is the fate of prophets of doom to see their prophecies belied by the facts, inasmuch as those prophecies are heeded and everything possible is done to avoid the catastrophes they predict. However that may be, in the future envisioned by Silver, "reprogeneticists" (i.e., reproductive geneticists) will replace the creative activity of God in Eden by taking over from evolution and modifying the essence of the human species. In doing so, they will complete the separation between procreation and sexuality that current technologies have already initiated. In claiming that reproductive technologies replace God, Silver rehearses an argument against this separation often advanced by the Vatican and other Churches who see it as both blasphemous and antinatural.

Thus, in one way or another, most of the projective (or futuristic) readings of the biblical myth bring it closer to the legendary stories of single divine male parenthood found in the most diverse ancient traditions. In particular, they see today's technologies replacing artificial—or miraculous—birth without a mother out of a male god. In this replacement, one can see a weakening of male power over procreation, or even its end. Gena Corea, however, in a work that reviews these myths, gives them a one-dimensional interpretation that sees technology as taking over in stripping women of procreative power, a dispossession even more definitive than the legal and symbolic one inflicted upon women under patriarchal systems.

As we have suggested, projecting origin myths into the future offers other possibilities. Let's explore some of these alternative paths.

Eden, or Overcoming the Nightmare

In the imagery of Eden before the fall and all the misfortunes that followed, we can see nostalgia for a golden age when the pain that today seems inherent in the human condition did not exist. Projected into the future, this imagery offers signs of new emotional and social ties. Men, women, and children will be required to enter into new relationships; their nature will be transformed, possibly humanized, by technology.

Therefore, let's start from the beginning.

The species *Homo sapiens*, like any species, has distinguishing or *specific* characteristics: erect posture, opposable thumbs, a large brain, reflexive articulate language, projective imagination, and the manufacture of tools, that is, the arts of technology and representation. Humans obviously share characteristics with other animal species, especially with mammals: carbon metabolism and biochemistry; the physiological functions of nutrition, excretion, circulation, and breathing; sensory relations; and viviparous sexual reproduction.

But the characteristics shared with other species are modulated by the characteristics proper to *Homo sapiens*. In particular, due to the phenomenon of neoteny, human reproduction implies a long period of training and rearing newborns, during which children are completely dependent on the adults for their survival. Besides, natural human sexuality is already partially separated from reproduction, since, unlike other mammals (with the exception of a few primates), it is not limited to particular periods of rut, that is to say, fertility. Finally, human sexuality is also different because, more often than not, it happens face-to-face. But above all, the art and technologies of the artificial have deeply modified the conditions of life and the environment of the human species and have brought it to the point where it distinguishes itself from all naturally existing beings. The arts of clothing, of fire, of agriculture and livestock, but also of the wheel and architecture (to say nothing of the arts of war and the sophistication in weaponry derived thereby) have all integrated into human nature more and more elaborate artificialities.

This brief summary of culture and the arts as constitutive of human nature shows that human nature is characterized by a "cultural evolution" that is different from the biological one and proceeds much faster. In other words, the hominization of *Homo sapiens*, which is also a humanization, is, at least in part, a gradual exodus out of animality.

It is not difficult to trace the major stages in sexual reproduction and what they have left us as legacy. Traversing and determining us, heterosexual impulses ensured the reproduction of the *species* from the beginning, and they still do. In this field as in others, freedom means knowing our determinism and accepting and directing it. To this strictly biological determinism, cultural evolution added familial structures focused on educating children; traditionally, these organizations ensure the reproduction of the *social order*. Here again, freedom means knowing these determinations, understanding their evolution, and, at the same time, accepting and guiding them. This is especially important now that we see a significant increase in forms of sexuality and family organization. This openness started long ago, namely with the separation of sexual enjoyment from sexual reproduction that is specific to the human species and to some primates. Over the past few centuries, it has accelerated with the opening of societies to each other and the discovery of diversity in family structures. It is in this context that we should place the dissociation of reproduction from sexuality that biotechnology brings into view.

The intrusion of technology into human reproduction may be seen as a continuation of this trend and another stage in humankind's gradual exodus out of animality. Far from being an "unnatural" regression, it may bear a more advanced humanization. Separated from reproduction (and death in the same breath), eros could focus on the unique experiences of meeting others and of openness to the plural world of modified states of consciousness, the "separate reality" that in the distant past distinguished the world of the sacred, before the established religions took over. At the same time, reproduction of the species and the social order will be maintained by the intervention of technology, without having to go through the battle between the sexes and patriarchal oppression. The biblical myth can be used in an optimistic projection of this sort, which, of course, does not eliminate the usual catastrophic perspectives. Let us recall that Genesis presents the painful obstetric labors of women, together with labor of men needed for subsistence, as the malediction accompanying their expulsion from Eden. Technology is progressively freeing mankind from the labor to survive, however. In parallel, technology is progressively freeing women from the pain of giving birth. As we have seen, the liberation of women from the yoke of masculine domination is also an effect of the technology that has freed them from the constraints of unwanted pregnancies and onerous domestic labor.

All this means that technology frees the descendants of Adam and Eve from the effects of the expulsion from Eden. It also means that now one can imagine the prelapsarian state as utopian existence, as life without suffering, as life filled with amorous friendship and evolutionary harmony with the environment. As we have seen, Eve's birth without a mother represents a separation within Adam's intrinsic bisexuality and the emergence of its dominant figures—the masculine and the feminine. These figures are capable of turning around, looking at one another face-to-face, and reuniting again in

an ecstatic knowledge. Their relationship to nonhuman nature is also different. They are in the "garden," planted with fruit trees serving as a source of
a vegetarian food obtained without violence against animals. They are there
not only to live by harvesting nourishment but "to work and maintain it"
(Genesis 2:15). This situation implies two things. On the one hand, work
already exists in Eden; however, it is work without pain, work in the sense of
an activity "in the service of" what is being worked. On the other hand, this
activity does not only conserve, it also transforms nature, which the prelapsarian Adam and Eve received as incomplete. In the garden, the man-woman
Homo faber is "associated with the Creator," following a talmudic expression,
thus prefiguring all the artificial transformations that technology will make
nature undergo in humanizing it. Note that the process that today we call
"cultural evolution" also concerns a transformation of human nature, as is
suggested by *two* pairs of Hebrew words for man and woman. In the first of
these pairs, '*adam*, "generic man," has '*adamah*, or the "earth," as its feminine, while in the second pair, '*ishah*, "woman," has as its opposite '*ish*, or
"particular man," which appears only with '*ishah* and will be related to Adam
only later. Everything happens as if the separation of the feminine and its
reunification with the residual masculine were part of an evolution from an
obscure, terrestrial human nature toward a human nature that is igneous,
burning with the heat and light of "fire" ('*esh*). (This is suggested by the
semantic pair '*ish*-'*ishah*, in which a "divine fire" ['*esh yah*] is inscribed.)

According to this system of interpretation, even the prohibition against
eating the fruit of the two trees—the tree of life and the tree of knowledge—is oriented toward future evolution. The prohibition is temporary and
pedagogical, awaiting a greater humanization that could incorporate without
harm the experience of eternity (i.e., the "tree of life"), as well as the timeless
true "knowledge" that accompanies it; without such delay, this incorporation
could be only approximate and ambivalent, good and bad at the same time,
"the tree of knowledge of good and evil." Even if Adam and Eve could not
have avoided transgressing, because transgression is part of their nature, their
"mistake" appears as an inversion of an ideal order of things in which intellectual precedes sexual maturity. This ideal order stands in contrast to the
natural order, in which puberty occurs long before an adult "age of reason,"
when sexuality becomes enlightened, at least in principle, by a relatively
developed intellect. (I have proposed a detailed analysis of traditional interpretations of the biblical myth elsewhere.[1])

---

1. [See Atlan 1999b, chap. 1.—Trans.]

Out of Eden and after the Fall, humankind has not lost everything. The process of humanization and the progressive overcoming of animality continues, urged on by the need to repair what has been broken. That is why technologically driven liberation from the economic and social necessities of slavery—the liberation that currently is phasing out hard labor—obviously leads toward humanization. Similarly, the liberation of women from the slavery in which they were kept because of physiological asymmetries in reproduction and the masculine domination that resulted certainly tends in the same direction and brings us closer to the Edenic utopia that the myth leads us to imagine. In this movement, sexual encounter becomes a value in itself, regardless of its possible procreative outcome, because it is above all an encounter between two parts of the same self or the same "flesh," as Genesis puts it: "This is why a man ['*ish*] will abandon his father and his mother and cleave to his wife ['*ishah*], and they will become one flesh" (Genesis 2:24). The mention of a father and a mother shows that this passage no longer speaks about Adam and Eve (who were born of the earth and of their separation from each other), but of all those who will later be born of a man and a woman. So everything happens as if by abandoning, if only for a time, the father-mother couple, the extraordinary founding event of the initial bisexual separation-reunion imprinted its seal onto "natural" sexuality, whatever implications this may have for the reproduction of the species. In this context, it is normal to distinguish the biblical injunction for a man to "cleave" to his wife from the earlier one "to be fruitful and to multiply" (Genesis 1:28).

Assuming the interpretative distance that the myth allows, we can see that the complete dissociation of sexuality from procreation that future techniques allow us to foresee does not necessarily imply the nightmarish "brave new world" of Aldous Huxley. There are more optimistic scenarios, where human life, free from the painful physiological and economic necessities imposed by nature, can develop to maximize the possibilities of aesthetic and intellectual enjoyment, celebration, ecstasy, and even moral surpassing that nature also offers.

Nonetheless, the best is no more guaranteed than the worst. Certain conditions must be met in order for the dream of a return to Eden not to become the nightmare of a new fall into barbarism and the war of all against all. A minimal condition would be a moral one: a disinterested, "maternal" compassion, a concern for others and justice would need to permeate human relationships, whatever forms of familial, political, economic, ideological, or religious organization future societies may assume. Hillel's "golden rule,"

"Do not do to your neighbor what you would hate to be done to you," should be applied to the relationship between generations. Before giving birth to a child in whatever way, shouldn't it be necessary to ask "Would I have appreciated or would have I hated being born this way?" An unbridled individualistic hedonism could pose the great danger of the future, made possible by the explosive combination of technical achievement and a savagely competitive economic liberalism.

Savage Economism and Hedonist Chaos, or the Overcoming of Eros by Eros?

Indeed, as soon as the functions of father and mother were to be reduced to the microscopic level of supplying laboratories with cells and artificial uteruses, the question of raising and educating children would become more acute than it has ever been. The generative relationship between parents and offspring would no longer necessarily be the norm. No longer a product of necessity (physiological in case of the mother, social and legal in case of the father), the relationship between parents and offspring would have to be either imposed or replaced by institutional means. Adults would need to guarantee the playful and intellectual emotional environment essential to the growth and development of children. To avoid falling into the automatic conditioning imagined by Huxley, in one way or another, adults would have to be motivated by something like the mixture of love and a sense of duty that thus far has characterized the family cell in all its forms. And this condition would have to exist in a world where, quite independently of procreation, eros would continue to animate and to inspire the emotions of the same adults.

The long period of maturation characteristic of the human species is at the origin of an obvious gap between early sexual maturity and a belated (sometimes very much so) intellectual and emotional one. As we have seen, the taboo against tasting the fruit of the trees of life and knowledge in the myth of Eden alludes to this gap. It results in a dissociation between the erotic emotions and the emotions that develop in relation to children, a dissociation that family institutions negotiate more or less successfully. In an ectogenetic society, this break risks becoming more pronounced and might lead adults to an almost exclusive pursuit of selfish pleasure at the expense of a necessarily altruistic parental affectivity.

The new feminine and masculine identities currently emerging in developed societies dominated by competitive economism are already exposed to the danger of hedonist chaos. But they also open new avenues in the progress toward a humanization that might increase the chances of overcoming eros with eros.

Where it has taken off, economic development has freed the workforce enslaved to the satisfaction of vital needs. Sexual liberation has begun to free women (and men) from the social constraints of patriarchy and religious puritanism. But the activation of desire is not yet freedom. On the contrary, it might be a source of even greater alienation because the enslavement of desire by desire would be even more difficult to escape. Ectogenesis would accentuate further the tension between the alternatives of totalitarian utopia, sweetened by a conditioning to an alienation disguised as happiness (Huxley's "brave new world"), or a fraternal utopia of men, women, and children who, even if not free, at least are engaged in an active pursuit of true freedom, not only political but inner as well.

It may seem paradoxical to imagine a new kind of fraternal love in a society where traditional families, along with the fraternities that characterized them, may disappear. Nonetheless, only an opening of eros, only a movement beyond a solitary enjoyment (even with a partner) closed on itself, will dissolve the tension between egotistical attachment to the pursuit of the pleasure of receiving and altruistic attachment to the pursuit of the pleasure of giving. Hormonal and neural structures that promote opposite affects—such as faithful attachment and individualistic detachment—pervade the human brain, as well as the brains of other animal species.[2] This fact could be a source of optimism . . . or pessimism. It all depends. Contradictory potential orientations, more or less selfless or individualistic, are already there, "imprinted" in cerebral structures. The dominance of an orientation depends on the relatively greater activation of one cerebral circuit rather than another. History and different social environments, fortuitous or planned events can shift the balance in one direction or another. Some rodents are known to be naturally monogamous, living as couples all their lives and raising their progeny together. Some neighboring species are naturally polygamous, because of the effect of "attachment hormones" that insure the dominance of the appropriate cerebral circuits (Cho et al. 1999). If we think about the possible uses of these mechanisms for the sake of psychological, chemical, and/or genetic

---

2. See Eisler and Levine 2002 and Insel and Young 2001.

conditioning, we can be pessimistic. But we can be optimistic in observing that these determinisms are not unidirectional. Unforeseen events can always change their course.

We have known this for a long time, even if today's neurophysiological findings give new connotations to this knowledge. Eros has a long history of conflicting, even opposed, impulses, emotions, feelings, and illuminations. In fact, the erotic experience offers the same ambivalences as sacred or mystical illuminations. It opens onto expanded states of consciousness and onto what some users of hallucinogenic plants in American Indian rituals have called "another reality"; it also presents the danger of alienated withdrawal into the addictive repetition of exhilaration, an exhilaration to be constantly sought after and less and less often found. The symbolic and sacral elements of human sexuality have always distinguished it from its animal counterparts. Like death or clothing, sexuality is always ritualized in some fashion. Its openness to the sacred is far from being limited to its reproductive function. From the sacred prostitutes of antiquity to the erotic explorations of sacred and modern art, including the celebrations of the Song of Songs and Indian tantrism, sacralized sexuality is found in all forms of sexual practice: heterosexual and homosexual, masculine and feminine. In all these cases, sexuality is torn between repetition of passages to the act that blur the boundary between eroticism and pornography (*porneia* originally meant prostitution markets) and the discovery of a love that is all the more profound because it is exclusive, the emergence of a couple whose sexual knowledge, in the proper sense of the word, is socialized by rite. This sacralization can result in a cult of amorous friendship, love sublimated without being de-sexualized; Plato's *Symposium* remains an archetype of such a relationship, that is, "platonic" love between the philosopher and his disciple. We find this pursuit of sublimation without de-sexualization in many other traditions as well, notably Indian, Jewish, and Muslim. Alan Watts celebrates the Indian and Chinese versions of this pursuit (Watts 1958). As Rachi says in his commentary on biblical sexual prohibitions, they are punctuated by a program of sacralizing the human, a program that responds to the sacred, which is divine: "Wherever it comes to a 'prohibition on nudity' [i.e., sexual prohibitions], it is a matter of the sacred."[3]

---

3. In Leviticus 19:2, the formula "Be sacred, for I am sacred" introduces bans on different sorts of action, such as incest and other forms of sexuality condemned by the Mosaic Law. The frequent translation as "Be holy, for I am holy" proceeds from a tendency

In his beautiful posthumous book, *The Sex of the Soul: The Vicissitudes of Sexual Difference in Kabbalah*, Charles Mopsik, a great scholar of Kabbalah, shows how some authors in the Jewish tradition renewed the biblical— Platonic—myth of the androgyne. In doing so, they elaborated a conception of human identity in which the separation between masculine and feminine gender is no longer cut and dried, certainly no longer essential:

> There is no soul and therefore no human being in the full sense that is not at the same time male and female. Sex is the separator which establishes a devastating dissociation between two halves made to be united. Sexuality as a desire for a union in love is the attempt to overcome the damages caused by this primary dissociation. . . . The common idea of the existence of a substantial gender identity attached to each was rejected in favor of the notion of an identity as a figure of the lack. To be man or woman is to be what is lacking to the other man or the other woman. The anatomical difference doesn't ground sexual identity, it is the inscription in the body of the organs lacking in the other, of which the other needs to be one in his/her body, to be himself/herself thanks to me. The psychological or character differences belong to the same logic: we differ only by what we complete. (Mopsik 2005: 32, trans. modified)

Exoteric religion sees the canonical biblical text that proclaims the unity of the divine, "Hear, O Israel, . . . YHVH is One" (Deuteronomy 6:4), as a profession of monotheistic faith. The writings we are examining now interpret this passage as alluding to the union of masculine and feminine divine principles, the "one flesh" that characterizes the unity produced by the human reunion of men and women after their separation in Genesis 2:24 (Mopsik 2005: 67–68). In his study, Mopsik draws on many texts that deal with this little-known aspect of what is called, perhaps misleadingly, "Jewish mysticism." He mentions how problems of infertility in couples in which sexual differences are blurred—because, for example, a master has been reincarnated as the spouse of another master, thus rendering her "masculine"— have been solved by forms of the homoparentality of souls, illustrating the psychological aspects of this essential bisexuality. With Jewish mysticism as his subject, Mopsik undertakes "the exploration of the construction of sexual difference in European thought which has dedicated a lot of energy to this question but still remains alien to the religious or philosophical discourse of the Christian West" (ibid. 66, trans. modified). This thought brings the gap

---

toward disincarnated spiritualization, as promoted by theologies of pure spirit, and leads to an image of the flesh as sin.

between sexuality and procreation down to the fundamental bisexuality of all humans. An obvious masculine or feminine dominance can shift within the same individual in relation to his or her deep complementarity; going beyond initial sterility, such shifts make possible fertile discrepancies in the second degree (ibid. 88). This transformation does not result in the reduction of these tensions and the disappearance of the difference between the sexes, however. Recognizing the constructed aspect of sexual difference doesn't lead to an undifferentiated fusion. Quite the contrary. We can already see this in the current increase in types of social expressions and especially in the emergence of new masculine and feminine identities.

In avowed homosexuality, homoparentality, and reconstructed families, sexual difference is multiplied, not erased. We can wager that this trend will continue to accelerate in the future. Instead of being limited to a simple difference within a couple, sexual difference will differentiate types of couples: male homosexuals, female homosexuals, and heterosexuals. The number of combinations can evidently be increased if one takes into consideration bisexual households comprising three or more members . . . In all these cases, genders will remain different, since they are already different within each individual. Sexual difference will always be there, but its appearance and physiological, psychological, and social functions will be modified.

These considerations (and many others of the same type) allow us to seek in ancient texts—including myths—a perspective on contemporary transformations that would not view them as an absolute aberration. What seems as an incredible novelty may then find a place in a human nature that is deeper and older than one tends to believe.

These reminders can help us avoid the trap of simplistic transposition and imagine new forms of eros in which one could transcend oneself both intellectually and emotionally. This future world, which today announces itself as a nightmare, may, on the contrary, bring tomorrow's humanity to what could turn out to be a new Eden.

From this point of view, it is instructive to read the new preface Huxley wrote to his *Brave New World* in 1946, right after the Second World War ended and fourteen years after the first edition. He observed that the totalitarian utopia of his book, in which art and science are used to solve what he called the "problem of happiness"—in other words, the "problem of making people love their servitude" (Huxley 2004: 12)—was a lot closer than one could have imagined fifteen years earlier. Originally, Huxley put it six hundred years into the future, but by 1946 it seemed "quite possible that this horror will be upon us within a single century" (ibid. 13). Among the early signs of the "deep, personal revolution in human minds and bodies" necessary for establishing "the love of servitude," he saw already in 1946 a trend toward a sexual promiscuity of the kind he portrayed in *Brave New World*, along with the "standardization of human product" by various techniques of conditioning. He predicted that:

> In a few years, no doubt, marriage licenses will be sold like dog licenses, good for a period of twelve months, with no law against changing dogs or keeping more than one animal at a time. As political and economic freedom diminishes, sexual freedom tends compensatingly to increase. And the dictator (unless he needs cannon fodder and families with which to colonize empty or conquered territories) will do well to encourage that freedom. In conjunction with the freedom to day-dream under the influence of dope and movies and the radio, it will help to reconcile his subjects to the servitude which is their fate. (Ibid.)

In this new preface, Huxley also reflects on his own evolution in relation to his novel. He now describes its author as an "amused, Pyrrhonic aesthete" (ibid. 6). As a result of this evolution, he says he would have imagined another end for his novel, so that it "would possess artistic and (if it is permissible to use so large a word in connection with a work of fiction) philosophical completeness which in its present form it evidently lacks" (ibid. 7). In the meantime, he had published an essay on *philosophia perennis* (Huxley 1944), and his later works focus on exploring the modified states of consciousness enabling a broadening and an elevation of the human mind.[4] Had he written his novel in 1946, he would have given it an ending different from the suicide of its hero, the Savage, caught between two forms of madness, that of a totalitarian techno-utopia and a primitive, pretechnological existence. A third possibility would be offered:

> If I were now to rewrite the book, I would offer the Savage a third alternative. Between the utopian and the primitive horns of his dilemma would lie the possibility of sanity. . . . Science and technology would be used, as though, like the Sabbath, they had been made for man, not (as at present and still more so in the Brave New World) as though man were to be adapted and enslaved to them. Religion would be the conscious and intelligent pursuit of man's Final End, the unitive knowledge of the immanent Tao or Logos, the transcendent God-head or Brahman. And the prevailing philosophy of life would be a kind of Higher Utilitarianism in which the Greatest Happiness principle would be secondary to the Final

---

4. See Huxley 2004.

End principle—the first question to be asked and answered in every contingency
of life being: "How will this thought or action contribute to, or interfere with, the
achievement by me and the greatest possible number of other individuals of man's
Final End?" (Ibid.)

We can certainly question the existence of such a "Final End"; we would
also have to give it explicit content (with, eventually, a universal vocation),
and we know that totalitarian ideologies or messianisms often pursue such a
goal. At the very least, one could imagine such a final end as a constructed
and personal one, seeking relative harmony with the ends of others. However
that may be, Huxley replaces the catastrophic projections of *Brave New World*
with a context that no longer excludes the possibility of salvation. He thus
concludes his preface in a fashion that is dark but not desperate:

> Indeed, unless we choose to decentralize and to use applied science, not as the end
> to which human beings are to be made the means, but as the means to producing
> a race of free individuals, we have only two alternatives to choose from: either a
> number of national, militarized totalitarianisms, having as their root the terror of
> the atomic bomb and as their consequence the destruction of civilization (or, if the
> warfare is limited, the perpetuation of militarism) or else one supranational totali-
> tarianism, called into existence by the social chaos resulting from rapid
> technological progress in general and the atomic revolution in particular, and
> developing, under the need for efficiency and stability, into the welfare-tyranny of
> Utopia. You pays your money and you takes your choice. (Ibid. 13)

Let's recall one last time that the role of catastrophic predictions (like
those offered by prophets of doom) is to encourage those who believe in
them to do everything to prevent their realization. Therefore, it is quite possible that a large part of what we envisage will not take place.

Perhaps all societies will not evolve on the American model of individualism and the all-powerful market. It is not certain that, in the near or more or
less distant future, American society itself will not mitigate the excesses of its
model and impose some restrictions on the unlimited desires of individuals,
even those who can pay.

The dual-parent family structure may not disappear, after all. It might
well be voluntarily preserved because of the emotional charms that many
still find in it and might continue to find there, despite the temptations of
individualism. It is conceivable that communities, religious or otherwise, may
jealously preserve this way of life, possibly even using reproductive technologies, including the artificial uterus and cloning, in the context of the parental,
fraternal, and family structure they have decided to keep.

Many types of biosocietal experience will be able to coexist. Some may prevail and survive, while others will disappear or transform themselves.

Finally, as the "second" Huxley suggested in 1946, a narrow path remains open despite everything—the path of a philosophical conversion to the search for freedom of the Spinozist type by means of knowledge and an understanding of the world, others, and oneself. Perhaps the future will see the emergence of societies like Castalia, which Hermann Hesse imagined in *The Glass Bead Game*. Or the future may hide a neo-platonic conversion toward the One, favored by a new technological Eden that has broken free from the constraints of animality, or conversions of the type Spinoza describes at the beginning of his *Treatise on the Emendation of the Intellect*, where he decides to embark on research that will lead, through the joy of knowledge, to the freedom and happiness of the sage we encounter at the end of his *Ethics*:

> I finally resolved to inquire whether there might be some real good having power to communicate itself, which would singly affect the mind, to the exclusion of all other apparent goods; whether, in fact, there might be anything of which the discovery and attainment would enable me to enjoy continuous, supreme, and unending happiness. . . . For the ordinary surroundings of life which are esteemed by men, as their actions testify, to be the highest good, may be classed under the three heads—Riches, Fame, and the Sexual Pleasure [*libidinum*]: with these three the mind is so absorbed that it has little power to reflect on any different good. . . . All the objects pursued by the multitude, not only bring no remedy that tends to preserve our being, but even act as hindrances, causing the death not seldom of those who possess them, and always of those who are possessed by them. . . . For I perceived that the evils were not such as to resist all remedies. . . . Especially after I had recognized that the acquisition of wealth, sexual pleasure, or fame, is only a hindrance, so long as they are sought as ends not as means; if they be sought as means they will be under restraint, and, far from being hindrances, will further not a little the end for which they are sought, as I will show in due time. (Spinoza 2002b: 3–4)

Translated by Gina Fisch

TWENTY-SEVEN

To Teach Virtue

(1991)

> Man is the measure of all things: of things which are, that they are, and of
> things which are not, that they are not.
>
> —Protagoras

Relative Relativism: The Measure of Things That Are Not

The conception of education I will defend here is different from the one we
inherited from both the classical world and the Enlightenment. Traditional
education rests on an act of faith in the liberating and educational value of
truth: training in critical pursuit of scientific truth, objective and disinterested, must automatically produce the liberation of men as individuals, love
of virtue, and an attachment to the moral values of their society. A testimony
to a kind of preestablished harmony between the liberation of individuals and
social cohesion instituted and unveiled by objective knowledge, this act of
faith is contradicted more and more regularly by facts. The relativistic conception that I am defending here proposes that the critical pursuit of truth is
only a component—an indispensable but not an exclusive one—of "apprenticeship in virtue" (and thus of the solution to the "question of education").
This conception rests on an act of faith as well, but of a different kind.

In this new conception of education, there is an incontestable show of faith in the future of educated children, in their ability to form opinions that are not too crazy in relation to what educators (the three unified powers[1]) consider as their reality, their truths, and their values; an act of faith in the ability of these children to renew these truths and these values, should some new constraints make it necessary, and to do so without completely breaking the continuity between generations. In short, it is an act of faith in the unknown future that represents the generation to come, an act all the more justified because education, while giving up totalitarian mastery, leaves room for some indetermination, the source of creativity and possible freedom. (Here one meets again the von Foerster theorem, now extended to the time of the generations.[2]) This act of faith, this show of confidence in the future, in education by opinion despite its changing character (or because of it, since this conception leaves room for play in the vagueness of the game and protects against the one-dimensional discourse of single and definitive truth), implies a vision of virtue to be taught as an *art* of living and not as an absolute knowledge of the truth and the sovereign good. In Hebrew, *amen* comes from the same root as *'oman*, a man of art; this common root designates faith, trustful compliance, fidelity, education through practice, and the creative capacity of the artist. It stands for Nietzsche's "yes to life," provided we understand it as a "yes to the future" rather than as founding a philosophy of Life, with its vitalist, pseudo-biological abstractions.

Even when our experience contradicts the Socratic tradition, the latter is still so powerfully present in our minds that many have difficulty renouncing an absolute belief in one and only one truth and accepting this kind of relativism. The critique most often directed at relativism is of a logical or moral order, with the latter resting on the former.

The moral critique of relativism condemns its inability to establish a universal norm, since relativism seems to allow each individual or each social group to justify any norm and its opposite. No superiority could be granted to "Thou shall not kill" over the morals of an assassin; inversely, one would not know, it is claimed, how to raise objections to torture or death camps or to justify such objections with reference to the prohibitions and limitations

---

1. [For an examination of the "three powers of speech," which should be kept separate: see Atlan 1991a: 231–39.—Eds.]

2. [See ibid., chap. 1.—Eds.]

that the Geneva Convention and humanitarian organizations such as the International Red Cross or Amnesty International try to instill. In answer to this critique, I would point out that, while theoretical relativism accepts the impossibility of grounding a system of norms or even a system of unique, universally acceptable ethical principles in theory, it does not thereby prevent establishing a practical universality by consensus, a type of universality that is certainly more difficult to formulate, but that is also more effective. This transcultural pragmatic universality of fact establishes itself by adhering to principles that eventually approach international agreements such as the Universal Declaration of Human Rights, the Helsinki Accords, and the Geneva Convention. This list can be expanded to include the prosecution of war crimes and crimes against humanity in Nuremberg after World War II.

As for the logical critique of relativism, it has no object as soon as one relativizes relativism itself, while recognizing it for what it is. In fact, this criticism always comes down to pointing out the contradiction between the fact of relativizing all belief and the attempt to justify this relativism to the exclusion of other attitudes, so that relativism appears as a privileged belief and the only one that is true. It is easy to answer this critique by showing that it is located on two different levels and that there is a difference between an "absolute relativism" (which, indeed, would be self-contradictory) and a relativism relativized because recognized as such.

One way, among others, to grasp this difference would be to return to the sources and to read a famous aphorism by Protagoras, while placing it in context: "Man is the measure of all things: of things which are, that they are, and of things which are not, that they are not."

Much ink has been spilt over this pronouncement, in many interpretations from the relativistic caricature Plato gives of this saying so as better to refute it and oppose to it his own alternative: "For us, the divine must be the measure of all things" (*Laws* 4.716c),[3] to more recent humanist interpretations, which were supposed to rid us of the gods, superstitions, and transcendent worlds not accessible to perception and to the measure of man, and to install man—more precisely, man in society—as the referential unit of all things, thanks to his conventional and artificial activity.

---

3. For example: "How things appear to me to be is how they are for me, how they appear to you to be is how they are for you" (*Cratylus* 386) or "But if he sees any one thing, he sees something that exists" (*Theaetetus* 161c).

Perhaps we haven't given sufficient weight to the negative part of Protagoras's proposition. I would like to suggest a relatively relativistic interpretation of it (and not an "absolutely relativistic" one, which is evidently incoherent and can easily be ridiculed in the manner of Plato or, more recently, Hilary Putnam [1981]). This interpretation starts from the assumption that, because we experience them, we know that there are things that men cannot measure and that are not to their measure; Protagoras knows this, too. It is on the basis of this knowledge or these observations that we say (with Protagoras): "Man is the measure of all things." This is why we cannot stop there and must now distinguish between "things that are" and "things that are not." Some things are accessible to man and, in that humans measure them, they "are," since "all" things—by definition—are in the measure of man. Other things are not to his measure and therefore are inexistent; this is still as a consequence of the assertion that "all" things are in the measure of man. They "are not." But the last part of the proposition tells us that these inexistent things are still in the measure of man, even in what they are not. It means simply that man *also measures that in which things are not to his measure*.

I suggest the following logical analysis of this proposition. (It would greatly surprise me if such an analysis has not already been suggested by one of the many exegetes who have addressed this question over the centuries.) The first part is a *definition* of what *all* things *are*—they are "in the measure of man." Then, a distinction is made between things "that are" and things "that are not." It is the same distinction as that between things that are in the measure of man and those that are not. It follows that the third part of the formula affirms that the things that are not (in the measure of man) are to his measure in what they are not (to his measure). He measures, for example, indirectly and by difference, that in which he cannot measure them. The second part, while appearing to repeat the first, acquires its meaning in relation to the third. Man measures—directly and positively—the things that he can measure. This direct measuring of these things, however, contrasts with the one done by man's measure—indirect and negative—of that in which he cannot measure the other things.

Insofar as Protagoras's man is concerned, to measure that in which things are not measurable amounts to measuring that in which—for him—they are not. This is so because for him only the things that he can measure exist. The advantage of this position is that it is not ingenuously or "absolutely" relativist. Things exist relative to the one who perceives and measures them in the broad sense that he assesses them from both quantitative and qualitative points of view. But I affirm this relativist proposition in a relativist *manner*. This means that it doesn't eliminate the possibility that things that we cannot measure also *are*. They are, but in another way of being for us, that is, as things that we cannot measure and that therefore *are* for us, but *negatively*.

This interpretation is coherent with other opinions of Protagoras in which relativism and agnosticism regarding the existence or nonexistence of the gods does not stop him from recommending not only virtue but piety and obedience to the laws of the city, including the laws of its gods.

Translated by Oleg Gelikman

TWENTY-EIGHT

The Center of the Universe and the Domain of Ethics

(1991)

Even if our image of man—which Michel Foucault has said is being effaced like an image written in sand—is changing, it does not easily disappear under the influence of new forms of knowledge. Today, even though man has freed himself from the enclosure of rational omnipotence in which humanist knowledge maintained him, and even though he has opened himself to natural determinations that, in making him exist, seem to obliterate him, he nonetheless remains present and irreducible as an unquantifiable totality. That is to say, not only as the imagined totality of past, present, and future human individuals, but also as the "whole" of man, which involves his unimagined "possible" beyond already real and describable potentialities, and also his capacity to negate the real, thereby making this "possible" possible both logically and empirically.

The dethroning of man from the center of the universe has thus been wrongly bemoaned, although it is vain to attempt to replace him with the help of renovated anthropomorphic constructions clothed in scientific jargon (such as those, reminiscent of Pierre Teilhard de Chardin, that postulate, in conformity with a hoaxlike "anthropic" principle, a growth in complexity from the *Big Bang* to the physicist who theorized it). Man, as a thinking being, was never dethroned from his unique and irreducible place . . . *in the world of thought*. Only because we operated as if the only reality that mattered was that of extension and matter was it important that man occupy a central place in that other world as well.

As a source of judgment—even if this can be judged illusory and delirious by the model of this or that theory—I remain at the center of the universe. Life in society, and the dependencies this implies (through, among other things, the use of language and speech), propel me, whether I like it or not, to extend this centrality to other human beings, everywhere I recognize the human form, the human body and face. Through this recognition-projection, I distinguish the nonhuman and establish, in the nonhuman animal, degrees of closeness or distance (degrees that are not always scientifically founded) from the whole of man, who is always at the center. Regardless of the off-center and alienated image provided by cosmology and the theory of evolution and completed by knowledge of the unconscious, biological, and social determinisms that traverse him, man, insofar as he thinks, judges, makes projects, and perceives himself as doing such things remains a center from which the universe projects itself. Better yet, as both center and periphery, he is able to reach the confines of the universe in time and space, and even in other dimensions his thought permits him to enter.

But of that man, a naturalist can speak only in a negative way, which is already a great deal. Scientific knowledge can trace his shadow by making precise what limits him, the physical, biological, and social constraints within which he is forced to negotiate the whole and the possible, what he can or cannot deny without putting himself in danger. But it has nothing to say about the positivity of man: who he is, what he does, or even whether he exists. On the contrary, the natural tendency of scientific knowledge is to act as if he did not exist, as if his limits could shrink to the point of annulling him as new determinations are progressively discovered. Current debates about the human person and the harm that biotechnologies are likely to inflict upon him perfectly illustrate our point: nothing in the biological sciences allows us to characterize in an objective ("scientific") manner the human person in its positive attributes. As a result, these sciences, which have produced techniques capable of modifying the biology of human beings, can tell us nothing concerning the good or evil, desirable or undesirable character of these modifications, if they are to be judged not from an instrumental point of view interior to logical technique itself but from the viewpoint of the interests of man.

To take Marie Curie's suggestive formulation: science is concerned with things, not with persons. The natural sciences make this evident: But what of the sciences of man? Insofar as man can be considered objectively, the object of these sciences is "man"—through the construction of an object of thought (and sometimes of experience) with a view to applying the scientific method to him. Men are thus also things. And, insofar as they are considered as persons, they are the ones who are concerned with science, and not science with them.

But if that is the case, who or what "is concerned" with them? Which activity, of both thought and practice, concerns men as persons?

The answer is all the rest, all that is not science, that is, all the philosophical and religious traditions of the West and the East, the mystical traditions, the ancient mythologies, and also the false modern sciences that pretend to replace these traditions. This separation between the science of things and the thought of persons is in fact quite recent. All these traditions included, in former times, the science of inanimate objects in great syntheses where everything—the thinking and conscious human being, the animal, the plant, and the mineral—had its place in a hierarchy of being that claimed to be perfectly intelligible. Modern science separated itself from all that and acquired its operational and theoretical efficacy thanks to this separation. Whence sprung the nostalgia for great syntheses, the search for lost unity.

In all of these nonsciences, we can—and some among us believe that we must—distinguish the rational from the irrational and, within the irrational, distinguish the conscious and lucid play of paradoxes (which implies a mastering of the rational) from confusion and the attitude that anything goes. It is clear that the entire philosophical tradition thinks itself to be rational and that a particular philosophy succeeds in inscribing itself in the tradition only insofar as it is rational. This is also true of certain intellectual currents that have attempted to restrict themselves to a rationality as rigorous as possible within the great mystical and mythological traditions. (Certainly, these rationalities differ from one another in that the rules of their games and their fields of application are different. They are, nonetheless, rationalities, in that they employ discursive language founded on as rigorous a use as possible of the meaning of words and phrases, even when these are metaphorical. These rationalities thus always lead to an infinite critical analysis imposed by the iron law of the principles of identity and noncontradiction.)

Out of this recent separation between science and everything else, over the course of the last few decades two nonscientific disciplines, fields of investigation, or ways of thinking that have man-person as object and subject at once—and that also lay claim to being rational—have imposed themselves: psychoanalysis and ethics. The former, already well established through numerous practices and theories, tends today to be divided between a rigorous reflection that in fact joins with the Western philosophical tradition (and thereby renews it), and an ensemble of pseudo-sciences that nourish modern mythologies of the irrational. As for the latter, while it has always been part of ancient traditions, philosophical and other, it strives today to become autonomous, essentially because of the *demand* for ethics and norms *coming from science*, which, it turns out, not only ignores persons but can eventually threaten them. Traditional ethics, borne up by philosophies and religions, does not respond to this demand in that it has not yet succeeded in accepting the autonomy of scientific knowledge; ethics continues to found itself on the representation of natural things, which modern science definitively rejected a long time ago. Even Kant, who seemed to have allowed philosophy—and at the same time morality—to take the turn toward a greater autonomy of science, is no longer of great help: although he takes into account the scientific revolution of the seventeenth century, that of Newtonian mechanics, he still falls within the tradition that precedes the second scientific revolution, that of twentieth-century biology, which allows us to treat as things (that is, inanimate objects "without a soul") not only celestial bodies but also living beings. We are thus brought back to the question of the purposiveness of nature, which has caused a permanent crisis in the question of natural law. For natural law to exist, the laws of men (judicial law) must have the attributes and the properties of the laws of nature. But for that, the laws of nature must appear to have the same function, the same purpose, as those of men: that is, they must express the order of things that allows the community of men who share these laws to exist—in other words, society in general and the individuals who compose it. It was necessary, then, that the laws of nature describe a nature ordered and having a purpose that coincides with men's experiences of life. Hence the evidence of the purposiveness of nature, indispensable for the transposition of the model of natural laws to that of the laws of men, through which the laws of men could claim for themselves a natural origin.

It is this evidence of the purposiveness of nature, however, that physico-chemical and mechanistic (thus resolutely causalist) biology has eliminated. At the same time, biotechnology appears to be threatening men as persons.

Whence the demand for an ethical reflection that would "concern" itself with the man-person while integrating the representation of things, including living things, abundantly supplied by today's science and technology, united as they are in the sociocultural complex that we call technoscience. If we do not want to take refuge in modern mythologies or in the irrational mysticisms of pseudo-science, then we have no choice but to renew ethical reflection on the basis of different traditions, philosophical and otherwise, while accounting for the rupture—taking into account the impossibility that follows, namely, founding an ethics (even indirectly and philosophically) on scientific knowledge. Hence the necessity of establishing new relations between ethics and truth, and thereby between education and truth.

But the task is not an easy one, as it requires protecting ourselves on all sides—not only from irrationality of all kinds but also from rational ideologies that forget that rational critique is reason's critique of itself, reason's delimitation of the domains of its legitimacy and the rules of its application. The temptation offered by both sides, as they renovate their forms and modernize their appearances, is strong. After the end of the "scientific" totalitarianisms of the twentieth century, do we not see on the horizon what risks becoming the ideology of the twenty-first century, with its risk of unidimensionality and of new totalitarianisms? On the one hand, the transformation of ecological science into ideology has already begun, with its political projects and its stonewalling, founded on projections at once apocalyptic and messianic, inspired, of course, by what science is supposed to teach us. On the other hand, religious fanaticism is resurging, whether in the extreme forms of the "Islamic" revolution or those, attenuated, of other religious ideologies whose intolerance feeds on sectarian particularism or a missionary universalism.

It is up to the philosopher to speak if we are to prevent the domain of ethics, of the person, and of the project from being abandoned to the obscurantism of religious fanaticism and the charlatanism of false science. Of course, the great mystical and religious traditions of East and West, rooted in their mythologies, have never ceased to speak from and for this at once centric and eccentric domain. But abandoned to themselves, they find it difficult not to withdraw into fanaticism and obscurantism, not knowing how to situate themselves and to resist the steamroller of technoscience. This is why they need the capacities of critical creativity that the philosophical tradition has developed. It is certainly no longer a question of imagining a great philosophical (in fact, mystical-scientific) synthesis, since we know today that it cannot but answer, albeit unwillingly, to the need for ideology and for a confusion of facts and of values, and that such a synthesis could not but be led astray by this need into totalitarianism and a new fanaticism.

For the philosopher, it is a matter of taking a stand. Attentive to objective sciences, he gains from them not certainties in matters of general and unifying theoretical knowledge but rather elements of knowledge as factual as possible, integrated into provisional theorizations, which resist as best they can the assaults of critique. He thereby learns all the more to exercise his own thought. It is in this way that, armed with critical reason, he becomes attentive to what escapes, by construction, the scientific—in other words, the life of man as center of a personal and impersonal universe.

If we assign this role to philosophy, philosophy is condemned to speak clearly of something that can be conceived in a partial and relative manner, and to measure itself against what the neo-positivists call non-sense, that is the mystical elements of language, for the earlier Wittgenstein, and, quite simply . . . philosophy for the later Wittgenstein. In the field where we thus situate it (at the opposite end of the place given to it by a neopositivism that would employ the model of an empirico-logical science), philosophy is no longer very far from non-Western philosophies that developed in parallel and at times in interaction with Greek philosophy, but in an autonomous way and in continuity with inspiration by the mystical traditions of the Near and Far East. But if philosophy wants to continue to differentiate itself from these mystical traditions, refusing to employ systematically their metaphorical and equivocal language (the "Unveil one cubit in order to hide two" of Kabbalists, the permanent paradox of "breaking the limits of reason" of Buddhists), then the only language still at its disposal is a *double language*. And one is at once surprised and perhaps reassured to observe that this is not an empty gesture: one finds examples in past philosophers (among them some of the greatest) where the use of a philosophical double language must be carefully distinguished from political doublespeak.

Translated by Danielle Dubois

TWENTY-NINE

Pleasure, Pain, and the Levels of Ethics

(1995)

When one reflects on the nature of moral judgment or the genealogy of ethics—and there are several ways of going about it—the most difficult question is still that of articulating the normative and the descriptive.

What I would like to say can be summarized like this: there is a first level of ethics that is first above all because on it this question of articulating the normative and the descriptive is immediately resolved, or at least so it seems: this is the level of avoiding pain and seeking pleasure. This level of pleasure and pain is also first because, as we will see, it is probably the most universal. But things are more complicated, because there are always several superimposed levels across which pleasure and pain are taken up and transformed. This superposition and integration of levels produces the richness and difficulty of moral problems. These problems can certainly not be reduced to pleasure and pain, but pleasure and pain cannot be eliminated from them.

Articulating the normative and the descriptive is a difficult problem because we must avoid falling into the so-called naturalistic fallacy, that is, the confusion of what is with what ought to be. We cannot maintain the objectivity of knowledge if what is is confused with what we would like it to be. But inversely, what we decide must be is not necessarily what is. If we discover that there are aggressive drives in us, that does not mean that we cannot or must not decide to resist them. By the same token, we decide that it is good to treat and heal diseases, even though we know that diseases are part of the reality of nature. On the contrary, our knowledge of what is—of the mechanisms of diseases, of our drives—can even help us to resist and thus to change what is according to what we judge should be, within the limits of certain constraints, of course.

And yet, once we have recognized that what is is different from what should be, the question of their articulation still poses itself, and in a different way at each level. First, it poses itself on the level of the moral judgment that makes us say, "This is good and we must seek it; this is evil and we must avoid it." Where does the normative character of moral judgments come from? (This is what distinguishes them from other judgments, judgments of reality about what does or does not exist, judgments of truth about what is true and false.)

But the question still poses itself, perhaps even more insistently, on other levels, notably on the level some call meta-ethics, which is concerned with judging moral judgments, with evaluating under which conditions a moral judgment is a good moral judgment and whether one system of values and norms is better than another. At this level, the question joins the question of moral relativism, for those who believe—wrongly, in my opinion—that the only alternative lies between a total relativism, in which everything is equivalent and for which the observation of the systems of values and norms in force in different cultures cannot extend beyond cataloging morals and customs, and universal ethics that would allow us to judge in what a given system is more "advanced" or "enlightened" than others, as we still like to say, under the influence of the Enlightenment.

I would like to defend the thesis that the origin of the normative character of ethics on all its levels can be found in our experience of pleasure and pain. Not, of course, in the raw experience of the immediate sensation of pain or the well-being that comes with satisfying a need or desire, but in that experience as it is modified, transformed, interpreted, and sometimes inverted through the prisms of our cognitive capacities, insofar as these cognitive capacities, when they are not dealing with data concerning pleasure and pain, are at the root of other sorts of judgments. It is in their effect on an aspect of pleasure and pain that our cognitive capacities intervene in moral judgment.

That is why I cannot but evoke briefly the still largely open questions concerning the nature and the mechanisms of our cognitive activities, even if these mechanisms are also at the root of the property particular to the human species that allows us to make judgments about good and evil. These mechanisms concern what we could call, by analogy to linguistic competency, the ethical competency of *Homo sapiens*. It comprises the mechanisms of memory, language, the capacity to represent and symbolize, and intentionality as a source of meaning and of the effects of sense, and also, more simply, of goal-oriented projects and activities.

Understanding the mechanisms of these activities of what we call the mind—or, more precisely, the brain-mind unit—is certainly pertinent to our task here, for in their absence, notably in animals, there is no ethical competency. Each of these activities, even those that seem the simplest to us, puts into play very complicated processes, processes that are studied by cognitive science, neurophysiology, psychology, linguistics, artificial intelligence, and so on. I myself work on preliminary elements of a physical theory of intentionality,[1] with the help of a model of how projects emerge in networks of automata in which no intentionality whatsoever is a priori supposed to exist. I try to understand how a purely causal sequence of brain states that produces a certain effect can be transformed into a goal-oriented procedure in which this effect, in its representation, ultimately sets off a repetition of the same sequence. This sequence then becomes the accomplishment of a project. That is a relatively simple and limited form of the emergence of intentionality, yet it is far from trivial, because outside humans, it is observed only in chimpanzees capable of using objects as tools.

I do not want to talk about that now, even though it could be pertinent to our task, because it concerns elements of ethical competency and not elements of performance, or rather performances, to stay within the linguistic analogy: the capacity to render moral judgments does not yet tell us much about the content of these judgments. Yet it is above all because of the content of moral judgments that the question of articulating the descriptive and the normative arises.

❀

---

1. See Atlan 1991a, chap. 2.

The idea that the origin of the normative character of ethics can be found in the experience of pleasure and pain is not new. We can find it in the philosophers of antiquity, not only in the Epicureans but also in Plato, and in Aristotle in particular. That is because, in fact, the experience of pleasure and pain is distinctive in that it is both cognitive and normative. A sensation or perception is quickly and automatically accompanied by a behavior aimed at avoiding pain or seeking pleasure. These already are complex phenomena— and that is why I would hesitate to speak of "primitive" facts—because, as we will see, there is no symmetry between pleasure and pain. On top of that, pain as sensation is already different from suffering at the neurophysiological level. Neurosurgical interventions and studies with Positron Emission Tomography brain imaging have shown the existence of neural pathways of suffering (i.e., roughly of a disagreeable character) different from the pathways of pain as a discriminating sensation (Willis 1995: 19–20, Craig et al. 1994: 770–73, Gybels and Sweet 1989). Nonetheless, at least for the individual and in the immediacy of his perception, his knowledge of a state of the world and of his body immediately determines a reaction of flight or avoidance, thus a kind of proto-judgment of what he *should* do. Once states of this sort have been committed to memory, reason, imagination, and language appropriate them and make them the object of a less immediate knowledge—a knowledge, however, that retains a trace of the normative character of pain that *should* be avoided and of pleasure that *should* be sought.

The ancients saw this clearly, especially the Epicureans, some of whom Cicero calls "subtle" because they are not content with "judging good and evil by the senses but suppose that one can also grasp by the intellect and reason [*animo etiam ac ratione*] that pleasure should be sought and pain avoided in and for themselves" (Cicero 1914: 35, trans. modified). But they also saw that things are not that simple, and we can find in Epicurus himself a description of the moral problem, which I believe to be exhaustive because it sums up, in a few sentences, all the difficulties at the root of its practically infinite complexity:

> And since pleasure is the first of natural goods, we do not for this reason choose any pleasure whatsoever, but often pass over many pleasures when a greater annoyance ensues from them. And often we consider pains superior to pleasures when submission to the pains for a long time brings us as a consequence a greater pleasure. While therefore all pleasure is good because it is naturally akin to us, not all pleasure is worth choosing; just as all pain is an evil and yet not all pain is to be shunned. It is, however, by measuring one against another, and by looking at the

conveniences and inconveniences, that all these matters must be judged. Sometimes we treat the good as an evil, and the evil, on the contrary, as a good. (Epicurus 1995: 655)

Everything has been said here if we emphasize two expressions that sum up two enormous problems, which generations of philosophers and moralists have not ceased to address.

One poses the question of accord with our nature: "all pleasure is good because it is naturally akin to us." The other concerns "measuring one against another." What is this wise consideration that "sometimes we treat the good as an evil, and the evil, on the contrary, as a good"? Who is this "we" who sometimes treat the good as an evil and evil as a good?

It could be the "we" of the philosophers, who decide rationally. But it could also be all of us, every single one of us, who in our behavior express an existing morality. I believe it is important to dwell on this distinction between a morality expressed in this or that moral philosophy and that, or those, practiced in fact in existing societies.

Let us first look at the case of the philosophers by noting that, in one way or another, they return to the question of accord with nature. We learn that, for the Epicureans, "pleasure and pain arise in every animate being; the one is favorable and the other hostile to that being's nature, and by means of pleasure and pain choice and avoidance are determined" (Diogenes Laertius 1995: 565). But this is not only about the Epicureans. For all naturalistic philosophies, the good conforms to nature in general, or to the nature of the living being or of human beings, and that is why we must choose it. Put otherwise, this is where the normative (what should be chosen) and the cognitive (what nature is) are identified. Knowledge of nature is at the same time knowledge of good and evil.

Obviously, this identity was destroyed by critical philosophy. But it is still necessary, though difficult, to come up with a possible articulation between the two orders.

Regarding the question of a possible knowledge of good and evil, within the framework of a study of the relations between ethics and cognition it is difficult not to evoke the biblical myth of the first man and of the tree of knowledge of good and evil, especially because in a moment we will evoke the role of myth in the genealogy of ethics. In the multitude of interpretations to which this myth has given rise, it is interesting to retain those of two philosophers, both rationalists, but very different: Maimonides and Spinoza. They disagree on many essential points, yet agree on the idea that a knowledge of good and evil different from a science of the true and the false is only a makeshift solution. In the paradisiacal state before the fall, rational knowledge of the true and the false rendered the knowledge of good and evil useless, even nonexistent.

For Maimonides, "it is the function of the intellect to discriminate between the true and the false—a distinction which is applicable to all objects of intellectual perception" (Maimonides 1956: 90). What is good and what is bad, like the beautiful and the ugly, by contrast, are not of the order of the intelligible and the necessary but of that of opinion and the probable.[2] When Adam had perfect and complete knowledge of all known and intelligible things, "he was not at all able to follow or to understand the principles of apparent truths" (ibid. 90). Good and evil as such did not exist for him, because for him only what reason made him understand as necessary existed. Only by losing this knowledge of all things as intelligible did he acquire a new state in which he knew things otherwise than rationally, contingently according to the opinion he had of them as beautiful or ugly, good or bad.

Spinoza, too, refers to the biblical myth, in part 4 of the *Ethics*: "If men were born free, they would form no conception of good and evil so long as they were free," for "a free man is he who is guided solely by reason" and "has only adequate ideas." Hence, he "has no conception of evil," for evil is only the product of partial and confused, inadequate ideas; in consequence, he has "no conception of the good (for good and evil are correlative)" (pt. 4, prop. 68; 2002a: 355). But he goes on to observe that this hypothesis (that human beings are born free) is impossible for humans, who are only part of nature and thus are subject to determinations external to them. This impossible hypothesis, he writes, seems "to be what Moses intended by his history of the first man." For Spinoza, it is from the perspective of this hypothetical but impossible originary state that Moses told "the story that God forbade the free man to eat of the tree of the knowledge of good and evil, saying that as soon as he should eat of it he would straightway fear death instead of desiring to live" (ibid., scholium, p. 355).

In any event, the knowledge of good and evil characterizes the human condition after the fall, for this approximate and conflictual knowledge replaces knowledge by means of the clear and distinct ideas of reason, the

---

2. [I have used "good and evil" for *le bien et le mal* and "what is good and what is bad" for *le bon et le mauvais.*—Trans.]

only possible source of freedom. For these philosophers, the knowledge of good and evil thus characterizes a state of alienation, given that human beings are born into the real world, as opposed to the intuition of a world of freedom and happiness that the rational knowledge of what is true would allow us to conceive.

Moreover, the role of science and philosophy is to grant us access, with the help of reason, to a knowledge of good and evil that is true because it flows from the "true" nature of the human. Thus a true knowledge of good and evil is conceivable. For Spinoza, this is the knowledge that reason teaches us as being useful or harmful to the nature of the human. This is not the place to discuss the criticisms of this utilitarianism, whose ethical character could be contested. Let us simply recall that, in the context of the *Ethics*, it turns out that what reason shows us to be the most useful for a human being are other human beings (see, e.g., pt. 4, prop. 35, corollaries 1 and 2; 2002a: 337). Be that as it may, we have here the idea of the possibility, theoretically, at least, of a natural foundation of ethics. But this obviously implies that we have adequate, that is, perfect and complete, knowledge of what the nature of the human is. Yet that is far from being the case. Instead, we have only religious or secular ideologies, each of which presents itself as the true knowledge of what nature in general and human nature in particular are. Above all, by admitting that we can acquire such a true knowledge of good and evil, such knowledge has no force in itself. Inasmuch as it is purely intellectual, disembodied knowledge, its possible truth could not be imposed on the ideas of good and evil we habitually come up with for ourselves on the basis of our experiences and our passions. This is another way of emphasizing the difficulty of articulating the descriptive and the normative: "true knowledge of good and evil cannot check an emotion by virtue of being true. But in so far as it is an emotion, if it be stronger than the emotion which is to be checked, to that extent only it can check an emotion" (pt. 4, prop. 14; 2002a: 328), that is, only to the extent that it can become interiorized and incorporated to the point of acquiring the force of an affect. Put otherwise, what is true does not constrain. It can be a norm only for itself and not for behavior.

This point is crucial: a content of knowledge can only be transformed into a norm if it is invested with an emotion that produces well-being or suffering, that is, a feeling of the agreeable to be sought or of the disagreeable to be avoided.

In the meantime, for life in society we can avail ourselves only of an inadequate knowledge of good and evil, which, for the philosopher, is merely the practice of a provisional morality passed on by the society in which he finds himself and toward which his personal ethics sometimes forces him to assume difficult attitudes. Put otherwise, the aspiration to a true knowledge of good and evil in which evil might disappear as such is a universal aspiration. For some scholars and philosophers, it is the aspiration to find in emotional life the same certitude as logical and mathematical truth can produce or, at the minimum, the sort of evidence the empirical sciences produce. But in reality, we are dealing with existing moralities.

✦

Thus, leaving the philosopher's theoretical reflection behind us, we find ourselves faced with multiple opinions concerning systems of values and norms inherited from our social and individual histories. It is by means of these norms and opinions that we render value judgments, deciding what is good and what is evil. Yet beyond the multiplicity of these value systems, we can still ask what is common to them and thus universal.

There remains, first of all, their very existence. The existence of systems of values and norms is a universal fact in that it characterizes all societies, even if the contents of these systems are not the same. And here we find once more our first level, which is common to all these systems because it is conditioned by the physiology of *Homo sapiens*, the experience of pleasure and pain, of what makes us feel good and of what makes us feel bad, however this experience may be described and, of course, integrated into a socially constructed and inherited system.

This first level of ethics, where good and evil blend with pleasure and pain, with what is good and what is bad, we can call the fundamental level or level zero: the fundamental level if we prefer to give it a positive connotation due to the fact of its physiological universality; level zero if we want to give it a negative connotation because of its apparent poverty in relation to the richness and complexity of moral problems. To the majority of modern philosophers, especially after Kant, the experience of pleasure and pain seemed too simple, too limited to have anything to do with the sophistication of the "problems" of conscience, liberty, and responsibility that are, if we may say so, the daily bread of ethical reflection.

Of course, at the crux of these considerations we once more find Spinoza, for he takes up this identification of good and evil with what is good and what is bad and makes it the foundation of his theory of the affects: "in so far as we perceive some thing to affect us with joy or sadness . . ., we call it what is good or what is bad; and so knowledge of good and evil is nothing other than the idea of joy or sadness."[3]

But here, what is good and what is bad already are no longer reduced to pleasure and pain but to what he calls the affects of joy and sadness, which, by the way, he defines in reference to an augmentation or diminution of the power to act. This definition allows him to enlarge the perception of pleasure and pain to include the perception of well-being and suffering, which are already less punctual phenomena and integrate a larger number of stimuli at a different level of organization, which is a "common gift." This gift is common, this time, not to all living beings, as in Epicurus, but to all beings endowed with reflexive consciousness and, in any event, to all human beings.

❦

From this first level on, things very quickly become complicated because of other characteristics proper to the human species, namely, our cognitive capacities of representation, memory, symbolization, intentionality, and so on. The experiences of pleasure and pain produced by the contingencies of each of us, individually and collectively, are committed to memory, and it is their content that then feeds our more or less rationalized imagination.

In other words, because of memory and the imagination, what we call good and bad is no longer reduced to the immediate perception of what makes us feel good and what makes us feel bad. The signification of the words *good* and *bad* is doubled twice by the imagination: once in time, where good or bad in the present is extended in duration; and once in space, where our subjective experience of what does us good or harm is projected onto others who resemble us and to whom we attribute the same experiences. It is thus extended to a collective good or evil.

This doubling of the things we call good and bad expresses the ambiguity of the words that name them. Insofar as the word *good* is concerned, this is the ambiguity between, on the one hand, what is good, a source of pleasure, well-being, and happiness, and, on the other hand, what is good by virtue of a reflexive or conditioned judgment, a moral judgment. The ambiguity is even more evident when "bad" as pain and "bad" as bad action are involved:

---

3. The text continues: "which [idea] necessarily follows from the emotion of joy or sadness," that is, the emotion itself in so far as we are conscious of it, put otherwise, in so far as we form the concept of it, in so far as we have the idea of its idea" (proof of pt. 4, prop. 8; 2002a: 326; prop. 8 states, "Knowledge of good and evil is nothing other than the emotion of pleasure or pain in so far as we are conscious of it"; trans. modified).

on the one hand, we have the bad that is suffering [*mal-souffrance*], penury, and pain, which tends to eliminate the search for the good of well-being [*du bien (être)*] and for the good of happiness [*du bon (heur)*]; on the other hand, we have the bad as the evil of wickedness [*mal-méchanceté*], that of bad actions, of vice and perversions, of hatred and injustice, from which the teaching of ethics is supposed to deliver us. The second bad sometimes joins the first, in cases where wickedness is suffering inflicted on another or on oneself in a more or less distant future.

Awareness of the distance between the good or bad that is perceived immediately in the present of sensation and the good or bad that is perceived more abstractly by a projection of the imagination onto others or onto oneself in a time other than the present constitutes what we call the "knowledge" of good and evil. The good and bad that are imagined, projected into the space of others, deferred in time, and not perceived immediately as what is good and what is painful cannot but be the object of a reflective judgment, a moral judgment. And moral judgments are marked by the ambiguity of what we call "good" and "bad," which unite their two aspects in the consciousness or representation we have of them.

This ambiguity is evident in the notions of joy and sadness, of happiness and unhappiness: in them we find individual experience in the present and its double doubling—by imagined extension to others and in time—inextricably blended. And this is where the *second level* of ethics institutes itself.

This second level can be characterized generally by individual or collective strategies in which the search for the good and the avoidance of the bad are deferred. For we know, or think we know, that a present good can be the source of bad in the future, whether for others or for ourselves. The second level of ethics is the level on which the search for pleasure and the avoidance of pain must undergo an imagined projection in time and space that, in practice, ends in deferring a desire.

In fact, everyone's desire to avoid suffering and to enjoy pleasure immediately is more or less forcibly deferred, according to circumstances and above all according to value systems, philosophies, and traditions. These, in the end, are all expressed in norms of behavior that indicate, for different sorts of desire, when, how, and how much their satisfaction must be deferred. It is not possible to remain in the present, and memory and the imagination suggest strategies for deferring satisfaction. The most difficult question, then, is how much one can and must defer. If immediately satisfying every desire instantaneously, as perhaps animals and infants do, is no longer possible, inversely, deferring the satisfaction of every desire indefinitely and in all circumstances, even through processes of repression and sublimation, is not possible, either. As King Solomon says in Proverbs: "Hope deferred makes the heart sick, but a desire fulfilled is a tree of life" (Proverbs 13:12).

A set of judgments on when, how much, and how to defer the satisfaction of desires constitutes this second level of ethics, which is universal in its existence but diversified in its contents. This law of desire to be satisfied or deferred is lived differently according to alternations and varieties of satisfaction and deferment. These alternations create a rhythm of their own that characterizes each ethics as a particular instance of the law of desire and its sublimation. Each instantiation of this type constitutes a particular definition of good and evil, generally produced by the history of each society and the history of its collective imaginary, in relation to the imaginary of individuals.

The second level of ethics, being a strategy in which the desire to enjoy pleasure and to avoid pain finds itself deferred, is thus already the level of reflective judgment, where good and bad are no longer immediately perceived as what is pleasurable and what painful. With the development in the adult of memory and new cognitive faculties, the instinctive drive of the animal and the baby is transformed by a knowledge at once rational and imaginative.

❊

This is where we encounter a *third level*. In fact, this is where we are now led to render judgment on moral judgment, on the criteria we use to decide on good and bad. This is, obviously, where the question of relativism arises. Are all criteria of judging good and bad equivalent? Is everything equivalent: gang morality, Nazi morality, the morality of human rights and that of the master and the slave, of "Thou shalt not kill" and the morality of the strongest, and so on? Or are there criteria that permit us to decide that one morality is better than another, and perhaps that another morality is immoral? The set of such criteria would constitute a universal ethics. We are certainly, therefore, on a third level, concerned with judging the value systems of the second level. And once more, here too it is convenient to distinguish between philosophy and existing moralities. We can discern roughly two possible attitudes. One is the attitude of the great tradition of Western moral philosophy, culminating in Kant and the Enlightenment, in which reason establishes the foundations of a universal ethics. The other attitude consists in asking how a universal ethics might be established, not by deduction from principles of reason but pragmatically by observing existing moralities. The first attitude postulates the universality of a rational ethics that teaches every human being what is good and bad. As we have seen, for the ancients this ethics already flows from a nature of the human, itself universal. This is why this philosophical tradition has always maintained a close relation with the sciences, the natural sciences as well as the human sciences, which are supposed to teach us what the nature of the human is.

Critical philosophy made a decisive turn away from the ancient philosophies, recognizing the impossibility of founding morality on pure reason and on the categories of natural science but nonetheless not renouncing the universality of practical reason as a possible source of a universal ethics. This turn was itself conditioned by the birth of modern science in the seventeenth and eighteenth centuries, by what is called the scientific revolution—which concerned above all physics and chemistry, more than the life sciences. That is why it is possible to show that the pursuit of the project of critical philosophy today puts into question the humanist philosophy founded upon the supposed existence of a suprasensible domain of human freedom, to the extent, that is, that this philosophy benefited from the vitalist and finalist climate of the era (Atlan 1991a).

Very schematically, within the framework of this philosophy, a necessary connection between the exercise of morality and the reality of our free will is taken for granted. In other words, our moral judgment on the second level is not a good moral judgment from the viewpoint of the third level unless it is rendered by a subject endowed with a voluntary consciousness that freely decides its future actions. Free will and a force of the individual will seem to be obvious conditions of access to the second level of ethics.

Nevertheless, this judgment concerning what a good moral judgment must be like is itself far from universal. It derives from our being accustomed to the idea that our moral judgment follows internal deliberations of a self with itself that precede a reflective choice between a good and a bad. But the inevitable character of these deliberations and of their eventual causal effect is evident neither from an anthropological nor even from a philosophical point of view. In particular, we discover—especially thanks to biology, which is both physicochemical and mechanistic—the implications of an absolute and machinelike natural determinism for our behavior. The scientific rationality of the classical age already presupposed this determinism, but at the time life and the human seemed to be able to escape it. We must therefore reactualize a difficult and forgotten representation of the world, which was overshadowed by the easy idea of an accountable responsibility that could be conceived only on the basis of free will.

It is as if we had to return to an ancient, Homeric view of the world, where humans are possessed by gods who use them and act through them, except that our behavior is no longer possessed and determined by the gods of a mythical pantheon, personalized in the image of the human, but by impersonal mechanisms, those of the chemical reactions in our genes' molecules and of the metabolism of our cells. One could try to gauge the respective advantages and downsides of these two modes of possession: on the one hand, possession by mythical personal yet somehow human gods, who are projections in whom we can recognize ourselves; on the other, possession by impersonal mechanisms over which technology, however, gives us some hold.

Whatever the case may be, today we have discovered a world of impersonal laws in which the difficult thing to understand, what constitutes the problem, is not the possible knowledge of an absolute and timeless determinism. On the contrary, each discovery of a new physical or biological mechanism, especially when it takes the form of a mathematical law or a timeless formal structure, allows us to experience such determinism, even if its field of application is local and limited. What constitutes a problem, however, in this mechanical representation of the things that govern us is our social and subjective experience of responsibility, which is supposed to depend on the free decrees of our will.

Contrary to what we often hear, the new science that tries to domesticate the uncertain and aleatory by confronting the complexity of the real does not renounce either the presupposition of causal determinism or the classical ideal of knowledge through causes. The indeterminations of current probabilistic science do not pave the way for free will any more than did the gaps in classical determinism. That is why, on the speculative and individual level— that is to say, on the level of the philosophers themselves or of those who read or listen to them—this criticism of Kantianism (Alexis Philonenko speaks of "the agony of Kantianism"; 1969: 335–36) may end up in a comeback for Spinoza, as if Spinoza, cast aside into the margins of the history of philosophy and sidelined for three centuries, had waited for scientific reason, on which he relied so much, to lead enough people partway down the road. In fact, only the *experience* of comprehending things by reason can lead to accepting Spinoza's solution to the problem of ethics. In that solution, intellectual knowledge itself serves as the criterion for judging good and evil, not in the content of its knowledge but in itself, by granting that the experience of the good and of what is good [*le bien-bon*] is the very fact of comprehension and that the experience of evil and what is bad [*le mal-mauvais*] is the obstacle to comprehension.

In the eighteenth-century German "Pantheism Controversy," Kant had to take sides against Spinoza, for Spinoza had drawn the consequences of rationalism (as apparently Kant had himself), for which algebraic geometry and the new physics served as models. But in so doing, Spinoza, unlike Kant, situated himself on the margins of the theologico-philosophical tradition of free will without renouncing an ethics of liberty through knowledge, one that was obviously much more difficult to conceive of and even more difficult to accept. In such a philosophy, free will does not found responsibility, but responsibility comes first—without need for guilt—in a world that is given to us as both determined and intelligible.

This ethics with Stoic accents, which Spinoza nonetheless insists on delimiting from Stoicism, is poorly adapted to the technologies of mass communication—for the moment at least. It requires a very distinctive pedagogy. Even though rational and universal in principle, it can be embodied only in singular individuals and can therefore be a matter only of a universal-singular belonging to the philosopher or the sage alone.

That's why, on this third level of ethics, we must return yet again to existing moralities, and we cannot avoid observing, between the individual and the universal, the reality of miscellaneous social groups in their concrete forms of behaving and existing. We cannot ignore the place where the systems of values and norms that orient the behavior of individuals are effectively and concretely elaborated. This place is always a particular social group with its culture, its language, and its history. And if we do not fall into a moral relativism in which everything is equivalent, we can seek and find common elements that can bring about contact, or a dialogue, between these existing moralities: for example, in a genealogical search for the source of the ethics that is concretely lived by this or that group, we always find traces a religion, that is, of a set of beliefs, myths, and ritual practices that the individuals of the group have interiorized through education—even if, later, especially in modern secular societies, individuals can to differing degrees exercise critical reflection and contribute to modifying accepted values and norms.

The interest of this genealogical research lies in that it does not dwell exclusively on observing the diversity of cultures, their moralities, and their religions, on cataloguing morals and customs alone, for one can try to identify behind this diversity common elements that may facilitate encounters and dialogue between different cultures and traditions. These reciprocal encounters, argumentation, and appreciation are also part of reality. They are not just pious wishes. They have themselves become realities as soon as individuals nourished by different traditions live together in pluralist societies, and also as soon as different societies, peoples, and nations are condemned to coexist by having renounced war as a habitual means of resolving conflict.

In other words, to say that a universal ethics that would impose itself on everyone simply by force of reason does not exist does not necessarily imply resignation to moral relativism. Another sort of universalism is possible, in which reason is a tool for dialogue and not a tribunal. Reason does not serve as the ultimate foundation, a sort of tabula rasa, but as a means of dialogue on the basis of different moral experiences. In this mode of proceeding, reason can construct a concretely universal ethics only out of existing moralities, by means of rationalization and rhetorical argumentation.

Contrary to appearances, this attitude is the only barrier that can, on the one hand, prevent a relativism of knowledge from falling into a nihilism in which "everything is equivalent" and, on the other, reject any and all abstract absolutes. No matter whether an absolute pretends to be rational and scientific or revealed, we know today that the result risks being the same: a totalitarian regime waging a (holy) war of extermination in which intolerance is wedded to belief. Only a dialogue between different cultures and traditions that seeks a compromise can give us hope for the birth of a universality founded on the recognition of others, a birth through progressive and empirical construction, without imposed revelation or excommunication. Obviously, no one and nothing can guarantee either the truth or the success of this mode of proceeding. There is always the risk that dialogue and negotiation will not carry the day and that conflict will be settled, once again as it always has been, by war.

But, as we shall see, this pragmatic way of proceeding can already claim some modest and partial successes. And it becomes possible to believe in its relative efficacy once we realize the nature of the framework within which dialogue can be undertaken. I'm not talking about a theoretical framework that would revert to posing principles that would be accepted by all—other than the simple desire or necessity to meet and to talk—because that would revert to the impasses of the abstract universal we have already discussed.

This, perhaps, is the reproach to be directed at philosophies of argumentation, such as Habermas's or Apel's, which have nonetheless advanced quite far in this direction. I am talking about the framework that necessarily constitutes the very practice of dialogue. This practice can only exist because the moral experiences produced and borne by the moralities of different societies are not totally dissociated from one another, even though they are different and sometimes irreconcilable. They cannot oppose one another unless they are not incommensurable, namely, in that they flow from the same problem of good and evil, a problem that remains concretely universal in so far as it remains reduced to the *first level* of the good and what is good and of evil and what is bad [*du bien-bon et du mal-mauvais*]. All agree in recognizing that suffering does harm and that pleasure does good. Argumentative and even rhetorical reason, too, can appropriate this first level and perhaps seek solutions on the basis of this common field of experience. In this search for a common denominator, a common finality might be more easily admitted when it is limited to the suppression of suffering and the search for well-being, without, obviously, forgetting that different cultures differ in their strategies for this search, taking into account the necessities of deferring the satisfaction of desire and of extending it to others.

In other words, the concern not to lift off too far from the first level of ethics, the only level that is universal in that it concerns the suffering or happiness of each individual in the present and in his or her singularity, already allows different systems of norms at least to meet on this common terrain. Then we can also understand, without necessarily providing justification, how the normative aspect of the second level of ethics for a certain social group derives from the particular history of this group's sensibilities. These sensibilities first express themselves in language and in words that designate what is good and what is bad. Depending on the language, words for the good and the bad of ethics, the just and the unjust, the generous and the cruel, the pure and the impure, the clean and the dirty, the perfumed and the nauseating are closer to or farther from those that designate well-being and suffering. The identity, pure and simple, that exists in Romance languages between words derived from *bonum* and *malum* that designate the first two levels is not always found in other languages: for example, in English, *good* and *bad* are transposed into *good* and *evil*, or *right* and *wrong*, which apparently no longer have anything to do with *pleasure* or *well-being* and *pain* or *suffering*. These verbal differences perhaps express what, with Freud, we could consider more or less significant degrees of a sublimation of the physiological into the sociocultural. But behind them, I would like to insist on a common element, which to me seems important in the history of different moral sensibilities, a common element that, like religion, is the cause of both rapprochement and opposition.

❃

On the second level of ethics, things are more complicated, not only because imagination and memory project pleasure and pain into time and space but also because of an additional phenomenon: the perception of the sphere of the *sacred* as objective reality. In the genealogy of ethics, in the sense of a set of forgotten sources, covered over by layers of new interpretation, we find, behind religion, the experience of enjoyments [*jouissances*] and sufferings of a different nature, related to what are called *modified states of consciousness*. The role of hallucinogenic plants in contemporary shamanic religions and their probable role in humanity's ancient religions is an element in this observation. But these modified states of consciousness, differently qualified as ecstatic, mystic, or otherwise, are not produced only by ingesting plants; they also characterize moments of so-called normal human life, even though they grant experiential access to another reality that many have not hesitated to call a separate reality, such as dreams, sexuality, a certain relation to danger and death in extreme situations lived as heroic (including, of course, war), and certain aspects of aesthetic experience, where everything is transfigured by the work of art. Even though these experiences can today be lived in a secular fashion and can be integrated into a positivist and even materialist view of the universe, they seem to be at the origin of what we call the sphere of the sacred in practically all religious traditions.

It is possible to maintain the hypothesis (Atlan 1986, chap. 8) that religions in general, and their sense of the sacred, have their source in this sort of perception. In this reading, one of the functions of myth and ritual, which is at their source and thus lies beyond their later symbolic rationalization, is to build a bridge between the two separate realities of habitual and modified states of consciousness. In fact, behind their diversity these experiences share the fact that they are not pleasurable or painful like other experiences, for they allow access to what appears as another reality, a world different from that of our habitual perception. Thus, if we admit something that is difficult to contest, namely, that the systems of values and norms in effect in societies inherit the mythical images, representations, and rituals of their religions, it is easy to find in the genealogy of ethics traces of these experiences of the sacred. Without neglecting the classic interpretation of the role of myth and ritual in the organization of traditional societies, when we take seriously both the psychic and the physiological reality of these experiences, these rituals appear as behavior that allows for building bridges, reunifying, tying together, for each individual, the pieces of his or her existence, in which he or she participates in two separate realities, two different temporalities, two distinct worlds, that of the sacred and that of everyday time. (The role of shamanic rituals in the treatment of disease is one of the more blatant examples. The hidden cause of a patient's suffering is found in a world to which only ecstatic experience allows access, and this is how the ritual takes effect.) This interpretation does not suppress others that emphasize, for example, the integrative function of myth and ritual, thanks to which social cohesion and politico-religious power are assured. But it allows us to understand the phenomenon of interiorization, that is to say, how individuals discover the force of this reality, both in the utmost depths of their being and in a way that cannot be dissociated from representations borne by the imaginary of the group or the rules of behavior that accompany these representations.

I know that many who are ready to recognize the origins of effective moralities in religion are not ready to recognize at the origin of religions experiences of the sacred that can be assimilated to dreams and hallucinoses. They would like to distinguish not only according to their histories but also according to their sources between religions that have preserved an explicitly lived relation to the sacred, in the manner of shamanic religions, which have only little relation to the written word, and religions considered to be more advanced, notably, the "great" religions, which have repressed and covered over the reality of these experiences with very rich literatures and philosophies. It is here that, in the West, the intellectual refinements of theology and, by way of many detours, via direct influences and reactive effects, philosophy and the sciences have their origin. By contrast, most shamanic religions have remained oral religions and, perhaps for this reason, have been considered to be primitive, in the pejorative meaning of the term. These distinctions are certainly pertinent from the viewpoint of the history of ideas and civilizations. But this takes away nothing from the fact that this dimension of ethics, rooted as it is in the experience of the sacred as objective reality and unusual as it may be for moral philosophy, cannot be ignored under the pretext that it constitutes an irrational dimension on the order of illusion and mystification.

Dreams are not illusions, for illusions are errors. For someone who knows he's dreaming, the dream, even the waking dream, is a lived experience. Certainly, its content is difficult to objectivize other than by means of interpretation, where finding an objective truth independent of the subjectivity of the interpreter is an illusion. But it is nonetheless a sense experience, despite the modification of perceptions of time and space that characterizes it. To the extent to which the experiences of dreams and hallucinoses, provoked and unprovoked, have played—and continue to play—a role in instituting the sphere of the sacred and the behavioral norms derived from it, it would be madness to ignore them out of some desire for universality, under the pretext that the nonobjectivizable contents of these experiences are often irreducible to one another and difficult to communicate, for the reality of these experiences seems to be a property of the species that is as universal as erect posture and reason itself.

It is important to recognize this dimension of ethics, for it considerably enriches and complicates our second level. On top of the imagination and memory that project pleasure and pain into time and space, these experiences of the sacred—through hallucinosis, dreams, or any other means—double good and bad into a double register that corresponds to these two modes of perception or two separate realities, if we should wish to call them that. The effects of this doubling are evident in the revealed and ritualized moralities of traditional societies. But they are evident in valorized and ritualized behavior in our own societies as well, even if here the meaning of the sacred is much toned down and has long been separated from its origin in dreams and hallucinations. (An extremely pathological and tragic, albeit unfortunately banal, example is that of drug addiction, where this enjoyment is sought even though it is also harmful and even fatal.) Distinctively, these experiences act both on the individual in his or her singularity and on collective representations, myths, and rituals. That is why they suggest a possible mechanism for individually interiorizing a group reality, which is also to say individually interiorizing the law and the norms of the group.

In this way, the first level of ethics, individual and immediate, is modulated on the second level by norms of seeking good and avoiding bad, where good and bad are perceived as both individual and collective, of a nature both sensorial and imagined on the basis of an experience of multiple realities. These norms, interiorized by the individuals of a group, make these individuals say "This is good," in the sense of "procuring good," and "this is bad," in the sense of "doing harm," even if these judgments sometimes stand in opposition to the immediate perception, here and now, of what is good or bad for this or that individual, whether for himself or for another. In different cultures, this individual interiorization of norms takes place in different ways according to the society to which individuals belong and in which they have grown up. To facilitate the encounter of cultures and different value systems, it is thus important to understand how the articulation of these norms with the immediate and individual level of pleasure and pain, which they hold in common, comes from the history and memory of particular collective experiences. In this, genealogical and ethnological inquiry—even applied to our own societies—can be fruitful in discovering the elements of these experiences behind language, social organization, the organization of time and space, and the myths and rituals that express the memory of these experiences.

In this way, the third level—where different systems of values and norms coexist—can, without being normative in the same way as a universalist moral philosophy, avoid being the field of a moral relativism where one could do nothing more than establish a catalogue of morals and customs. This third level, rather than being the level of a meta-ethics, as it is sometimes called, can thus be the level of an ethics of argumentation, where different systems of values and norms enter into dialogue. The only rule in the dialogue is the necessity of resolving practical problems of coexistence without, insofar as possible, having recourse to war.

Success is never guaranteed, but neither is it as utopian as one might think. Several factors can contribute to success, and the relatively recent experience of ethics commissions in pluralist societies, where different systems of values and norms coexist, testify to this. One of these factors, as we have seen, is the presence in all individuals of a common denominator provided by experiences of pleasure and pain, even if they are covered over by the strata of the group's history and culture.

Another, very important factor is a particularity of our systems of communication and argumentation, where we attempt to answer *why?* questions in terms of causes, reasons, or motivations.

It is a remarkable property that appears as a theoretical weakness in many of our explicatory systems but becomes a force, an advantage, in our practice of intersubjectivity.

❖

It is a phenomenon that, in different contexts, has been analyzed under the name of *underdetermination of theories*, first by P. Duhem (1914) in epistemology, then by W. V. O. Quine (1960: e.g., 23 and 78) in epistemology, as well as in his works on translation, a phenomenon we find today in our modeling of dynamic systems by means of networks of automata (Atlan 1989: 247–53, Atlan 1991a: 130 and 163). Very schematically, every dynamic system (e.g., a network of neurons, or of cells of the immune system, or of metabolic reactions in a cell) has the property of converging toward one of its stable states or one of its attractors, taking a multitude of possible pathways. These pathways differ among themselves not only because of the fact that their initial conditions, which can result from the contingencies of history or from preceding determination, are different, but also because of the fact that the network can have different structures. The structure of a network depends on the way in which its elements are connected with each other. But it turns out that different connections, which express different models or theories, can nonetheless produce identical attractors and thus predict the same observation when these observations, as is often the case, are reduced to stable states (or even generally reduced to attractors, whether stable, oscillating, or chaotic).

This can be demonstrated quickly and by means of relatively simple examples. We can show that in the overwhelming majority of systems constituted of elements observable in only a few discrete states (states of activity of a neuron or of a cell population in the immune system, etc.), the number of observable states of the network is much smaller than the number of possible connection structures. In fact, the number of observable states varies as a power n (n being the number of connected elements) and the number of possible connection structures varies as a power n^2.[4] The result is that, in the

---

4. The number of possible states of a network of n elements is 2^n, 3^n, . . . , q^n if each element, respectively, can find itself in 2, 3, . . . , q states. The number of connection structures is 2^{n^2}, 3^{n^2}, . . ., p^{n^2} if the number of values possible for the force of each connection can be, respectively, 2 (e.g., 1 or 0, i.e., existing connection or nonexistent connection), 3 (e.g., 1, 0, −1, if a connection can be inhibiting), or p (for a connection whose "weight" can vary and take p possible values). In order for these two numbers to be of the same order of magnitude and for there thus to be no underdetermination, it must be the case that $q^n \geq p^{n^2}$ or $q \geq p^n$. This condition is fulfilled only in controlled experimental situations with systems broken down into subsystems, where n is small and where the elements can be observed in quasi-continuous fashion (i.e., where q approaches infinity).

majority of cases, hundreds or hundreds of thousands of different models (their number increases exponentially with n) predict the same observed facts. When we are dealing with natural systems, such as biological systems studied *in vivo*, once we have discovered the structure and function of the elements that constitute these systems (cells, cell populations, populations of molecules, etc.), the difficulties of theorization thus do not derive from a lack of theories but, on the contrary, from too great a number of correct theories that predict the same observations—without our being able to determine which one is more "true" than another.

This, to be sure, is a weakness of theory: but the same phenomenon, when we apply it not to models but to natural systems (e.g., to the neural networks that constitute our brains rather than to the formal neural networks we use to simulate them), appears, on the contrary, as a strength of their dynamic. Applied to the functioning of our brains, the property of converging toward one and the same final state by way of different pathways and despite different structures perhaps explains a remarkable property of our intersubjectivity: it is much easier to agree on conclusions than on the way to get there. This may seem curious and even shocking to our deductive and rationalizing way of thinking. Yet we often experience it. The deliberations of ethics commissions furnish us with striking examples. It happens rather often that, confronted with a particular question, such as whether a given practice is morally acceptable, people spontaneously react in the same way, even before they present the reasons for their position. The difficulties arise when it comes to analyzing and justifying this characteristic—morally acceptable or not—which seems at first sight almost obvious. All members of the French National Advisory Committee on the Ethics of the Life and Health Sciences, for example, immediately agreed to judge the use of aborted human embryos in the cosmetics industry to be unacceptable. But it was very difficult for individuals with opposed conceptions of the nature of the person and the status of the embryo to agree on a minimal expression of the reasons for this rejection. In other words, confronted with a given situation—all the more so when this situation is a particular, circumscribed case—we first react through our sense of what is good or bad. But when it happens that our conclusions are the same as others' (even if only because the number of options is limited), it is easy to see that it might be better not to ask why and not to analyze our motivations too much, for then disagreements, sometimes irreducible, appear at the level of beliefs and principles.

Thus, a relative misunderstanding, or what Wittgenstein calls "the vagueness of natural language" (in opposition to the precision of formal logical and mathematical languages) sometimes allows dialogue to come about. While general agreement on a priori concepts and principles is out of reach, it is thanks to such misunderstandings that a de facto universal ethics can be constructed, piece by piece, in a procedure characterized by hesitation, from the bottom up rather than, deductively, from the top down. In this process, in which we find ourselves in agreement, locally, on this or that practical conclusion even though our interlocutor adheres to a system of values different from our own, we can be led to relativize our own motivations, which we believe to be those of our own ethics, without devalorizing them. In other words, misunderstanding about motivations sometimes allows dialogue to move forward, but beyond that, the fact of advancing makes it possible to take into account the misunderstanding itself and to relativize, if not suppress, opposition between beliefs.

❀

This is where the importance of the first level of ethics, which has never really disappeared, can reappear and where its weight in the argumentation can lead more quickly to new agreement.

More and more, we witness to these practices when we observe how, case by case, problems of coexistence between different cultures are handled [*se traitent*] when dialogue and negotiation are accepted as the only means of resolving conflict. I have mentioned the experience of ethics commissions in pluralist societies. On another scale, we have the example of great shifts in public opinion [*mouvements d'opinion*], where a large consensus is built up on a moral basis, which is then said to be humanitarian. What remains universal and allows for agreement of a moral sort—besides, of course, the convergence of interests that leads to renouncing war—is the invocation of good and bad on the first level by means of the immediate experience, the spectacle, of suffering or well-being. We are present at someone else's suffering through the media, and this is what triggers the shifts of opinion in which we participate. These movements, to be sure, are ambiguous, in part because there is an obvious asymmetry between good and bad, and that constitutes the limitations of these large consensuses.

It is much easier to agree on what we perceive as pain or suffering, than it is to agree on our images of happiness [*bonheur*]. That is why these very large agreements rest on a minimal ethics we could call a "moral of indignation," which is in fact the most immediately and universally accessible ethical level. We are much more disposed spontaneously to combat an evil we feel or witness than we are to seek a happiness we have difficulty imagining, given that the search for well-being is much more individual and extensive in time. It is easier to agree on a minimal ethics of protection against visible pain and suffering than on a more elaborate ethics of accomplishing good and seeking happiness. That is why the moral of indignation in the face of suffering can most immediately and most easily constitute a consensus.

But, of course, it has disadvantages, too: it is a somewhat infantile moral of the here and now, a moral of sentiment produced by images in the instant of their perception, where one image chases another, a moral of the "underdog," where the one who suffers is necessarily right, even if, over the course of time, he is wrong. As such, this moral of indignation lends itself to being manipulated in ways familiar from the mass media, which play, for example, on the infantile perversion of emotional blackmail. But in the end, that is better than nothing. It is a first level we'd be wrong to neglect, for avoiding suffering, one's own future suffering imagined at the same time as the suffering of others, constitutes at the very least a meeting place where dialogue can begin.

This attitude has already led to a sort of new, de facto universalism, where pragmatic agreement can be attained between peoples whose traditional criteria of good and bad are quite different. The Universal Declaration of Human Rights is the best example of this and perhaps still the only one. It is now more and more obvious that different cultures conceive of the notion of human rights in ways different from those imagined by the successive authors of this declaration during the French Revolution and then after World War II—but this means that nearly all nations on the planet have reached an agreement despite, or rather thanks to, these misunderstandings. The condemnation of torture, following on the condemnation of slavery, the condemnation of racism, the notion of crimes against humanity, and the protection of the environment on a global scale, are so many achievements on the path to a universal moral, achievements due much more to the horror that evils inflicted have concretely inspired, once they have been presented to the largest possible number of people, than to the persuasive force of a "natural" ethics founded on an image of the supreme good, allegedly imposed by reason. This way of proceeding does not necessarily suppress the philosophical universal—such as, for example, the Kantian imperative never to use another human being solely as a means—but it does radically transform its origin. It is no longer necessarily a question of maxims of practical reason flowing from an image of human freedom that imposes itself on reason, but of rules of behavior that are accepted as minimal means for preserving concrete individuals from bodily and emotional pain.

Translated by Nils F. Schott

Agassi, J. 1969. "Unity and Diversity in Science." *Boston Studies in the Philosophy of Science* 4: 463–519.

Aho, A. V., J. E. Hopcroft, and J.D. Ullman. 1974. *The Design and Analysis of Computer Algorithms*. Reading Mass.: Addison Wesley.

Ait-Si-Ali, S., S. Ramirez, F.-X. Barre, F. Dkhissi, L. Magnaghi-Jaulin, J. A. Girault, P. Robin, M. Knibiehler, L. L. Pritchard, B. Ducommun, D. Trouche, and A. Harel-Bellan. 1996. "Histone acetyltransferase Activity of CBP Is Controlled by Cycle-dependent Kinases and Oncoprotein E1A." *Nature* 396: 184–86.

Anonymous. 1982. *Doctrine de la nondualité* (advaita-vâda*) et Christianisme: Jalons pour un accord doctrinal enter l'Église et la Vwedânta, par un moine d'Occident*. Paris: Dervy-Livres.

———. 1998. "Adult Cloning Marches On: New Results on Cloning Technology Increase the Urgency for Regulations to Ensure its Responsible Use." *Nature* 394: 303.

———. 1999a. "Symposium: Human Primordial Stem Cells . . . Ethical Considerations." *Hastings Center Report* 29(2): 30–48.

———. 1999b. "First Principles in Cloning." *The Lancet* 353: 81.

Anscombe, G. E. M. 1957. *Intention*. London: Basil Blackwell.

Aristotle. 1998. *Nicomachean Ethics*. Trans. David Ross. Oxford: Oxford University Press.

Artavanis-Tsakonas S., K. Matsuno, M. E. and Fortini. 1995. "Notch Signaling." *Science*. 268: 225–32.

Ashby, W. R. 1958. "Requisite Variety and Its Implications for the Control of Complex Systems." *Cybernetica* 1(2): 83–99.

———. 1962. "Principles of the Self-Organizing System." In *Principles of Self-Organization*, ed. Heinz von Foerster and George W. Zopf, Jr., 255–78. New York: Pergamon.

———. 1967. "The Place of the Brain in the Natural World." *Currents in Modern Biology* 1 (2): 95–104. Amsterdam: North Holland.

Atlan, H. 1968. "Applications of Information Theory to the Study of the Stimulating Effects of Ionizing Radiation, Thermal Energy, and Other Environmental Factors: Preliminary Ideas for a Theory of Organization." *Journal of Theoretical Biology* 21: 45–70.

———. 1970. "Rôle positif du bruit en théorie de l'information appliquée à une défi-nition de l'organisation biologique." *Annales de physique biologique et médicale* 1: 15–33.

———. 1972 (1992). *L'organisation biologique et la théorie de l'information*. Paris: Hermann.

———. 1974. "On a Formal Definition of Organization." *Journal of Theoretical Biology* 45: 295–304.

———. 1978. "Le principe d'ordre à partir du bruit: L'apprentissage non dirigé et le rêve." In *L'unité de l'homme*, ed. E. Morin and M. Piatelli-Palmerin, 469–75. Paris: Seuil.

———. 1979. *Entre le cristal et la fumée*. Paris: Seuil.

———. 1982. "Temps biologique et auto-organisation." In Atlan, *Temps de la vie et temps vécu*, 24–35. (Paris: CNRS).

———. 1983. "L'emergence du nouveau et du sens." In *L'Auto-organisation, de la physique au politique*, Colloque de Cerisy, ed. P. Dumouchel and J.-P. Dupuy, 115–38. Paris: Seuil.

———. 1986. *À tort et à raison*. Paris: Seuil.

———. 1987a. "Self-Creation of Meaning." *Physica Scripta* 36: 563–76.

———. 1987b. "Le milieu naturel et la personne humaine face aux biotechnologies." In *Le jaillissement des biotechnologies*, ed. P. Darbon and J. Robin. Paris: Fayard–Fondation Diderot.

———. 1989. "Automata Networks Theories in Immunology: Their Utility and Their Underdetermination." *Bulletin of Mathematical Biology*. 51(2): 247–53.

———. 1991a. *Tout, non, peut-etre: Education et verité*. Paris: Seuil.

———. 1991b. "L'intuition du complèxe et ses formalisations." In *Les théories de la complexité: Autour de l'oeuvre de H. Atlan*, Colloque de Cerisy 1984, ed. F. Fogelman-Soulie, 9–42. Paris: Seuil.

———. 1992a. "Personne, espèce, humanité?" In *Patrimoine génétique et droits de l'humanité*, ed. F. Gros and G. Huber, 52–63. Paris: Odile Jacob.

———. 1992b. "Self-Organizing Networks: Weak, Strong and Intentional; the Role of Their Underdetermination." *La Nueva Critica* 1–2: 19–20, 51–70.

———. 1992c. Preface to M. Idel, *Le Golem*. Trans. C. Aslanoff. Paris: Le Cerf.

———. 1993. *Enlightenment to Enlightenment: Intercritique of Science and Myth*. Trans. Lenn J. Schramm. Albany: State University of New York Press.

———. 1994a. "Intentionality in Nature: Against an All Encompassing Evolutionary Paradigm; Evolutionary and Cognitive Processes Are Not Instances of the Same Process." *Journal for the Theory of Social Behavior* 24(1): 67–87.

———. 1994b. "Le projet 'Génome humain' et la transmission du savoir biologique." *Alliages* 18: 27–36.

————. 1995a. "DNA: Program or Data? (or: Genetics Is Not in The Gene)." *Bulletin of the European Society for the Philosophy of Medicine* 3(3), CD-ROM 1.0.1.a, b, E.

————. 1995b. "Projet et signification dans des réseaux d'automates: Le role de la sophistication." In *L'intentionnalité en question: Entre phénoménologie et recherches cognitives*, ed. D. Janicaud, 261–88. Paris: Vrin.

————. 1995c. "Le plaisir, la douleur et les niveaux de l'éthique." *International Journal of Bioethics* 6: 53–64.

————. 1997. "Transfert de noyau et clonage: aspects biologiques et éthiques." *L'aventure humaine* 8: 5–18.

————. 1999a. *La fin du "tout génétique"? Vers de nouveaux paradigmes en biologie*. Paris: INRA Éditions.

————. 1999b. *Les étincelles de hasard*. Vol. 1: *Connaissance spermatique*. Paris: Seuil.

————. 2000. "Self-Organizing Networks: Weak, Strong, Intentional." In *Functional Models of Cognition*, ed. Arturo Carsetti. Dordrecht: Kluwer Academic Publishers.

————. 2003. *Les étincelles de hasard*. Vol. 2: *Athéisme de l'écriture*. Paris: Seuil.

————. 2005. *L'uterus artificiel*. Paris: Seuil.

————. 2007. "Stem Cell Research: From What Stage May a Cell or Cellular Artifact Be Qualified as an Embryo?" *Journal international de bioéthique* 18, no. 4 1–6.

————. 2008. "From French Algeria to Jerusalem." In *Religion: Beyond a Concept*, ed. Hent de Vries, 339–53. Fordham: Fordham University Press.

————. 2011. *Le vivant post-génomique ou qu'est-ce que l'auto-organisation?* Paris: Odile Jacob.

Atlan, H., M. Augé, M. Delmas-Marty, V. Droit, and N. Fresco. 1999. *Le clonage humain*. Paris: Seuil.

Atlan, H., E. Ben Ezra, F. Fogelman-Soulié, D. Pellegrin, and G. Weisbuch. 1986. "Emergence of Classification Procedures in Automata Networks as a Model for Functional Self-Organization." *Journal of Theoretical Biology* 120: 371–80.

Atlan, H., and M. Botbol-Baum. 2007. *Des embryons et des hommes*. Paris: Presses Universitaires de France.

Atlan, H., and C. Bousquet. 1994. *Questions de vie*. Paris: Seuil.

Atlan, H., and Irun R. Cohen, eds. *Theories of Immune Networks*. New York: Springer, 1989.

Atlan, H., F. Fogelman-Soulie, J. Salomon, and G. Weisbuch. 1981. "Random Boolean Networks." *Cybernetics and Systems* 12: 103–21.

Atlan, H., and M. Koppel. 1990. "The Cellular Computer DNA: Program or Data?" *Bulletin of Mathematical Biology* 52(3): 335–48.

Aurobindo Ghose. 1955. *The Life Divine*. Pondicherry: Sri Autobindo Ashram.

————. *The Message of the Gita: With Text, Translations, and Notes*. Ed. Anilbaran Roy. Pondicherry: Sri Autobindo Ashram.

Barkai, N., and S. Leibler. 1997. "Robustness in Simple Biochemical Networks." *Nature* 387: 913–17.

Bateson, G. 1980. *Men and Grass: Metaphors and the World of Mental Process.* West Stockbridge, Mass.: Lindisfarne Press.

Bennett, C. H. 1988. "Logical Depth and Physical Complexity." In *The Universal Turing Machine: A Half-Century Survey*, ed. R. Herken, 227–58. London: Oxford University Press.

Beurle, R. L. 1962. "Functional Organization in Random Networks." In *Principles of Self-Organization*, ed. Heinz von Foerster and George W. Zopf, Jr., 291–314. New York: Pergamon.

Bichat, X. 1827. *Physiological Researches on Life and Death.* Trans. F. Gold. Boston: Richardson and Lord.

Bienenstock, E. 1985. "Dynamics of Central Nervous Systems." In *Dynamics of Macrosystems*, ed. J. P. Aubin, D. Saari, and K. Sigmund, 3–20. Berlin: Springer.

Bonnefoy, Yves. 1989. "'Image and Presence': Inaugural Lecture at the Collège de France." Trans. John T. Naughton. In Bonnefoy, *The Act and the Place of Poetry: Selected Essays*, ed. John T. Naughton, 156–72. Chicago: University of Chicago Press.

Bosk, C. L. 1999. "Professional Ethicist Available: Logical, Secular, Friendly." *Daedalus* 128(4): 47–68.

Bray, D. 1995. "Protein Molecules as Computational Elements in Living Cells." *Nature* 376: 307–12.

Bray, D., R. B. Bourret, and M. I. Simon. 1993. "Computer Simulation of the Phosphorylation Cascade Controlling Bacterial Chemotaxis." *Molecular Biology of the Cell* 5: 469–82.

Bunge, M. 1982. "Is Chemistry a Branch of Physics?" *Zeitschrift für allgemeine Wissenschaftstheorie/Journal for General Philosophy of Science* 13(2): 209–23.

Campbell, K. H. S., J. McWhir, W. A. Ritchie, and I. Wilmut. 1996. "Sheep Cloned by Nuclear Transfer from a Cultured Cell Line." *Nature* 380: 64–66.

Cazenave, M., ed. 1984. *Science and Consciousness: Two Views of the Universe, Edited Proceedings of the France-Culture and Radio-France Colloquium, Cordoba, Spain.* Trans. A. Hall and E. Callander. Oxford: Pergamon Press.

Chaitin, G. J. 1975. "A Theory of Program Size Formally Identical to Information Theory." *Journal of Applied Computation and Mathematics* 22: 329–40.

Chalmers, D. J. 1995. "The Puzzle of Conscious Experience." *Scientific American* 12: 62–68.

Cho, M. M. et al. 1999. "The Effects of Oxytocin and Vasopression on Partner Preferences in Male and Female Prairie Voles (*Microtus ochrogaster*)." *Behavioral Neurosciences* 113: 1071–79.

Churchland, P. M., and P. S. Churchland. 1990. "Could a Machine Think?" *Scientific American* 262: 32–37.

Cibelli, J. B., S. L. Stice, P. J. Golueke, J. J. Kane, J. Jerry, C. Blackwell, F. A. Ponce de Leon, and J. M. Robl. 1998. "Cloned Transgenic Calves Produced from Nonquiescent Fetal Fibroblasts." *Science* 280: 1256–58.

Cicero, Marcus Tullius. 1914. *De finibus bonorum et malorum.* Trans. Harris Rackam. London: Heinemann.

Clothier, C., et al. 1992. *Report of the Committee on the Ethics of Gene Therapy.* London: HMSO.

Comité consultatif national d'éthique pour les sciences de la vie et de la santé (CCNE). 1988. *Recherche biomédicale et respect de la personne humaine.* Paris: La Documentation française.

———. 1997a. "La constitution de collections de cellules, tissus et organes humains et leur utilisation à des fins thérapeutiques ou scientifiques." Brief no. 52–53. *Cahiers du CCNE* 12: 10–16.

———. 1997b. "La constitution de collections de cellules embryonnaires humaines et leur utilisation à des fins thérapeutiques ou scientifiques." Brief no. 53. *Cahiers du CCNE* 12: 6–9.

———. 1997c. "Réponse au président de la République au sujet du clonage reproductive." Brief no. 54. *Cahiers du CCNE* 12: 17–39.

Connolly, W. E. 2002. *Neuropolitics: Thinking, Culture, Speed.* Minneapolis: University of Minnesota Press.

Corea, G. 1985a. *The Mother Machine: Reproductive Technologies from Artificial Insemination to Artificial Wombs.* New York: Harper Collins.

———. 1985b. *The Hidden Malpractice: How American Medicine Mistreats Women.* New York: Harper & Row.

Coustou, V., C. Deleu, S. Saupe, and J. Begueret. 1997. "The Protein Product of the Het-s Heterokaryon Incompatibility Gene of the Fungus *Podospora anserina* Behaves as a Prion Analog." *Proceedings of the National Academy of Science* 94: 9773–78.

Cowan, J. D. 1965. "The Problem of Organismic Reliability." In *Progress in Brain Research*, vol. 17, *Cybernetics and the Nervous System*, ed. N. Wiener and J. P. Schade, 9–63. Amsterdam: Elsevier.

Craig, A. D., M. C. Bushnell, E. T. Zhang, and A. Blourquist. 1994. "A Thalamic Nucleus Specific for Pain and Temperature Sensation." *Nature* 372: 770–73.

Crescas, Hasdai. 1998. *The Light of the Lord.* Trans. in Warren Zev Harvey, *Physics and Metaphysics in Hasdai Crescas.* Amsterdam: J. C. Gieben.

Daly, M. 1978. *Gyn/Ecology.* Boston: Beacon Press.

Damasio, A. 2003. *Looking for Spinoza: Joy, Sorrow, and the Feeling Brain.* New York: Harvest Books.

Dancoff, S. M., and H. Quastler. 1953. "The Information Content and Error Rate of Living Things." In *Information Theory in Biology*, ed. H. Quastler, 267–73. Urbana: University of Illinois Press.

Davidson, D. 1970. "Mental Events." In *Experience and Theory*, ed. L. Foster and J. Swanson Amherst: University of Massachusetts Press, 79–81. Reprinted in Davidson. 1980. *Essays on Actions and Events.* New York: Oxford University Press.

———. 1999. "Spinoza's Causal Theory of the Affects." In *Desire and Affect: Spinoza as Psychologist*, ed. Y. Yovel, 95–111. New York: Little Room Press.

Davis, M., ed. 1965. *The Undecidable*. New York: Raven Press.

Davy, M. M. 1981. *H. Le Saux, Swami Abhishiktananda, le passeur entre deux rives*. Paris: Cerf.

Delaisi de Parseval, G., and A. Janaud. 1983. *L'enfant à tout prix*. Paris: Seuil.

Dennett, D. C. 1978. *Brainstorms: Philosophical Essays on Mind and Psychology*. Montgomery, Vt.: Bradford Books.

Detienne, M. 1984. "Le mythe, plus ou moins." *L'infini*, no. 6, 27–41.

———. 1986. *The Creation of Mythology*. Trans. Margaret Cook. Chicago: University of Chicago Press.

Detienne, M., and J.-P. Vernant. 1974. *Les ruses de l'intelligence: La Mêtis des Grecs*. Paris: Flammarion.

Diogenes Laertius. 1995. "Epicurus." In Diogenes Laertius, *Lives of Eminent Philosophers*. Trans. R. D. Hicks. Cambridge, Harvard University Press.

Dreyfus, H. L. 1994. *What Computers Still Can't Do: A Critique of Artificial Reason*. Cambridge: MIT Press.

Dubnau, J., and G. Struhl. 1996. "RNA Recognition and Translational Regulation by a Homeodomain Protein." *Nature* 379: 694–99.

Duhem, P. 1914. *La théorie physique, son objet et sa structure*. 2nd ed. Paris: M. Riviere & Cie. Reprint, Paris: Vrin, 1981. Trans. P. Wiener as *Aim and Structure of Physical Theory* Princeton: Princeton University Press, 1954.

Dumay, Jean-Michel. 2001. "Punir les fous?" *Le Monde*, November 4, 2001.

Dupuy, J.-P. 1994. *Aux origines des sciences cognitives*. Paris: La Decouverte.

Eigen, M. 1971. "Self-Organization of Matter and the Evolution of Biological Macromolecules." *Die Naturwissenschaften* 58: 465–523.

Eigen, M., and P. Schuster. 1979. *The Hypercycle: A Principle of Natural Self-Organization*. Heidelberg: Springer.

Eisler, R., and D. S. Levine. 2002. "Nurture, Nature, and Caring: We are Not Prisoners of Our Genes." *Brain and Mind* 3: 9–52.

Eldredge, N. 1995. *Reinventing Darwin*. New York: J. Wiley and Sons.

Eliade, Mircea. 1958. *Patterns in Comparative Religion*. Trans. Rosemary Sheed. New York: Sheed and Ward.

Eliaschow, S. 1977. *Léchem chébo veha'hlama: Sefer hada'at*. Jerusalem: Hahayim Vehashalom.

Elkana, Y. 1981. "A Programmatic Attempt at an Anthropology of Knowledge." In *Science and Cultures*, ed. E. Mendelsohn and Y. Elkana, 1–76. Dordrecht: Reidel.

Epicurus. 1995. "Letter to Menoeceus." Trans. R. D. Hicks. In *Diogenes Laertius, Lives of Eminent Philosophers*, 528–677. Cambridge: Harvard University Press.

Falke, J. J., R. B. Bass, S. L. Butler, S. A.. Chervitz, and M. A. Danielson. 1997. "The Two-Component Signaling Pathway of Bacterial Chemotaxis: A Molecular View of Signal Transduction." *Annual Review of Cell and Developmental Biology* 13: 457–512.

Faye, J.-P. 1974. *Migrations du récit sur le peuple juif.* Paris: Belfond.

Feynman, Richard P. 1963. *Feynman Lectures on Physics.* Reading, Mass.: Addison-Wesley.

———. 1967. *The Character of Physical Law.* Cambridge: MIT Press.

———. 1969. "What Is Science?" *The Physics Teacher* 7(6): 313–20.

Fisher, E. 1979. *Woman's Creation.* New York: McGraw Hill Co.

Fodor, J. A. 1975. "Two Kinds of Reductionism." In Fodor, *The Language of Thought,* 1–26. New York: Thomas Y. Crowell.

———. 1981a. "The Mind-Body Problem." *Scientific American* 244: 124–32.

———. 1981b. *Representations.* Cambridge: MIT Press.

Foerster, Heinz von. 1960. "On Self-Organizing Systems and Their Environments." In *Self-Organizing Systems,* ed. M. C. Yovitz and S. Cameron, 31–50. New York: Pergamon.

Foucault, Michel. 1994. *The Order of Things,* New York: Vintage.

Fox, R. C. 1999. "Is Medical Education Asking Too Much of Bioethics?" *Daedalus* 128(4): 1–25.

Fox Keller, E. 1995. *Refiguring Life: Metaphors of Twentieth-Century Biology.* New York: Columbia University Press.

Friedman, M. 1983. *Foundations of Space-Time Theories.* Princeton: Princeton University Press.

Fryer, C. J., and T. K. Archer. 1998. "Chromatin Remodeling by the Glucocorticoid Receptor Requires the BRG1 Complex." *Nature* 393: 81–91.

Gabirol, Solomon Ibn. 1970. *Livre de la source de la viee (Fons vitae).* Introd. Jacques Schlanger. Paris: Aubier-Montaigne.

Gardner, A. R., and B. Gardner. 1969. "Teaching Sign Language to a Chimpanzee." *Science* 165: 664–72.

Gearhart, J. 1998. "New Potential for Human Embryonic Stem Cells." *Science* 282: 1061–62.

Gurdon, J. B. 1962. "The Development Capacity of Nuclei Taken from Intestinal Epithelial Cells of Feeding Tadpoles." *Journal of Embryology and Experimental Morphology* 10: 622–40.

Gurdon, J. B., R. A. Laskey, and O. R. Reeves. 1975. "The Developmental Capacity of Nuclei Transplanted from Keratenized Skin Cells of Adult Frogs." *Journal of Embryology and Experimental Morphology* 34: 93–112.

Gurdon, J. B., and V. Uehlinger. 1966. "'Fertile' Intestine Nuclei." *Nature* 210: 1240–41.

Gybels, J. M., and W. H. Sweet. 1989. *Neurosurgical Treatment of Persistent Pain.* Basel: Karger.

Haken, H. 1975. *Synergetics: An Introduction.* Heidelberg: Springer.

Hardt, M. and Ant. Negri. 2000. *Empire*. Harvard University Press.

Hayles, M. K. 1991. *Chaos and Order: Complex Dynamics in Literature and Science*. Chicago: University of Chicago Press.

Hesse, H. 2004 [1943]. *The Glass Bead Game*. Trans. Richard and Clara Winston. New York: Picador.

Hinton, G. E., and J. A. Anderson, eds. 1981. *Parallel Models of Associative Memory*. Hillsdale, N.J.: Erlbaum.

Hofstadter, D. R. 1979. *Gödel, Escher, Bach*. New York: Basic Books.

Holliday, R. 1987. "The Inheritance of Epigenetic Defects." *Science* 238: 163–70.

Hopfield, J. J. 1982. "Neural Networks and Physical Systems with Emergent Collective Computational Abilities." *Proceedings of the National Academy of Sciences* 79: 2554–58.

Hori, V. S. 2004. "Le Zen et le soi." *Théologiques* 12: 1–2, 125–33.

Hume, David. 1999 [1748]. *An Enquiry Concerning Human Understanding*. Ed. Tom L. Beauchamp. New York: Oxford University Press.

Husserl, E. 1947. *Méditations cartésiennes*. Paris: Vrin. Trans. Dorion Cairns as *Cartesian Meditations*. The Hague: Martinus Nijhoff, 1960.

———. 1950. *Idées directrices pour une phénoménologie*. Trans. P. Ricoeur. Paris: Gallimard.

Huston, N. 1986. "Le féminisme comme éthique." *Le Genre Humain* 14: 161–74.

Huxley, Aldous. 1944. *The Perennial Philosophy*. New York: Harper & Row.

———. 2004. *Brave New World and Brave New World Revisited*. New York: HarperCollins.

———. 2009. *The Doors of Perception and Heaven and Hell*. New York: Harper & Row.

Huxley, T. 1884. "Evolution in Biology." In Huxley, *Science and Culture, and Other Essays*. New York, Appleton and Co.

Idel, M. 1988. *Kabbalah, New Perspectives*. New Haven: Yale University Press.

———. 1990. *Golem: Jewish Magical and Mystical Traditions on the Artificial Anthropoid*. Albany: State University of New York Press.

Insel, T. R., and L. J. Young. 2001. "The Neurobiology of Attachment." *Nature Review of Neurosciences* 2: 129–36.

Israel, Jonathan. 2001. *Radical Enlightenment: Philosophy and the Making of Modernity 1650–1750*. Oxford: Oxford University Press.

Israel, Jonathan. 2009. *Enlightenment Contested: Philosophy, Modernity, and the Emancipation of Man 1670–1752*. Oxford: Oxford University Press.

Jacob, F. 1993. *The Logic of Life*. Trans. Betty E. Spillman. Princeton: Princeton University Press.

Jambet, C. 1983. *La logique des Orientaux: Henry Corbin et la Science des formes*. Paris: Seuil.

Jeannerod, M. 1992. "The Where in the Brain Determines the When in the Mind." (With commentary by D. C. Dennett and M. Kinsbourne, "Time and the Observer.") *Behavioral and Brain Sciences* 15: 212–13.

Jonas, H. 1965. "Spinoza and the Theory of Organism." *Journal of the History of Philosophy* 3: 43–57.

Kant, I. 2007. *Critique of Judgment.* Trans. J. C. Meredith. Oxford: Oxford University Press.

Katchalsky, A. 1971. "Biological Flow-Structures and Their Relation to ChemicoDiffusional Coupling." *Neurosciences Research Program Bulletin* 9(3): 397–413.

Kato, Y., T. Tani, Y. Sotomaru, K. Kurokawa, J. Kato, H. Doguchi, H. Yasue, and Y. Tsunoda. 1998. "Eight Calves Cloned from Somatic Cells of a Single Adult." *Science* 282: 2095–98.

Kauffman, S. A. 1970. "Behaviour of Randomly Constructed Genetic Nets." In *Towards a Theoretical Biology,* ed. C. H. Waddington, 3: 18–37. Edinburgh: Edinburgh University Press.

———. 1984. "Emergent Properties in Random Complex Automata." *Physica* 10 D: 145–56.

———. 1993. *The Origin of Order.* New York: Oxford University Press.

Kinzler, K. W., and B. Vogelstein. 1996. "What's Mice Got to Do with It?" *Nature* 382: 672.

Kolmogorov, A. N. 1965. "Three Approaches to the Quantitative Definition of Information." *Problems of Information Transmission* 1: 1–7.

Kono, T. 1997. "Nuclear Transfer and Reprogramming." *Reviews of Reproduction* 2: 74–80.

Kook, A. I. Hacohen. 1964. *Orot Hakodesh.* Vol. 2. 2nd ed. Jerusalem: Mossad Harav Kook.

———. 1969. *Olat Reiya.* Vol. 1. 3rd ed. Jerusalem: Mossad Harav Kook.

Koppel, M. 1988. "Structure." In *The Universal Turing Machine: A Half-Century Survey,* ed. R. Herken, 435–52. Oxford: Oxford University Press.

Koppel, M., and H. Atlan. 1991a. "Les gênes: Programme ou données? Le rôle de la signification dans les mesures de complexité." In *Les théories de la complexité: Autour de l'oeuvre d'Henri Atlan,* ed. M. Milgram, V. Havelange, and F. Fogelman Soulié, 188–204. Paris: Seuil.

———. 1991b. "An Almost Machine-Independent Theory of Program Length Complexity, Sophistication and Induction." *Information Sciences* 56: 23–33.

Koppel, M., H. Atlan, and J.-P. Dupuy. 1991. "Complexité et aliénation: Formalisation de la conjecture de von Foerster." In *Les théories de la complexité: Autour de l'oeuvre d'Henri Atlan,* ed. M. Milgram, V. Havelange, and F. Fogelman Soulié, 410–21. Paris: Seuil.

Kritzman, L., ed. 2007. *Columbia History of Twentieth-Century French Thought*. New York: Columbia University Press.

Laing, R. D. 1967. *The Politics of Experience*. New York: Pantheon.

Lederberg, Joshua. 1966. "Experimental Genetics and Human Evolution." *Bulletin of the Atomic Scientists* 23: 4–11.

Leibniz, G. W. 1991. *De l'horizon de la doctrine humaine: La restitution universelle*. Trans. M. Fichant. Paris: Vrin.

Lewontin, R. 1992. "The Dream of the Human Genome." *New York Review of Books*, May 28, 1992, pp. 31–40.

Libera, A. de. 1990. *Albert le Grand et la philosophie*. Paris: Vrin.

Libet, B. 1985. "Unconscious Cerebral Initiative and the Role of Conscious Will in Voluntary Action." *Behavioral and Brain Sciences* 8: 529–66.

———. 1992. "Models of Conscious Timing and the Experimental Evidence." With commentary by D. C. Dennett and M. Kinsbourne, "Time and the Observer." *Behavioral and Brain Sciences* 15: 213–15.

Lichenstein, D., and H. Atlan. 1990. "The 'Cellular State': The Way to Regain Specificity and Diversity in Hormone Action." *Journal of Theoretical Biology* 145: 287–94.

Lilti, A. 2009. "Comment écrit-on l'histoire intellectuelle des Lumières? Spinozisme, radicalisme, et philosophie." In *Annales HSS* (Jan–Feb. 2009), 171–206.

Linden, E. 1974. *Apes, Men and Language*. New York: Saturday Review Press / E. P. Dutton & Co.

Louzoun, Y., and H. Atlan. 2007. "The Emergence of Goals in a Self-Organizing Network: A Non-Mentalist Model of Intentional Actions." *Neural Networks* 20(2007): 156–71.

Lucas, Philippe. 1990. *Dire l'éthique: Éthique biomédicale, le débat*. Paris: Actes Sud/INSERM.

MacIntyre, A. 1975. "How Virtues Become Vices: Values, Medicine and Social Context." In *Evaluation and Explanation in the Biomedical Sciences*, ed. H. T. Engelhardt, Jr., and S. F. Spicker. Dordrecht: Kluwer Academic Publishers.

Maimonides, M. 1912. *The Eight Chapters of Maimonides on Ethics*. New York, Columbia University Press.

———. 1956. *The Guide to the Perplexed*. Trans. Michael Friedländer. New York: Dover.

Marshall, E. 1998. "A Versatile Cell Line Raises Scientific Hopes, Legal Questions." *Science* 282: 1014–15.

Mayr, E. 1961. "Cause and Effect in Biology." *Science* 134: 1501–6.

McCulloch, W. S., and W. Pitts. 1943. "A Logical Calculus of the Ideas Immanent in Nervous Activity." *Bulletin of Mathematical Biophysics* 5: 115–33.

Miklos, G. L. G., and G. M. Rubin. 1996. "The Role of the Genome Project in Determining Gene Function: Insights from Model Organisms" *Cell* 86: 521–29.

Mopsik, C. 2005. *The Sex of the Soul: The Vicissitudes of Sexual Difference in Kabbalah.* Los Angeles: Cherub.

Morin, E. 1973. *Le paradigme perdu: La nature humaine.* Paris: Seuil.

Morowitz, H. J. 1955. "Some Order Disorder Considerations in Living Systems." *Bulletin of Mathematical Biophysics* 17: 81–86.

Morrey, D. 2005. *Jean-Luc Godard.* Manchester: Manchester University Press.

Moyn, S. 2010. "Mind the Enlightenment" in *The Nation*, May 31, 2010.

National Bioethics Advisory Commission [USA]. 1997. *Cloning Human Beings: Report and Recommendations.* Rockville, Md.

Neher, André. 1986. *Jewish Thought and the Scientific Revolution of the Sixteenth Century: David Gans (1541–1613) and his Times.* Translated by David Maisel. Oxford: Oxford University Press.

———. 1987. *Faust et le Maharal de Prague: Le mythe et le réel.* Paris: Presses Universitaires de France.

Neumann, J. von. 1956. "Probabilistic Logics and the Synthesis of Reliable Organisms from Unreliable Components." In *Automata Studies*, ed. C. E. Shannon and J. McCarth, 43–98. Princeton: Princeton University Press.

———. 1966. *Theory of Self-Reproducing Automata.* Ed. A. W. Burks. Urbana: University of Illinois Press.

Neumann, J. von, and O. Morgenstern. 1944. *Theory of Games and Economic Behavior.* Princeton: Princeton University Press.

Newell, A. 1990. *Unified Theories of Cognition.* Cambridge: Harvard University Press.

Nicolis, G., and I. Prigogine. 1977. *Self-Organization in Non-Equilibrium Systems: From Dissipative Structures to Order Through Fluctuations.* New York: J. Wiley & Sons.

Oppenheim, P., and H. Putnam. 1958. "The Unity of Science as a Working Hypothesis." In *Minnesota Studies in the Philosophy of Science*, ed H. Feigl, G. Maxwell, and M. Scriven, 2: 3–36. Minneapolis: University of Minnesota Press.

Oster, D. 1981. *Monsieur Valéry.* Paris: Seuil.

Pagels, Elaine. 1988. *Adam, Eve, and the Serpent.* New York: Random House.

———. 1989. *The Gnostic Gospels.* New York: Random House.

Paulson, W. R. 1988. *The Noise of Culture: Literary Texts in a World of Information.* Ithaca, N.Y.: Cornell University Press.

Peden, K. 2009. "Reason Without Limits: Spinozism as Anti-Phenomenology in Twentieth-Century French Thought." Ph.D. Thesis, University of California, Berkeley.

Pennisi, E. 1997. "Opening the Way to Gene Activity." *Science* 275: 155–57.

Penrose, R. 1989. The *Emperor's New Mind.* Oxford: Oxford University Press.

———. 1995. *Shadows of the Mind: A Search for the Missing Science of Consciousness.* Oxford: Oxford University Press.

Philonenko, A. 1969. *L'œuvre de Kant.* Vol. 1. Paris: Vrin.

Pico della Mirandola, G. *On the Dignity of Man.* Trans. Charles Glenn Wallis. Indianapolis: Bobbs-Merrill, 1965. 19.

Pirsig, Robert M. 1974. *Zen and the Art of Motorcycle Maintenance.* New York: William Morrow.

Plato. 1929. *Cratylus, Parmenides, Greater Hippias, Lesser Hippias.* Trans. Harold North Fowler. Cambridge: Harvard University Press, 1926.

————. 2000. *Laws.* Trans. Benjamin Jowett. Amherst, N. Y.: Prometheus Books.

————. 2008. *Theaetetus.* Trans. Benjamin Jowett. Rockville, Md.: Serenity Publishers.

Poliakov, L., D. F. Green, and J.-C. Pala, eds. 1972. *Les Juifs et Israël vus par les théologiens arabes, extraits des procès-verbaux de la 4e Conférence de l'Académie de recherches islamiques.* Geneva: Éditions de l'Avenir.

President's Commission for the Study of Ethical Problems in Medicine and Biomedical and Behavioral Research. 1982. *Splicing Life: A Report on the Social and Ethical Issues of Genetic Engineering with Human Beings.* Washington, D.C.

Prigogine, Ilya, and Isabelle Stengers. 1984. *Order out of Chaos: Man's New Dialogue with Naure.* Boulder, Colo.: New Science Library.

Putnam, Hilary. 1973. "Reductionism and the Nature of Psychology." *Cognition* 2: 131–46.

————. 1981. *Reason, Truth and History.* Cambridge: Cambridge University Press.

————. 1986. "Meaning and Our Mental Life." In *The Kaleidoscope of Science,* 17–32. Dordrecht: Reidel.

Quine, W. V. O. 1960. *Word and Object.* Cambridge: MIT Press.

————. 1972. "Methodological Reflections on Current Linguistic Theories." In *Semantics of Natural Language,* ed. D. Davidson and Harman, 442–54. Dordrecht: Reidel.

Renard, J.-P., S. Chastant, P. Chesné, C. Richard, J. Marchal, N. Cordonnier, P. Chavatte, and X. Vignon. 1999. "Lymphoid Hypoplasia and Somatic Cloning." *Lancet* 353: 1489–91.

Rivera-Pomar, R., D. Niessing, U. Schmidt-Ott, W. J. Gehring, and H. Jackle. 1996. "RNA Binding and Translational Suppression by Bicoid." *Nature* 379: 746–49.

Robinson, I., L. Kaplan, and J. Bauer, eds. 1990. *The Thought of Moses Maimonides: Philosophical and Legal Studies/La pensée de Maimonide: Études philosophiques et halakhiques.* Lewiston, N. Y: Mellen.

Rosenberg, C. E. 1999. "Meanings, Policies, and Medicine: On the Bioethical Enterprise and History." *Daedalus* 128, no. 4: 27–46.

Roustang, François. 1976. *Un destin si funeste.* Paris: Minuit.

Rutherford, S. L., and S. Lindquist. 1998. "Hsp90 as a Capacitor for Morphological Evolution." *Nature* 396: 336–42.

Salthe, S. N. 1993. *Development and Evolution: Complexity and Change in Biology.* Cambridge: MIT Press.

Sartre, J.-P. 1954. *Réflexions sur la question juive*. Paris: Gallimard.

Schimmel, G., and H. Atlan. 1996. "Henri Atlan: The Frontiers of Science." *UNESCO Courier*, March 1996.

Schnieke, A. E., A. J. Kind, W. A. Ritchie, K. Mycock, A. R. Scott, M. Ritchie, I. Wilmut, A. Colman, and K. H. S. Campbell. 1997. "Human Factor IX Sheep Produced by Transfer of Nuclei from Transfected Fetal Fibroblasts." *Science* 278: 2130–33.

Scholem, Gershom. 1954. *Major Trends in Jewish Mysticism*. New York: Schocken.

———. 1965. *On the Kabbalah and Its Symbolism*. Trans. by Ralph Manheim. New York: Schocken.

———. 1974. *Kabbalah*. New York: Quadrangle/New York Times Book Co.

———. 1987. *Origins of the Kabbalah*. Ed. R. J. Zwi Werblowsky. Trans. Allan Arkush. Philadelphia: Jewish Publication Society/Princeton: Princeton University Press.

Schrödinger, E. 1945. *What Is Life?* Cambridge: Cambridge University Press.

Searle, J. R. 1983. *Intentionality: An Essay in the Philosophy of Mind*. Cambridge: Cambridge University Press.

———. 1990. "Is the Brain's Mind a Computer Program?" *Scientific American* 262: 26–31.

Serres, M. 2007. *The Parasite*. Trans. Lawrence R. Schehr. Minneapolis: University of Minnesota Press.

Shannon, C. E., and W. Weaver. 1949. *The Mathematical Theory of Communication*. Urbana: University of Illinois Press.

Shanon, B. 1991. "Réflections sur la complexité et la cognition humaine." In *Les Théories de la Complexité: Autour de l'Oeuvre d'Henri Atlan*, ed. M. Milgram, V. Havelange, and F. Fogelman Soulié. Paris: Seuil.

———. 1993. *The Representational and the Presentational*. New York: Simon & Schuster.

Shanon, B., and H. Atlan. 1990. "Von Foerster's Theorem on Connectedness and Organization: Semantic Applications." *New Ideas in Psychology* 8, no. 1: 81–90.

Siddheswarananda, Swami. 1977. "Introduction à l'étude desouvrages védantiques." In *Comment discriminer le spectateur du spectacle?*, trans. M. Sauton. Pais: Maisonneuve.

Silver, L. M. 2002. *Remaking Eden: How Genetic Engineering and Cloning Will Transform the American Family*. New York: Perennial-Harper Collins.

Simon, H. 1969. *The Science of the Artificial*. Cambridge, MIT Press.

Smith, Pierre. 1974. "La nature des mythes." In *L'unité de l'homme*, ed. E. Morin and M. Piatelli-Palmerini, 248–63. Paris: Seuil.

Solter, D. 1998. "Dolly is a Clone—and No Longer Alone." *Nature* 394: 315–16.

Sperber, Dan. 1975. *Rethinking Symbolism*. Trans. Alice L. Morton. Cambridge: Cambridge University Press.

Spinoza, B. de. 2002a. "The Ethics." In *Complete Works*, trans. Samuel Shirley, 213–383. Indianapolis: Hackett.

———. 2002b. "Treatise on the Emendation of the Intellect." In *Complete Works*, trans. Samuel Shirley, 1–31. Indianapolis: Hackett.

St. Johnston, D., and C. Nüsselein-Volhard. 1992. "The Origin of Pattern and Polarity in the Drosophila Embryo," *Cell* 68: 201–19.

Strohman, R. C. 1997. "Epigenesis and Complexity: The Coming Kuhnian Revolution in Biology." *Nature Biotechnology* 15, no. 3: 194–200.

Tardieu, Michel. 1974. *Trois mythes gnostiques: Adam, Éros et les animaux d'Egypte dans un écrit de Nag Hammadi (II,5)*. Paris: Institut d'études augustiniennes.

Taylor, M. (2001). *The Moment of Complexity: Emerging Network Culture*. Chicago: University of Chicago Press.

Thomson, J. A., J. Itskivitz-Eldor, S. S. Shapiro, M. A. Waknitz, V. S. Swiergel, V. S. Marshall, and J. M. Jones. 1998. "Embryonic Stem Cells Derived from Human Blastocysts." *Science* 282: 1145–47.

Turing, A. M. 1952. "The Chemical Basis of Morphogenesis." *Philosophical Transactions of the Royal Society* 237: 37–72.

UNESCO. 1997. *Universal Declaration on Human Genome and Human Rights*. Article 11.

Vigée, Claude. 1960. *Les artistes de la faim*. Paris: Calmann-Levy.

———. 1982. *L'extase et l'errance*. Paris: Grasset.

Vries, H. de. 1910. *Intracellular Pangenesis, Including a Paper on Fertilization and Hybridization*. Trans. Charles Stuart Gager. Chicago: Open Court.

Wakayama, T., A. C. F. Perry, M. Zucotti, K. R. Johnson, and R. Yanagimachi. 1998. "Full-Term Development of Mice from Enucleated Oocytes Injected with Cumulus Cell Nuclei." *Nature* 394: 369–74.

Watts, Alan W. 1951. *The Wisdom of Insecurity*. New York: Pantheon.

———. 1958. *Nature, Man, and Woman*. New York: Pantheon.

———. 1966. *The Book, on the Taboo Against Knowing Who You Are*. New York: Vintage.

White, J. 1972. *The Highest State of Consciousness*. New York: Doubleday.

Wickner, R. B. 1994. "(URE3) as an Altered URE2 Protein: Evidence for a Prion Analog in *Saccharomyces cerevisiae*." *Science* 264: 566–69.

Wiener, Norbert. 1948. *Cybernetics,or Control and Communication in the Animal and the Machine*. New York: J. Wiley and Sons.

———. 1964. *God and Golem, Inc.: A Comment on Certain Points Where Cybernetics Impinges on Religion*. Cambridge: MIT Press.

Willis, W. D. 1995. "Cold, Pain, and the Brain." *Nature* 373: 19–20.

Wilmut, I., A. E. Schnieke, J. McWhir, A. J. Kind, and K. H. S. Campbell. 1997. "Viable Offspring Derived from Fetal and Adult Mammalian Cells." *Nature* 385: 810–13.

Winograd, S., and J. D.Cowan. 1963. *Reliable Computation in the Presence of Noise.* Cambridge: MIT Press.

Wittgenstein, Ludwig. 1961. *Tractatus Logico-Philosophicus.* Trans D. F. Pears and B. F. McGuiness. London: Routledge and Kegan Paul.

———. 1965. *The Blue and Brown Books.* New York: Harper and Row.

———. 1979. *Notebooks 1914–1916.* 2nd ed. Ed. G. H. von Wright and G. E. M. Anscombe. Trans. G. E. M. Anscombe. Oxford: Basil Blackwell.

———. 2001. *Philosophical Investigations.* 3rd ed. Trans. G. E. M. Anscombe. Oxford: Basil Blackwell.

Young, L. E., K. D. Sinclair, and I. Wilmut. 1998. "Large Offspring Syndrome in Cattle and Sheep." *Reviews in Reproduction* 3: 155–63.

Zvonkin, A. K., and L. A. Levin. 1970. "The Complexity of Finite Objects and the Development of the Concepts of Information and Randomness by Means of the Theory of Algorithms." *Rassiar Mathematical Surveys,* 25 (6): 82–124